BERNOULLI'S
FALLACY

———

BERNOULLI'S FALLACY

—

Statistical Illogic
and the Crisis of
Modern Science

AUBREY CLAYTON

COLUMBIA UNIVERSITY PRESS
NEW YORK

Columbia University Press gratefully acknowledges the generous support for this book provided by a member of the Publisher's Circle.

Columbia University Press
Publishers Since 1893
New York   Chichester, West Sussex
cup.columbia.edu

Library of Congress Cataloging-in-Publication Data
Names: Clayton, Aubrey, author.
Title: Bernoulli's fallacy : statistical illogic and the crisis of modern science / Aubrey Clayton.
Description: New York : Columbia University Press, 2021. | Includes bibliographical references and index.
Identifiers: LCCN 2021004250 (print) | LCCN 2021004251 (ebook) | ISBN 9780231199940 (hardback) | ISBN 9780231199957 (pbk.) | ISBN 9780231553353 (ebook)
Subjects: LCSH: Bernoulli, Jakob, 1654–1705—Influence. | Probabilities—Philosophy—19th century. | Probabilities—Philosophy—20th century. | Mathematical statistics—Philosophy. | Binomial distribution. | Law of large numbers.
Classification: LCC QA273.A35 C53 2021 (print) | LCC QA273.A35 (ebook) | DDC 519.2—dc23
LC record available at https://lccn.loc.gov/2021004250
LC ebook record available at https://lccn.loc.gov/2021004251

Cover design: Noah Arlow

Disclaimer: The views expressed are solely those of the author and do not reflect the views of his employer, Moody's Analytics, or its parent company, Moody's Corporation, or its affiliates.

Dedicated to Jameel Al-Aidroos, of blessed memory.

CONTENTS

PREFACE

ince this book risks being accused of relitigating old arguments about statistics and science, let us first dispense with the idea that those arguments were ever settled. The "statistics wars" never ended; in some ways they have only just begun.

Science, statistics, and philosophy need each other now as much as ever, especially in the context of the still unfolding crisis of replication. Everyone regardless of ideology can likely agree that something is wrong with the practice of statistics in science. Now is also the right time for a frank conversation because statistical language is increasingly a part of our daily communal lives. The COVID-19 pandemic has, sadly, forced statistical terms like "test sensitivity," "specificity," and "positive predictive value" into our collective lexicon. Meanwhile, in other recent examples, (spurious) statistical arguments were a core component of the allegations of electoral fraud in the 2020 U.S. presidential elections, and (non-spurious) statistical arguments are central to the allegations of systemic racial bias in the U.S. criminal justice system. The largest stories of our time—in public health, education, government, civil rights, the environment, business, and many other domains—are being told using the rhetorical devices of statistics. So the recognition that statistical rhetoric might lend itself to misuse makes this an urgent problem with an ethical dimension. On that we can probably also agree.

What to do about it is another matter. In science, several proposed methodological changes (discussed in the following) have gained support as potential solutions to the replication crisis, but there are no clear winners yet. The reason consensus is hard to come by is that there are unresolved *foundational* questions of statistics lurking within these debates about methods. The discussions happening now can, in fact, be seen as a vibrant remixing of the same philosophical issues that have colored the controversies about statistics since the 1800s. In short, assessing whether a proposed change successfully fixes a problem requires one to first decide *what the problems are*, and such decisions reveal philosophical commitments about the process by which scientific knowledge is created. When it comes to such foundational questions, we are not all on the same page, for reasons explored in this book.

Because statistical methods are a means of accounting for the epistemic role of measurement error and uncertainty, the "statistics wars" (at least on the frequentist versus Bayesian front) are best described as a dispute about the nature and origins of probability: whether it comes from "outside us" in the form of uncontrollable random noise in observations, or "inside us" as our uncertainty given limited information on the state of the world. The first perspective limits the scope of probability to those kinds of chance fluctuations we can, in principle, tabulate empirically; the second one allows for probability to reflect a degree of confidence in a hypothesis, both before and after some new observations are considered. Unfortunately for the conflict-averse, there is no neutral position here.

As a snapshot of the ways these philosophical commitments are now playing themselves out in practice, consider that much of the current debate about statistical and scientific methods can be organized into three categories of concerns:

Where does the hypothesis come from and when? If a particular hypothesis, representing a concrete prediction of the ways a research theory will be borne out in some measured variables, is crafted after peeking at the results or going on a "fishing expedition" to find a version that best suits the available data, then it may be considered a suspicious product of "post hoc theorizing," also known as hypothesizing after results are known ("HARKing"), taking advantage of "researcher degrees of freedom," the "Texas sharpshooter fallacy," "data dredging," the "look-elsewhere" effect, or "p-hacking." Various proposals to combat this include the pre-registration of methods, that is, committing to a certain rigid process of interpreting the data before it has been gathered, sequestering the "exploratory" phase of research from the "confirmatory" one, or correcting for multiple possible comparisons, as in the Bonferroni correction (dividing the threshold for significance by the number of simultaneous hypotheses being considered) or others like it.

What caused the experiment to begin and end, and how did we come to learn about it? Subcategories of this concern include the problem of "publication bias," or the "file drawer problem," and the problem of "optional stopping." If, say, an experimenter conducting a trial is allowed to keep running the experiment and collecting data until a favorable result is obtained and only then report that result, there is apparently the potential for malfeasance. Attempts to block this kind of behavior include making publication decisions solely on the basis of the pre-registered reports—that is, based purely on the methods—to encourage publishing negative results, and for the "stopping rule" to be explicitly specified ahead of time and adhered to.

Is there enough data? Small samples are the bane of scientists everywhere, for reasons ranging from lack of resources to the phenomena of interest being inherently rare. In the standard statistical framework, this creates a problem of low power, meaning a high chance of failing to find an effect even though it's present. It also, perversely, means that an effect—if it *is* found—is likely overstated and unlikely to be replicable, a paradox known as the "winner's curse." A different but related dynamic is at work when statistical models with many parameters are "overfit" to the available data. Too few data points are asked to carry too heavy a load, and as a result the model may look good when evaluated on a training data set and yet perform miserably elsewhere. Apart from simply collecting larger data samples (easier said than done), the emerging best practice recommendations are to facilitate collaboration by sharing resources and materials, incentivize replication studies and meta-analyses, reserve some amount of data for "validation" or "out-of-sample testing" of any model being fit, and perform power analyses to determine how large a sample is needed to find a meaningfully sized effect with high probability.

All three appear at a glance to be legitimate causes for concern, and the proposed solutions appear to be sensible countermeasures, *but only if the standard (non-Bayesian) mode of doing statistical inference is assumed as a given.* As we'll see, Bayesian statistics provides natural protection against these issues and, in most circumstances, renders them non-issues. The safeguard, missing completely from the standard template, is the *prior probability* for the hypothesis, meaning the probability we assign it before considering the data, based on past experience and what we consider established theory.

Upon further reflection, it becomes apparent that many of these proposed limitations are at odds with common sense about the way hypotheses are usually formed and tested in light of the evidence. For example, a doctor measuring a child's height might report that measurement including some error margins, indicating the best guess for the true value and a probability distribution around it, but strictly speaking that hypothesis would be made after the results were known—violating the rules established to protect us from concern (1). Is the doctor now obligated to measure the child again as a "confirmatory" analysis? Many of the prototypical examples of basic statistics like surveying a population by random sampling would not pass such scrutiny. Nor would, say, the application of probability to legal evidence. If a suspect to a crime is identified based on evidence gathered at the scene, can their probable guilt also be established by that same evidence, or must all new evidence be gathered? Is a reviewer for a scientific journal, in the course of critiquing a paper, allowed to conjure up alternative explanations to fit the reported data, or must those alternatives also be pre-registered?

The problem of publication bias—the bulk of concern (2)—is exacerbated by the standard statistical methods because they unnaturally level the playing field between all theories, no matter how frivolous and outlandish. So, of course, among the theories that meet a given standard of publication-worthiness, the more surprising and counterintuitive ones (and those least likely to actually be substantive) will tend to get more attention. Requiring more surprising hypotheses to meet a higher standard of evidence would realign publication incentives and remove much of the chaff. Similarly, as we'll see, "optional stopping" is a source of anguish for standard methods only because they're sensitive to the possibility of what *could* have happened but did not. Bayesian inferences are based only on what was actually observed, and the experimenter's plans for other experiments usually do not constitute relevant information.

Similarly, the problems in (3)—low statistical power and overfitting of models—are only problems if the answer from a statistical procedure is interpreted as final. In the Bayesian mode, hypotheses are never definitively accepted or rejected, nor are single estimates of model parameters taken as gospel truth. Instead, uncertainty can change incrementally as more data is collected; a single observation can be useful, two observations more so, etc. The more simultaneous questions being answered, in the form of many adjustable dials in a given model, generally speaking the more data will be required to reduce that uncertainty to manageable levels, but all along the way we will be reminded where we "are," inferentially speaking, and how much further we have to go. Reserving data to use for validating models is a waste of perfectly good data from which we could have learned something.

All of the above and more is possible if we're simply willing to let probability mean *uncertainty* and not just *frequency* of measurement error. So, first we need to get over that philosophical hump. To put the case another way: compared with Bayesian methods, standard statistical techniques use only a small fraction of the available information about a research hypothesis (how well it predicts some observation), so naturally they will struggle when that limited information proves inadequate. Using standard statistical methods is like driving a car at night on a poorly lit highway: to keep from going in a ditch, we could build an elaborate system of bumpers and guardrails and equip the car with lane departure warnings and sophisticated navigation systems, and even then we could at best only drive to a few destinations. Or we could turn on the headlights.

So, while the floor is open for a discussion of "questionable research practices," it's the perfect time to question whether some of those practices are questionable, or whether there might be a better way to think about the whole scheme. To that end, contained in this book are suggestions concerning probability and statistical inference that at first will appear heretical to anyone trained in statistical orthodoxy but that might, with some meditation, seem more and more sensible. The common theme is that there would be no need to continually treat the symptoms of statistical misuse if the underlying disease were addressed.

I offer the following for consideration:

Hypothesizing after the results of an experiment are known does not necessarily present a problem and in fact is the way that most hypotheses are ever constructed.

No penalty need be paid, or correction made, for testing multiple hypotheses at once using the same data.

The conditions causing an experiment to be terminated are largely immaterial to the inferences drawn from it. In particular, an experimenter is free to keep conducting trials until achieving a desired result, with no harm to the resulting inferences.

No special care is required to avoid "overfitting" a model to the data, and validating the model against a separate set of test data is generally a waste.

No corrections need to be made to statistical estimators (such as the sample variance as an estimate of population variance) to ensure they are "unbiased." In fact, by doing so the quality of those estimators may be made worse.

It is impossible to "measure" a probability by experimentation. Furthermore, all statements that begin "The probability is . . ." commit a category mistake. There is no such thing as "objective" probability.

Extremely improbable events are not necessarily noteworthy or reason to call into question whatever assumed hypotheses implied they were improbable in the first place.

Statistical methods requiring an assumption of a particular distribution (for example, the normal distribution) for the error in measurement are perfectly valid whether or not the data "actually is" normally distributed.

It makes no sense to talk about whether data "actually is" normally distributed or could have been sampled from a normally distributed population, or any other such construction.

There is no need to memorize a complex menagerie of different tests or estimators to apply to different kinds of problems with different distributional assumptions. Fundamentally, all statistical problems are the same.

"Rejecting" or "accepting" a hypothesis is not the proper function of statistics and is, in fact, dangerously misleading and destructive.

The point of statistical inference is not to produce the right answers with high frequency, but rather to *always* produce the inferences best supported by the data at hand when combined with existing background knowledge and assumptions.

Science is largely not a process of falsifying claims definitively, but rather assigning them probabilities and updating those probabilities in light of observation. This process is endless. No proposition apart from a logical contradiction should ever get assigned probability 0, and nothing short of a logical tautology should get probability 1.

The more unexpected, surprising, or contrary to established theory a proposition seems, the more impressive the evidence must be before that proposition is taken seriously.

■ ■ ■

Of course, I'm far from the first person to write down such theses and nail them to the door of the church of statistics. What follows will not be anywhere near a comprehensive account of the history or current status of the statistics debates, nor the multitude of unique perspectives that have been argued for. No two authors, it seems, have ever completely agreed on the foundations of probability and statistics, often not even with themselves. Roughly speaking, though, the version presented here descends most closely from the intellectual lineage of Pierre-Simon Laplace, John Maynard Keynes, Bruno de Finetti, Harold Jeffreys, Leonard Jimmie Savage, and Edwin Jaynes.

I have tried wherever possible, though, to avoid taxonomizing the camps of people who have contributed to that vast body of literature because I don't think it adds much to the story. If anything, reducing someone's position via shortcuts like "[X]ive [Y]ist [Z]ian" and so on can convey the wrong idea since, depending on context, the meanings of [X], [Y], and [Z] may imply shifting and contradictory conclusions as they've been used differently over centuries. There is, in particular, a danger in allowing various subcamps to be labelled "objective" and "subjective," as these have historically been used mostly as ad copy to sell the different schools of thought.

For our purposes, most of the nuanced differences between the positions staked out by these camps are irrelevant, anyway. What concerns us is one essential question: Is it possible to judge hypotheses based *solely* on how likely or unlikely an observation would be if those hypotheses were true? Those who answer in the affirmative, whether they be Fisherians, neo-Fisherians, Neyman-Pearsonians, equivalence testers, "severe testers," and so on, commit the fallacy that is our subject. Those who answer in the negative have at least avoided that trap, though they may fall into others.

Nor should this book be interpreted as a work of original scholarship. All the main ideas presented here exist elsewhere in higher resolution detail. I've attempted as much as I can to provide the necessary references for an interested reader to follow any thread for more detail. (The answer, in the majority of cases, is Jaynes.)

Consider this, instead, a piece of wartime propaganda, designed to be printed on leaflets and dropped from planes over enemy territory to win the hearts and minds of those who may as yet be uncommitted to one side or the other. My goal with this book is not to broker a peace treaty; my goal is to win the war. Or, to use a less martial metaphor, think of this as a late-night infomercial for a genuinely amazing product, pitched at those researchers cursed with sleepless nights spent crunching statistical analyses and desperately hoping for a "$p < 0.05$" ruling from an inscrutable computer judge. Over a black-and-white video of such a researcher grimacing and struggling to get their appliance to simply function while the complex, clunky machinery of significance tests, power analyses, and multiple comparison corrections finally falls apart in a chaotic mess, a voice asks knowingly, "Are you tired of this always happening to you?" Our exasperated scientist looks up at the camera and nods. "Don't you wish there were a better way? Well now, thanks to Bayesianism, there is . . ."

ACKNOWLEDGMENTS

The Venn diagram of people who made this book possible by (1) offering friendship and encouragement, (2) reading half-finished versions to help clarify my thoughts, and (3) applying their professional skills to transform the work from an unfocused mess of ideas I was animated about into a finished product would look like three circles with a large union and elements in every possible intersection.

First, I'm profoundly grateful to Eric Schwartz, Lowell Frye, and Miranda Martin from Columbia University Press, and to Joan Brookbank for her enthusiastic representation. Craig Turnbull and Adam Reich provided invaluable feedback on multiple drafts and pointed out where the holes were. All remaining mistakes are, of course, my fault (except for those that occur randomly, which are nobody's fault). For their guidance in crafting the raw materials into a viable book proposal, I thank Ben Adams, Lorin Rees, Cathy O'Neil, and Jay Mandel.

Before this project existed, its immediate ancestors were a series of articles and blog posts for *Nautilus* and a similar recurring feature in the *Boston Globe*, all about probability, statistics, and current events. For taking the chance on there being people hungry for this material, and for helping make it palatable to them, I have my editors, Brian Gallagher, Kevin Berger, Brian Bergstein, and Bina Venkataraman, to thank. For encouraging me to write more, I thank George Gantz and Nassim Taleb.

Before any of the main concepts were written in public, they were taught in a course about probability and logic at the Harvard Extension School. For giving me an opportunity to test these ideas with inquisitive and thoughtful students—plus access to the Harvard libraries—I thank Andy Engelward, Mark Lax, and the staff of the Division of Continuing Education.

Before that material was taught officially, it was taught unofficially in an informal reading course and a series of online lectures about Jaynes's *Probability Theory: The Logic of Science*. For donating their valuable time to struggle with this difficult book and tell me what was difficult about it, my enormous thanks go to Sam Thompson, Lilli Thompson, Dev Kumar, Michael Waddington, Laura Grother, Summer Block, and Crawford Crews. A special second helping of thanks goes to Travis Waddington, my confidant and erstwhile co-author, who read with careful attention and helped me through many iterations of writing these things down and every time made the writing better. My gratitude also goes out to Will Kurt—my Jaynes study partner—and to Kevin S. Van Horn, G. Larry Bretthorst, and the whole of the online "Jaynes community."

Numerous other friends provided sympathetic ears and much-needed pep talks while these generations of projects were painfully begetting other projects. There's no way to name everyone, but a partial list would include: Graeme Wood, Louisa Lombard, Teresa Sharpe, Zach White, Charless Fowlkes, Nick Eriksson, Nirit Eriksson, Daniela Lamas, Emily Oldshue, Ezra Zuckerman Sivan, Lisa Wasserman Sivan, Jessica Ullian, Jordan Piel, Zoe Piel, Ian Abrams-Silva, Kelsey LeBeau, and Jameel Al-Aidroos, to whose memory this book is dedicated.

I could not have undertaken any of these endeavors without a vast support network of loving people, especially Lyn Willis, Karl Keyton, Marilyn Tapper, Ronna Tapper-Goldman, Jim Goldman, Adam Goldman, Melissa Tapper Goldman (my first and best reader on any new project), and Naftali Clayton.

I chose probability and statistics as subjects to study in depth because they are infuriatingly and fascinatingly counterintuitive. But the nature of such pursuits is that you will inevitably find yourself feeling hopelessly adrift, and it becomes essential to have certain "fixed point" references to get reoriented and find your way back to shore. For me, the main reference has been Jaynes, who saw the truth and preserved it, but there are many others. Stephen Stigler's historical documentation has been invaluable as has been the work of Richard McElreath, Eric-Jan Wagenmakers, Stephen Ziliak and Deirdre McCloskey, Persi Diaconis and Brian Skyrms, and Sandy Zabell.

Going even further up the family tree, I owe a debt of appreciation to Steve Evans, through whom I am mathematically descended from David George Kendall, Maurice Stevenson Bartlett, John Wishart, Karl Pearson, and Francis Galton.

And, of course, I owe my existence, my happy upbringing, and my love of mathematics to my mother and father, both excellent role models of math educators, through whom I am also mathematically descended from Hubert Spence Butts, Felix Klein, Carl Friedrich Gauss, and, along other branches, from Pierre-Simon Laplace and Jacob Bernoulli.

BERNOULLI'S FALLACY

INTRODUCTION

The methods of modern statistics—the tools of data analysis routinely taught in high schools and universities, the nouns and verbs of the common language of statistical inference spoken in research labs and written in journals, the theoretical results of thousands of person-years' worth of effort—are founded on a logical error. These methods are not wrong in a minor way, in the sense that Newtonian physics is technically just a low-velocity, constant gravitational approximation to the truth but still allows us successfully to build bridges and trains. They are simply and irredeemably wrong. They are logically bankrupt, with severe consequences for the world of science that depends on them. As some areas of science have become increasingly data and statistics driven, these foundational cracks have begun to show in the form of a reproducibility crisis threatening to bring down entire disciplines of research. At the heart of the problem is a fundamental misunderstanding of the quantification of uncertainty—that is, probability—and its role in drawing inferences from data.

To clarify before we go any further, I do not intend the preceding statement to be interpreted as "antiscience" or even as an indictment of the establishments of science: its journals, labs, university departments, etc. Despite all its problems, science remains, in the words of Carl Sagan, a "candle in the dark." It is the best organizing principle humanity has for disciplined and skeptical thinking, under constant threat both from moneyed interests resistant to its discoveries and from the forces of superstition and pseudoscience. What I hope to achieve, more than anything, is to inspire scientists to turn that skeptical eye inward and examine critically the received wisdom of how statistical arguments are constructed—*in order to make those arguments stronger and better*. My goal is to nudge science toward improvement for its own sake—to make the candle brighter, not to snuff it out.

There are also whole swaths of science to which these criticisms will not be very damaging, if at all—areas like particle physics that are dominated either by theory or by experimental methods that don't rely heavily on statistical techniques. Here is another opportunity for growth. Since probability and statistics are the unified language of uncertainty—a common ingredient in all observational disciplines—it seems natural that statistical techniques, properly reconstituted, should play a role in all research. As those methods currently stand, though, scientists are right to avoid them if possible; the research areas, especially in social science, that have used them most are the ones now paying the greatest costs in irreproducible results.

However, I don't want to risk understating the size of the problem or its importance. The problem is enormous; addressing it will require unwinding over a century of statistical thought and changing the basic vocabulary of scientific data analysis. The growth of statistical methods represents perhaps the greatest transformation in the practice of science since the Enlightenment. The suggestion that the fundamental logic underlying these methods is broken should be terrifying. Since I was first exposed to that idea almost fifteen years ago, I've spent nearly every day thinking, reading, writing, and teaching others about probability and statistics while living with the dual fears that this radical proposal I've committed myself to could be wrong and that it could be right.

I first got a feeling something was amiss when as a graduate student I tried to use statistics to make money gambling on basketball. I was about halfway toward finishing my PhD in mathematics at the University of California, Berkeley, when I realized I had no idea what I was doing. My research area was the mathematical theory of probability, basically an amped-up version of the problems we all did in high school where you draw colored marbles from a jar. I'd studied probability at the graduate level for a couple of years by that point and was full-to-brimming with theoretical knowledge. I loved the subject. On the walls of my apartment bedroom, I taped up sheets of notebook paper on which I had written the statement of every theorem from my first-year graduate textbook and an outline of the proof of each one. I gave talks at departmental seminars and was making slow but steady progress toward solving the problem that would be the centerpiece of my dissertation, a particularly nasty question about a random process living in an infinite-dimensional space. I thought I knew a lot about probability.

At a New Year's Eve party at the end of 2005, I ran into my friend Brian, who had left the math program at Berkeley to work for a company that made video poker machines. His job was to be in charge of the math—specifically, the probabilities. Gaming regulations require machines to conform to rigid rules about how often they pay out so companies can't cheat the players by, say, having the machines deal a royal flush only half as often as they should. Brian was responsible for testing the computer code inside the machines to make sure they worked properly. Mathematically, it was pretty trivial, but it was the kind of job where you had to be absolutely sure you were right.

Brian also liked to gamble. This was at the height of the Texas hold 'em boom in the mid-2000s. Suddenly it seemed like everyone was playing poker online and on their dining room tables. I was particularly frustrated by the conventional wisdom that anyone with a math background could get rich playing poker. I had dreams of dropping out of school and making easy cash playing poker professionally. "It's all just probability," everyone said. But despite my probability expertise, I was always a pretty lousy player. *I can prove that a unique optimal strategy for this situation exists, but I have no idea what it is*, I would think as I watched my chip stack slowly diminish until it disappeared. Usually on the receiving end of my chips were people like Brian, who somehow could do the mental calculations that gave him an advantage.

That New Year's Eve, though, instead of poker, he wanted to talk about basketball. Brian also liked betting on sports, and he noticed what he thought was a chance to make money in the online NBA betting markets. There are two kinds of bets you can place on any basketball game: which team will win and whether the total score of the game, adding the scores of both teams together, will be greater or less than an amount called the *over/under*. You can also sometimes *parlay* multiple bets together, meaning you win if all the predictions you make come true. Usually people use a parlay to bet on more than one game at the same time, but Brian noticed you could also parlay the two possible bets, winner and over/under, on the *same* game. So, if the Dallas Mavericks were playing the Los Angeles Lakers, you could bet on, say, the Mavericks winning *and* the total score being less than 200. The problem, as well as the opportunity, was that the betting websites always figured the odds for parlay bets as though the component parts were *independent*, meaning they assumed the probability of both things happening was the product of the two separate probabilities, like rolling a 6 twice in a row with a fair die. To Brian, that didn't feel right. He knew a thing or two about basketball, and it seemed to him the strengths and weaknesses of the teams would play out in a way that could affect these bets. In theory, this could allow for some advantageous bets because the parlay payoffs might be greater than they should be, on average.

Brian had acquired several years' worth of game data and asked if I could look it over to see if he was right. I was happy to try (I was still fantasizing about dropping out of grad school and getting rich quickly, somehow), but I realized pretty quickly I didn't know how to actually answer the question. As an undergrad, I had minored in statistics, probability's twin-sister discipline, and remembered enough that I could identify what the standard tools were for this kind of problem. But the more I thought about it, the more I realized I didn't really understand what those tools were doing—or even why they were the standard tools in the first place. I came up with *an* answer, even though I couldn't be sure it was the *right* one, and Brian and I made a tidy profit betting on the NBA that year.

The basketball problem kept nagging at me. Did our success conclusively prove our theory was correct, or had we just gotten lucky? The statistical arguments supporting the theory still didn't make much sense to me. The nature of the association we were trying to exploit was *probabilistic*. It wasn't *guaranteed* a given team would win if the game was low-scoring; we just needed it to be *more probable* than the bookies gave it credit for being. But how could I tell from the data whether something was actually more probable? I could maybe see if a certain combination of outcomes had happened *more often* than expected, but there was always noise in the data. How could I tell if the results I saw were just the product of chance variation and not truly evidence of a real effect?

This is, it should be said, exactly the kind of question the standard techniques of statistics are designed to answer—and not just for sports betting. Since the 1920s, these have been the go-to methods of analyzing data in every conceivable situation. Techniques like *significance testing* are supposed to tell you whether a phenomenon you observe in the data is real or just the product of chance. They never answer the question definitively one way or the other, but there are certain rules of thumb. If a deviation (in whatever sense is appropriate for the problem at hand) as big as or bigger than the one you've seen should happen only with probability 5 percent, then the general advice is to reject chance as an explanation.

To my mathematical brain, these tests always felt a little off. Why should I care about how often I *might have* observed the result I actually did observe? I could see it staring back at me from my spreadsheet, so I knew for a fact it had happened. Where did this magical 5 percent threshold of significance come from? And if my data did pass the threshold and I rejected the chance hypothesis, so what? What I cared about was whether I could make money gambling. The conventional explanations I found in my stats textbooks were never ultimately that satisfying and seemed to be deliberately obscured by jargon about Type I and Type II error rates and so on. My training had told me not to trust any idea I couldn't derive for myself from first principles, and no matter how hard I tried, I just couldn't *prove* the statistics tests were right. It turns out I wasn't alone.

My real problem was that, in all the probability calculations I had ever done, the probabilities had been given from the start, which is typical of the way the subject is usually taught. What I understood as probability theory was a set of rules for manipulating probabilities once given, not a way of establishing them in the first place. There was always a fair coin or a pair of dice or, in the more abstract mathematical setting, a probability measure. The questions were then all about the consequences of those assumptions: If a die is fair, what's the probability you'll observe a run of three 6s in a row sometime in the next 10 rolls? If five couples are randomly split up between two tables at dinner, what's the chance of any given couple sitting together? And so on. But I had no way to recognize a probability when it wasn't handed to me. What is the probability I'll be in a car accident or get skin cancer? What was the probability of the 1989 Loma Prieta earthquake? What is the probability Bigfoot is real? The more I thought about it, the more I realized I didn't know what probability really *meant*. This was a pretty destabilizing epiphany for someone who had already spent a good chunk of his life studying probability.

That spring I went searching for any books that could shore up my philosophical understanding of probability, and a chance recommendation led me to *Probability Theory: The Logic of Science* by Edwin Jaynes. The book answered the questions that had been bothering me about the meaning of probability, but it also did so much more. It fundamentally rewired the ways I thought about statistics, uncertainty, and the scientific method. It was like the world went from black-and-white to color. I read it over and over and begged my friends to read it too. I would occasionally ask my girlfriend (a nonmathematician) to put down whatever she was doing so I could read her a new favorite passage. I started to drop Jaynesisms like "desiderata," "state of knowledge," and "ad hockery" in casual conversation. It got a little weird.

Jaynes's book occupies a kind of cult niche in the world of probability. It's pitched at a pretty advanced level, dotted throughout with intimidating terms like "autoregressive model of first order"[1] and "inverse Laplace transform."[2] The preface opens with this: "The following material is addressed to readers who are already familiar with applied mathematics, at the advanced undergraduate level or preferably higher."[3] And that's the version he *intended*. Tragically Jaynes died in 1998 before he could finish the book, and the version published posthumously, with minimal editing, by his student G. Larry Bretthorst has numerous gaps, errors, and long, florid digressions. This is a typical snippet from early in the book, a section on the fundamentals of probability (only chapter 2!):

> Therefore everything depends on the exact way in which $S(1-\delta)$ tends to zero as $\delta \to 0$. To investigate this, we define a new variable $q(x, y)$ by
>
> $$\frac{S(x)}{y} = 1 - \exp\{-q\}.$$
>
> Then we may choose $\delta = \exp\{-q\}$, define the function $J(q)$ by
>
> $$S(1 - \delta) = S\left(1 - \exp\{-q\}\right) = \exp\{-J(q)\},$$
>
> and find the asymptotic form of $J(q)$ as $q \to \infty$.[4]

As a result, not many people have actually read Jaynes's book cover to cover. But those of us who have tend to be somewhat evangelical about it because in between the advanced-undergraduate-level-plus math sections is some of the most strikingly clearheaded and revolutionary thinking on the subject of probability ever committed to print.

You might have heard of the feud in the world of statistics between two camps: the "frequentists" and the "Bayesians." As these ideological team identifications go, Jaynes was a hardcore Bayesian, maybe the most Bayesian who ever Bayesed. Jaynes's essential argument comes down to this: *modern, frequentist statistics is illogical.* He starts with the basic, yet surprisingly thorny, question, the same question I found myself asking: What is probability? He then proceeds to demolish the traditional (non-Bayesian) answer—that the probability of an event is just how often you'll see it happening over many repeated trials—and shows how that mistaken idea has led all of 20th-century statistics down a ruinous path.

To understand the flaw in modern statistics, and what's so ruinous about it, we'll need to travel back to the 1600s and see what problems first motivated mathematicians to develop probability theory. In particular, we'll need to consider one of the first questions of what we would now recognize as statistics, a problem asked and, supposedly, answered by one of the founding fathers of probability, the Swiss mathematician Jacob Bernoulli.

■ ■ ■

Bernoulli (1655–1705) was a brilliant mathematician from a family of geniuses. You'll still find his name sprinkled around in any calculus or differential equations textbook today. His most important work, though, was in probability, which he grappled with for about the last 20 years of his life. He was still writing his great book on the subject, *Ars Conjectandi* (*The Art of Conjecturing*), when he died. Published posthumously in 1713 by his nephew, Nicolaus, the book is widely regarded as the foundational work in the field, clarifying ideas that had been gradually coming into focus over the preceding decades thanks to the efforts of other great mathematical pioneers like Christiaan Huygens, Blaise Pascal (Pascal's triangle), and Pierre de Fermat (Fermat's last theorem).

Before Bernoulli, people had mostly needed probability to solve problems about dice throwing and other games of chance. The problem Bernoulli was mainly concerned with, though, was one of probabilistic *inference*. This is a typical example of the kind of problem Bernoulli considered:

> Suppose you are presented with a large urn full of tiny white and black pebbles, in a ratio that's unknown to you. You begin selecting pebbles from the urn and recording their colors, black or white. How do you use these results to make a guess about the ratio of pebble colors in the urn as a whole?

At the time of his writing, the basic concepts of probability—the kinds of combinatorial calculations for things like dice and card games we'd now assign as problems to high school students—had begun to be formally worked out only recently. So taking on this challenge was ambitious, and he struggled with the problem for years. The main insight of the solution he ultimately landed on was this: if you take a *large enough* sample, you can be very sure, to within a small margin of absolute certainty, that the proportion of white pebbles you observe in the sample is close to the proportion of white pebbles in the urn. That is, Bernoulli developed what we'd now understand as a first version of the Law of Large Numbers. Stated in concise modern mathematical language, Bernoulli's result would be

> For any given $\varepsilon > 0$ and any $s > 0$, there is a sample size n such that, with w being the number of white pebbles counted in the sample and f being the true fraction of white pebbles in the urn, the probability of w/n falling between $f - \varepsilon$ and $f + \varepsilon$ is greater than $1 - s$.

Let's consider each term to unpack the contents of this theorem. Here, the fraction w/n is the ratio of white to total pebbles we observe in our sample. The value ε (epsilon) captures the fact that we may not see the true urn ratio exactly thanks to random variation in the sample: perhaps the true ratio is 3/5, and in a sample of 50 pebbles, we get, say, 32 white instead of 30. But we can be pretty well assured that, for large samples, we'll be pretty close on one side or the other of the true value. The value s reflects just how sure we want to be. No matter how large a sample we take, there will always remain the remote chance we'll be unlucky and get a sample outside our epsilon-sized tolerance of the true urn proportion, but we can be *almost* certain this won't happen. We can, for example, set $s = 0.01$ and be 99 percent sure. Bernoulli called this "moral certainty," as distinct from absolute certainty of the kind only logical deduction can provide.

So there are three moving parts to the calculation: how big a sample we take, how close we want to be to the absolute truth, and how certain we want to be to find ourselves within that margin. Wrangling all those terms was what made the problem so difficult.

Stated in words, then, Bernoulli's strategy for the inference question is this: "Take a large enough sample that you can be morally certain, to whatever degree that means to you, the ratio in the sample is within your desired tolerance of the ratio in the urn. Whatever you observe is therefore with high probability a good approximation of the urn's contents." He even helpfully tabulated values of how big a sample you would need to take in order to achieve various levels of moral certainty (99 percent, 99.9 percent, and so on). As a theorem, Bernoulli's Law of Large Numbers was technically sophisticated for its day, requiring clever mathematical innovations to prove and inspiring multiple generalizations over the years. As a proposed solution to his inference problem, it was conceptually appealing, powerful, elegant, and wrong.

Or, more generously speaking, it was incomplete.

The exact ways in which Bernoulli's answer was lacking will take some time for us to unfold, but the essential idea is that despite the apparent linguistic symmetry, there is a crucial difference between these statements:

The sample ratio is close to a given urn ratio with high probability.
  and
The urn ratio is close to a given sample ratio with high probability.

The former refers to a *sampling* probability, which Bernoulli and his contemporaries had grown adept at handling in games of chance (the probability of throwing a certain number with a pair of dice, say), while the latter involves an *inferential* probability, which we use when testing hypotheses, forecasting one-time events, and making other probabilistic claims about the world. Sampling probabilities lend themselves to a frequency-based interpretation; the probability of something measures how *frequently* it happens. Inferential probabilities require something more subtle; the probability of a statement depends on how much *confidence* we have in it. Sampling probabilities go from hypothesis to data: Given an assumption, what will we observe, and how often? Inferential probabilities go from data to hypothesis: Given what we observed, what can we conclude, and with what certainty? Sampling probabilities are fundamentally *predictive*; inferential probabilities are fundamentally *explanatory*.

Pictorially, sampling probabilities could be represented like this:

Hypothesis → Data
(Given a hypothetical mix of pebbles in the urn equal to *f*, the
probability of getting a data sample of *w* white out of *n* is . . .)

Inferential probabilities look like this:

Data → Hypothesis
(Given an observed data sample of *w* white out of *n*, the
probability of the mix of pebbles in the urn being *f* is . . .)

The problem with Bernoulli's answer to his question is that *the arrow is pointing in the wrong direction*. He wanted to answer a question about the probability of a hypothesis, but he did so by thinking only about the probability of an observation.

The confusion of the two—and the general idea that one can settle questions of inference using only sampling probabilities—is what I call *Bernoulli's Fallacy*. Although it may seem like a minor academic nitpick at first, the implications of this confusion are profound and go to the very core of how we think about and quantify uncertainty in settings from the lofty to the mundane.

∎ ∎ ∎

Generations of mathematicians, statisticians, and scientists of various disciplines have fallen prey to this fallacy over the ensuing centuries, and to this day, many would deny it is even a fallacy at all. The debate over the interpretation of probability—in particular, what kinds of probability statements are allowed and how those statements are justified—has gone on for almost as long as probability has existed as a mathematical discipline, with wide-ranging implications for statistical and scientific practice. For an argument about math, this debate has been surprisingly heated. Probability has a unique way of igniting confusion, frustration, and turf wars among otherwise staid, analytically minded people in the world of math. You may even be experiencing some strong feelings about probability right now. Witness, for example, the borderline hate mail *Parade* magazine columnist Marilyn vos Savant received, much of it from PhD mathematicians and statisticians, when she first published her (correct) solution to the probability puzzle called the Monty Hall Problem. Or go check out the furious online arguments over probability brainteasers like the Boy or Girl Paradox.

It is into this uncomfortable territory that we will plunge headlong in this book (we'll settle both the Monty Hall Problem and the Boy or Girl Paradox), and we will take sides.

For starters, we'll need to clarify what probability means. We'll argue for an interpretation of probability that fell out of fashion with the ascendancy of the frequentist school, which now underlies statistical orthodoxy, but that nevertheless has been present since probability's earliest days. This new/old interpretation has seen something of a comeback in recent years thanks in large part to authors like Jaynes. The main idea is this: *probability theory is logical reasoning, extended to situations of uncertainty*. We'll show how this flexible definition includes both the sampling and the inferential types of probability in Bernoulli's problem and how it allows us to both solve the problem and describe how Bernoulli's attempt was only part of a complete answer.

This goes far beyond drawing pebbles from urns or solving brainteasers, though. Once you learn how to spot Bernoulli's Fallacy—the misguided belief that inference can be performed with sampling probabilities alone—you'll start to see it almost everywhere. To illustrate the problems of Bernoulli's Fallacy in the real world, we'll explore some areas of daily life involving probabilistic reasoning, primarily the evaluation of legal evidence and medical diagnoses. Maybe someday you'll be on a jury, and you'll hear arguments like the one presented against Sally Clark, wrongfully convicted in 1999 of murdering her two infant children because the prosecution claimed the probability of a family having two children die of SIDS was astronomically low. That's a sampling probability, not an inferential one. The question *should* have been this: Given that the two children died, what's the probability they died of SIDS? Or maybe you'll go to the doctor and receive a positive test result for a dangerous disease, only to be told (misleadingly) that if you didn't have the disease, the chance is very small you'd test positive. Again, that's a sampling probability, but what you should care about is the inferential probability of having the disease *given* that you tested positive.

Using probability in these real-world situations is very much in the spirit of Bernoulli, whose goal was to formalize the kinds of conjecturing needed to make legal or other decisions on the basis of inconclusive evidence (that is, nearly all decisions). The first example he gave of a probabilistic inference in part IV of his book (containing the Law of Large Numbers) was not the contrived urn-drawing problem we saw earlier but one of determining the guilt or innocence of a suspect accused of murder based on the available facts of the case. As a generalized logic of uncertainty, we find probability well suited to the task. But we'll describe the ways an incomplete understanding of probability has flummoxed even the experts who should know better, resulting in other cases of wrongful convictions and misguided advice.

Most disturbing of all, though, once you can recognize Bernoulli's Fallacy in the wild, is that you'll start to see that it forms the basis for all of frequentist statistics—and therefore a good portion of modern science. Anyone trained in statistics these days will immediately see the flawed reasoning in the preceding legal and medical examples. For example, to correctly determine the probability of having a disease, given that you tested positive for it, we would also need to know the rate of incidence of the disease in the population. This allows us to weigh the relative chances of the positive result being a true positive (you actually have the disease and the test confirmed it) versus a false positive (you don't have the disease and the test was mistaken). If the disease in question is rare, the positive results from even a highly accurate test will usually be false positives. Forgetting to account for this fact is such a well-known statistical mistake it has a name: *base rate neglect*.

In the language of probability, the base rate determines what's known as the *prior* probability. Generally a prior probability is the degree of confidence we have in a proposition, such as a suspect being guilty or a patient having a disease, before considering the evidence. The problem with statistical methods in science is that they commit the same conceptual mistake as base rate neglect. The standard methods as used almost everywhere give no consideration to the prior probability of a scientific theory being true before considering the data. In fact, they don't even allow for that *kind* of probability at all. Instead, they evaluate a theory only on how probable or improbable it makes the observed data—the equivalent of caring only about the accuracy of the test and not about the incidence rate of the disease—so they commit Bernoulli's Fallacy. We'll see this play out, theoretically and practically, many times.

■ ■ ■

How did we get here? If the standard methods are so illogical, how did they ever become the standard? There are three reasonably possible explanations:

1. The problems mentioned previously are really not problems at all, and the standard statistical techniques are perfectly logically coherent.
2. The users and advocates of the standard methods have been unaware of these problems and have simply been reasoning incorrectly all this time.
3. The people responsible for shaping the discipline of statistics were aware of these issues but chose the standard techniques anyway because they offered certain advantages their alternatives could not.

As we'll see, all three theories are true to a greater or lesser extent, and these explanations have mutually reinforced each other over the years like three strands in a braid of mathematics, philosophy, and history. For simple situations, like Bernoulli's urn-drawing problem, the inferences supported by the standard methods are, in fact, totally reasonable. Numerically, they happen to match almost exactly the conclusions that a logical approach would produce under certain conditions, including the absence of any strong prior information. Intuitively this makes sense: if you don't have much of any background knowledge about a situation, then your reasoning process may ignore what little information you do have and still produce reasonable judgments. The problem, as we'll see, is that this condition does not always hold true, and the problems start to arise when we try to extend the techniques from simple situations to more complex ones.

The apparent success of the methods for simple problems, then, has at various critical junctures given cover for philosophical and rhetorical arguments about the meaning of probability. As Bernoulli's example illustrates, the arguments behind the standard statistical methods can, if expressed in the right kind of sloppy and sophistic language, sound extremely convincing. It's only with great care and precision that we can (and will!) dissect these arguments to see their flaws. But anyone, particularly a scientist in a rush to interpret their data, could be forgiven for not taking the time to wade through the details. They may reasonably decide to either accept what they've been taught as standard because it's always been that way or jump to the end and confirm the orthodox methods give the right answer for a simple thought experiment. If, in the latter case, their canonical example of a probabilistic inference is something like Bernoulli's urn, they'll see the methods apparently confirmed. If, as I hope will be true for anyone who's read this book, they think of an example like the Sally Clark case or the test results for a rare disease, their confidence in the correctness of statistical methods may be shaken.

The central conflict in this book is between two historical schools of thought, one that commits Bernoulli's Fallacy and one that avoids it. It would be a mistake to present the latter as perfectly complete and without its own difficulties, though. As we'll experience firsthand, avoiding Bernoulli's Fallacy requires us to think about probability and inference in an entirely different way from what is usually taught, but since at least the 1800s, this alternative approach has been subjected to harsh criticism that in some cases has only recently been answered and in other cases is still unanswered. The main point of concern is the question of how exactly to turn background knowledge into a numerical probability. What is the exact probability you would give to a proposition like "Tall parents tend to have tall children" or "Smoking causes lung cancer"? Different starting points would seem to produce different conclusions based on the data. Even the question of what probability answer to give when we have no information at all turns out to be much harder than it should be. An outsized philosophical concern with these problems was another key factor in the victory of the standard school of statistics. From a certain perspective, Bernoulli's Fallacy may have appeared the lesser of two evils. (It's not.) However, it would also be a mistake to divorce that choice from its historical context or to pretend that statistics exists somehow outside of history; the debate has been judged not by neutral third-party observers but by people motivated to prefer one answer over the other because it supported their agenda.

So to truly understand what the methods are now and how they got to be that way, we'll need to look at where they came from. Historically, probability problems were first confined to games of chance and scientific fields such as astronomy. Over time, innovative thinkers like the 19th-century Belgian scientist Adolphe Quetelet widened the scope of probability's influence by applying it to questions involving real live people, with all the conceptual messiness their complex lives entailed. It was at this historical moment that Bernoulli's Fallacy hit the mainstream. The use of probability methods to answer questions of social importance provoked extreme resistance and legitimate theoretical concerns among scientists and philosophers of the day. In response, a new idea emerged that probability could mean only frequency, an interpretation that would appear to ground all of probabilistic inference in empirically measurable fact. Bernoulli's Law of Large Numbers, and the method of inference he claimed to make possible, seemed to support this interpretation. We'll discuss the mathematical tools making possible the leap from "hard" to "soft" sciences, how they set the scene for the new discipline of social science, and what the demands of this new enterprise dictated for the meaning of probability.

We'll trace the development of statistical methods in the 19th and 20th centuries to understand how the modern discipline of statistics became what it is today. We'll see how, by insisting sampling probability is the *only* valid kind of probability, the titans of that field—most notably, the English trio of Francis Galton, Karl Pearson, and Ronald Fisher—embedded Bernoulli's Fallacy deep in the heart of what we now call statistics. Galton is particularly well-known today for inventing some of the basic ideas of statistics, including correlation and regression, as well as for being among the first to prioritize data collection as integral to science. Pearson and Fisher were both incredibly influential and are jointly responsible for many of the tools and techniques that are still used today—most notably, significance testing, the backbone of modern statistical inference.

Here, again, we'll see the continuation of a theme: the higher the stakes became, the more statistics became reliant on sampling probabilities (the predictive facts, measurable by frequencies) instead of inferential probabilities (the explanatory statements, dependent on judgment). Galton, in addition to being a statistical pioneer, famously coined the term *eugenics* and was an early advocate of using evolution to shape humanity's future by encouraging breeding among the "right" people. Pearson and Fisher were also devotees to the cause of eugenics and used their newly minted statistical tools with great success to support the eugenics agenda. For these early statisticians, the proper function of statistics was often to detect significant differences between races, like a supposed difference in intelligence or moral character. "Objective" assessments of this kind were used to support discriminatory immigration policies, forced sterilization laws, and, in their natural logical extension, the murder of millions in the Holocaust. Sadly the eugenics programs of Nazi Germany were linked in disturbingly close ways with the work of these early statisticians and their eugenicist colleagues in the United States.

I want to make clear what exactly I hope to accomplish by telling the stories of statistics and the eugenics movement at the same time. It may seem that in doing so I am unfairly judging people who lived over a century ago by modern standards of inclusiveness and egalitarianism. But that's not my goal. It's likely any intellectual of that time and place held views that would sound abhorrent to present-day ears, and my intention is not to dismiss the work of these statisticians simply because they were also eugenicists. It would be impossible to study the great works of history without engaging with authors who were not "pure" by our standards. Meanwhile, if we ignored that intellectual context and focused only on their abstract ideas instead, we would sacrifice valuable understanding. As Pearson himself once wrote, "It is impossible to understand a man's work unless you understand something of his character and unless you understand something of his environment. And his environment means the state of affairs social and political of his own age."[5]

The main reason this matters when it comes to Galton, Pearson, and Fisher, though, is that their heinous attitudes were not at all incidental to their work. In many ways, eugenicist ideas animated their entire intellectual projects, as we'll see. What's most important of all, though, is how a desire for authority—understandable as it was given the circumstances—affected their statistical philosophy. The work of Galton, Pearson, and Fisher represents just the extreme end point of the relationship between high-stakes social science and discomfort over any idea of probability as something other than measurable frequency, a connection that's existed for as long as both have been around.

Like any human institution, statistics is and was largely a product of its times. In the late 19th and early 20th centuries, scientific inquiry demanded a theory free from even a whiff of subjectivity, which led its practitioners to claim inference based *solely* on data without interpretation was possible. They were mistaken. But their mistake was so powerfully appealing, and these first statisticians so prolific and domineering, that it quickly took hold and became the industry standard.

■ ■ ■

As a result of the sustained efforts of these frequentist statisticians, decades' worth of scientific research that relies on the standard statistical methods has, since the 1930s, been built on an unstable foundation. Since that time, the accepted way to test a hypothesis using experimental data has revolved entirely around the question, How likely would this (or other) data be if the hypothesis were true? without any consideration given to the prior probability for the hypothesis—that is, the probability before considering the data.

While it may sound like good objective science to measure all theories with the same yardstick, meaning the sampling probability of the data, this practice has all but guaranteed the published literature is polluted by false positives, theories having no basis in reality and backed by data that only passed the necessary statistical thresholds by chance. The best way to avoid this would have been to require that extraordinary claims be supported by equally extraordinary evidence, but this has not been the standard approach. We'll see some disturbing examples, like Cornell University professor Daryl Bem's study, published in a prestigious psychology journal in 2011, showing college students have slight ESP abilities but only when thinking about pornography. Because Bem followed all the same rules and used all the same statistical tests as those doing more normal research, the journal saw no option but to publish his paper.

So scientists now have to choose: either accept this result and others like it as legitimate, or call all other published research using the same methods into question. As difficult as it is, many have chosen the latter. Prompted by incidents like Bem's publication, people have begun reexamining and attempting to replicate the established results in their own fields. What they've found hasn't been pretty. In what's been called the *crisis of replication*, something like half of all results tested so far have failed the replication test. Several of the celebrated theories in fields such as social psychology are going up in flames.

Here are just a few of the recent high-profile casualties:

- The 1988 study by Strack, Martin, and Stepper, claiming that forcing people to smile by holding a pen between their teeth raises their feeling of happiness.[6] A replication attempt in 2016 involving 17 labs around the world and close to 2,000 participants found no significant effect.[7]
- The 1996 result of Bargh, Chen, and Burrows in "social priming," claiming that, for example, when participants were unwittingly exposed to words related to aging and the elderly, they left the lab walking more slowly—that is, they adopted stereotypically elderly behavior.[8] A 2012 replication attempt by researchers at the Université Libre de Bruxelles and the University of Cambridge found no effect.[9]
- Harvard Business School professor Amy Cuddy's 2010 study of "power posing," the idea that adopting a powerful posture for a couple of minutes can change your life for the better by affecting your hormone levels and risk tolerances.[10] The effect failed to replicate in a 2015 study led by Eva Ranehill at the University of Zurich and has since been debunked by several meta-analyses.[11]

- The 2014 study of "ego depletion" by Chandra Sripada, showing that the ability to perform tasks requiring willpower was inhibited if the tasks were preceded by other willpower-demanding tasks.[12] A meta-analysis of 116 studies and a replication effort by 23 labs with 2,000 participants found no support for the effect.[13]

But psychology isn't the only field in crisis. Ongoing replication projects have shown the same problem in fields from economics to neuroscience to cancer biology, and the costs, in both time and money, of chasing statistical ghosts in these other research areas may be much greater. A widespread feeling of panic is emerging among scientists across disparate specialties who fear that many published results in their fields are simply false and that their statistical methods are to blame. A recent large survey, for example, showed that 52 percent of scientists in biology, chemistry, earth science, medicine, physics, and engineering considered their fields to be in "a significant crisis" of reproducibility, and among the most popular suggested remedies was "better statistics."[14] They are right to be afraid.

My basketball problem is just like the problems of scientists today. They have some data, sometimes in abundance and full of rich complexity, and they're sifting through that data for evidence of some probabilistic effect. Statistical orthodoxy tells them what tools to use, but the methods don't follow a coherent logical process—so whether a result is deemed publication worthy is largely a random decision. It's time to give up on those methods. What I ultimately came to understand about my own problem, and what may come as a mixed blessing for these scientists, is that *statistics is both much easier and much harder than we have been led to believe.* All that's needed is a radical overhaul of how we think about uncertainty and statistical inference. That unfortunately may provide little comfort, but until the logical fallacies at the heart of statistical practice are resolved, things can be expected only to get worse.

Many apologists for the discipline will object: "There's no need for such a reinvention of statistics; we already have a handle on what's causing the replication failures, and the problem is solvable by simply using the existing tools in a more disciplined and scrupulous manner." However, before we consider other possible reforms, we must first address the task of fixing the logic of statistical methods themselves. Without this necessary (but not sufficient) correction, I argue that we can expect future replication crises no matter what other improvements are made at the margins.

This isn't just about protecting scientists from the embarrassment of a failed replication, though. We all have a stake in scientific truth. From small individual decisions about what foods to eat or what health risks to worry about to public policies about education, health care, the environment, and more, we all pay a price when the body of scientific research is polluted by false positives. Given infinite time and money, replication studies could perhaps eventually sort the true science from the noise, but in the meantime, we can expect to be constantly deceived by statistical phantoms. The alternative, an approach to statistics that properly contextualizes experimental evidence against prior understanding, will help keep these phantoms in check.

I certainly didn't come up with the idea of Bernoulli's Fallacy. For years, it has been the problem in statistics that made many people vaguely uneasy, but few have criticized it out loud. It is the reason that so much of the standard statistics curriculum is awkward and unnatural. Sampling probabilities are attractive entities for educational purposes because they can appear to be backed up by experimental frequencies; there's always a "right answer." But they're only a part of the complete inferential picture. Requiring all probability statements to refer only to frequencies means avoiding the most sensible probability question in science, How likely is my theory to be true based on what I have observed? The route from data to inference that bypasses this basic question—which is expressly disallowed in the standard way of thinking—must therefore be full of twists and turns. If statistics never quite made sense to you before, Bernoulli's Fallacy may be the real reason why.

This book is the story of a mistake: a wrong idea of the relationship between observation and probabilistic inference so tempting it has ensnared some of history's greatest mathematical minds. I hope that by naming the problem, we can recognize it for the fallacy it is and get to work on the difficult task of cleaning up the mess it's caused. In the coming decades, as we face the existential threats science has been warning us about for years, science will need defending. As it stands now, though, much of science needs defending from itself.

The story is simultaneously about people calling out that mistake over and over when it's cropped up in various new guises, only to see their concerns dismissed and the resistance stamped out. This time around, though, there are reasons to be optimistic, in part because the replication crisis has drawn so much attention. The revolution is coming; in many domains, it is already well underway. I predict that eventually all the concepts we now think of as orthodox statistics will be thrown on the ash heap of history alongside other equally failed ideas like the geocentric theory of the universe. As we go through our story, we'll get several glances at what the discipline of statistics will look like in the future, made ever more tractable with advances in computing power.

It's a testament to Bernoulli's foresight that he saw the larger implications of probabilistic thinking far beyond its humble origins in resolving gambling disputes. We have reason to suspect he was right, even though his proposed solution fell short of the mark. It's only taken 300 years, and counting, for us to begin to figure out how it works.

1

WHAT IS PROBABILITY?

Some of the Problems about Chance having a great appearance of Simplicity, the Mind is easily drawn into a belief, that their Solution may be attained by the mere Strength of natural good Sense; which generally proving otherwise, and the Mistakes occasioned thereby being not unfrequent, 'tis presumed that a Book of this Kind, which teaches to distinguish Truth from what seems so nearly to resemble it, will be looked upon as a help to good Reasoning.

~Abraham de Moivre (1718)

Defining *probability* is more difficult than it might seem at first. Maybe you feel sure you already know what it is. If so, here's a warm-up exercise to make you less sure: Why is the probability of a coin flip coming up heads 50 percent? Maybe you answered that it's because, over a long series of flips, about half of them will come up heads—in which case here are two follow-up questions: (1) Have you ever actually flipped a coin enough times to convince yourself that what you think would happen will actually happen? (2) What about something like the weather forecast? Does the claim that there is a 30 percent chance of rain tomorrow mean that over a long series of tomorrows about 30 percent of them will be rainy? What about the probability that you'll live to be 100 or that the New York Mets will win the World Series? What about the probability that there is life on other planets?

Returning to the coin flips, maybe you said the answer is 50 percent because that's how sure you feel about the prospect of a coin coming up heads. What if your best friend said they were 99 percent sure a coin flip would come up heads and wanted to bet their life savings on it? Would you try to talk them out of it? What if a casino offered you a million dollars to compute the right payoffs for a new video poker machine? Would you go by gut feel?

Or maybe you said it's because there are two possibilities and heads is one of them—ergo the chance is one out of two. Okay, so what's the probability the sun will rise tomorrow? Also 50 percent?

Or maybe you think the coin-flip probability is 50 percent because you have no information to prefer one outcome over the other, so you must assign them equal chances. How do you know what information you have? What would you say is the probability that the number $2^{2^{61}-1} - 1$ is prime? Don't you think that, with the information you "have," you could eventually figure it out for sure?

If you find yourself feeling more confused the more you think about it, you're certainly not alone. Attempting to answer the question, What exactly do we mean by probability anyway? has frustrated mathematicians and philosophers for a long time, and as we'll see, that frustration has had ripple effects throughout scientific and statistical practice. To begin to get our heads around it, we'll start in ancient times and see how different people have tried to solve the puzzle of probability in various ways through the centuries and why their attempts were ultimately unsatisfying. That will bring us all the way back to the present day, where the question still isn't completely answered, but we can assemble the pieces we collect along the way into something that looks kind of like a start.

THE CLASSICAL ANSWER

Advantages: *Intuitive, works well for dice and card games.*
Disadvantages: *Doesn't work so well for anything else.*

Ideas of fate, fortune, or luck and methods of generating chance outcomes—either for purposes of divination or just for fun—have existed in various cultures around the world since the beginning of recorded history. The Greeks had Tyche (called Fortuna by the Romans), the god of luck, whose responsibilities included safeguarding the city. The *I Ching* describes a method of fortune-telling in which the user gathers stalks of the yarrow plant, converts these to a number, and consults a corresponding prophetic hexagram. Ancient Egyptians played a dice game called Hounds and Jackals, and the Indus Valley civilization had a similar game, Gyan Chauper, which survives today in slightly modified form as Chutes and Ladders. The Romans had their *aleae*, rudimentary tetrahedral dice made from the ankle bones of sheep, as memorialized in Julius Caesar's declaration upon crossing the Rubicon: "Alea iacta est." (The die is cast.) The Jewish holiday Purim (from the Akkadian *pur* meaning "fate") celebrates the story of the book of Esther, in which the Jewish people were saved from genocide on a date that had been chosen randomly. The biblical book of Leviticus refers to the high priest casting lots (in Hebrew, *goralot*) to decide the fate of a sacrificial goat or make other priestly decisions, and the Gospels also describe Roman soldiers casting lots (in Greek, *kleron* meaning "portion") to see who would get Christ's cloak after the Crucifixion.

Our word *probability*, from the Latin root *probare* meaning "to test," is therefore related to several other English words involving testing, including *probe*, *proof*, *approve*, and *probity*. The connection between probability and relative soundness or believability was present from the earliest times. Aristotle's *Rhetoric* described "the Probable" (in Greek, *eikos*, from *eoika* meaning "to seem") as "that which happens generally but not invariably."[1] The context for this was his classification of the arguments one could use in a courtroom or legislative debate, where perfect logical deductions may not be available. He called this form of argument an *enthymeme*, to be distinguished from the purely logical form of argument called the *syllogism*, which links together a set of assumed premises to reach deductive conclusions, as in this famous example:

All men are mortal.
Socrates is a man.

∴ Socrates is mortal.

Often, Aristotle observed, pure logic won't suffice on its own because some links in the chain are missing or uncertain. But he offered the advice that "we should also base our arguments upon probabilities as well as upon certainties,"[2] and this could still constitute an appeal to good reason, since "the true and the approximately true are apprehended by the same faculty; it may also be noted that men have a sufficient natural instinct for what is true, and usually do arrive at the truth. Hence the man who makes a good guess at truth is likely to make a good guess at probabilities."[3]

A standard of ancient and medieval legal proceedings was that testimony or evidence, even if not providing absolute certainty, could be (in Latin) *probabilis* meaning "credible and generally agreed upon." The Talmud even suggested that some doubt was a *necessary* component of justice because a feeling of absolute certainty meant a judge had not properly examined all the arguments to find the holes in them. In a capital case before the lesser Sanhedrin, the rule was that if all 23 judges voted to convict, then the suspect was *acquitted*. Probable opinions were those we humans hold to be true within the limits of our own fallible judgment. It was in this sense that Cicero, writing about the nature of the gods, advised that probability "should guide the life of a wise person."[4]

In ancient times, chance was purely qualitative, likely for two main reasons: (1) arithmetic was still carried out mostly using Roman numerals, which made probability calculations awkwardly difficult, and (2) the only things people had to gamble with were sheep ankle bones and other unpredictably shaped objects, which didn't behave reliably enough to bother with numerical probabilities. But by the Renaissance, when gaming equipment was more reliably crafted and real money was on the line, the *mathematics* of chance started to become a topic of concern. What should be the fair odds for a bet on a dice or card game? Even educated people found they were unable to answer the basic questions that naturally presented themselves.

The French gambler Antoine Gombaud, who gave himself the nickname of Chevalier de Méré, had a particular interest in what became known as the *problem of points*. Here's the basic setup:

> You and I are playing a game with equal chances of winning each round (such as flipping a fair coin), with the result that each time I win I earn 1 point and each time you win you earn 1 point. We start at 0–0 and play until one of us gets to some number of points, say 5. The winner gets $100. However, suppose at some intermediate point, say with you leading by a score of 3–1, the game is *interrupted*. How should we divide up the stakes?

On the one hand, it seems unfair to you if we simply call off the game and divide the stakes equally. After all, you are significantly ahead and well on your way to victory. On the other hand, it seems unfair *to me* if I'm made to forfeit entirely; it's still possible I could come back to win. So what should we do? Various gamblers had encountered this problem and attempted a fair division of the stakes by different means but without a clear mathematical reason to justify the division.

Gombaud posed this question to his friend, the mathematician and philosopher Blaise Pascal, who in 1654 was able to work out a rigorous solution in a series of letters with fellow member of the Académie Parisienne and mathematical superstar Pierre de Fermat. In the process, the two of them effectively created probability theory. The main idea of their solution was to divide the stakes according to what we would now refer to as each player's *expected value*. That is, we could *imagine* playing the game out to its conclusion (considering each of the remaining sequences of wins and losses to be equally likely), tally up the scenarios in which each player won the full prize, and finally divide up the stakes according to how many scenarios led to a victory for each player. In modern language, we would say the outcomes (you win) and (I win) were given a probability distribution, and the weighted average payoffs were given to each of us as a fair value of our position in the game.

The word *chance*, from the Latin *cadentia* meaning "a case, an instance," thus had something of a legal connotation. An event or claim was deemed to be probable or likely according to the number of "cases" for it, as though each possible outcome were an argument or piece of evidence in a courtroom. A fair judgment for an uncertain claim was the one given according to the weight of evidence in favor of each claimant. Fermat, in particular, would have found this way of thinking familiar, since he was previously a lawyer and magistrate of the court at Toulouse. The legal analogy, within an all-encompassing view that probability should apply to all reasoning, was apparently on Jacob Bernoulli's mind when he wrote *Ars Conjectandi*, which includes multiple suggestions for the application of probability to legal decision-making. He began chapter 2 of book IV with this preamble discussion:

> The art of conjecturing or stochastics is defined as the art of measuring the probability of things as exactly as possible, to be able always to choose what will be found the best, the more satisfactory, serene and reasonable for our judgements and actions. This alone supports all the wisdom of the philosopher and the prudence of the politician. Probabilities are estimated both by the number and the weight of the arguments which somehow prove or indicate that a certain thing is, was, or will be. As to the weight, I understand it to be the force of the proof.[5]

He then considered the many different forms of argument that might be presented in an example of a suspect accused of murder:

> Titius is found killed in the street. Maevius is charged with murder. The accusing arguments are: 1) He is known to have hated Titius (an argument from a *cause*, since this very hate could have incited to murder. 2) When questioned, he turned pale and answered timidly (this is an argument from the *effect* since it is possible that the pallor and fright were caused by his being conscious of

the evil deed perpetrated). 3) Blood-stained cold steel is found in Maevius' house (this is an *indication*). 4) The same day that Titius was killed, Maevius had been walking the same road (this is *circumstance* of place and time). 5) Finally, Cajus maintains that the day before Titius was killed, he had quarrelled with Maevius (this is a *testimony*).[6]

For these reasons, the theory of probability in the beginning days was known as the *doctrine of chances*—most prominently in the title of French mathematician Abraham de Moivre's seminal textbook on the subject, *The Doctrine of Chances: Or, a Method of Calculating the Probability of Events in Play* (published in 1718). The conventional definition of *probability* was what we would now call the "classical" one:

$$P[\text{event } A] = \frac{\text{\# ways event } A \text{ can occur}}{\text{\# possible outcomes that can occur}}$$

In fact, this persists as the definition given in most introductory probability textbooks today.

The main calculation required to solve any probability problem under the classical definition, then as now, was counting the possibilities for the numerator and denominator of this fraction. Pascal, in particular, was helped by the triangle that now bears his name, tabulating the number of combinations of any n objects taken r at a time, which is a regular ingredient for such computations. For example, if we were computing the probability of getting three heads in a sequence of five coin flips, we might at some point need to know how many such sequences there are, depending on the order in which the flips occur. Pascal's formula gives us the answer:

$$\binom{5}{3} = \frac{5!}{3!(5-3)!} = 10$$

However, a subtle difficulty arose in some problems where the accounting of possible outcomes was not so clearly apparent. For example, around 1620 Ferdinando dei Medici, the Grand Duke of Tuscany, became bothered by a particularly thorny problem of dice probabilities and sought help from a beneficiary of his patronage, who, luckily for the duke, was Galileo Galilei. The problem was to determine whether 9 or 10 was more probable as a sum of three six-sided dice. It seemed from one way of counting that they were equally likely, since each could be written as a sum in six distinct ways (ordering the terms from greatest to least):

$$9 = 6 + 2 + 1 = 5 + 2 + 2 = 5 + 3 + 1 = 4 + 3 + 2 = 4 + 4 + 1 = 3 + 3 + 3$$
$$10 = 6 + 2 + 2 = 6 + 3 + 1 = 5 + 3 + 2 = 5 + 4 + 1 = 4 + 3 + 3 = 4 + 4 + 2$$

The duke, who was both a compulsive gambler and an extremely observant person, had noticed 10 seemed to come up slightly more often but was at a loss to explain why. Galileo's insight was that counting outcomes according to the numbers they contained was the *wrong way of counting*. Instead, he argued, one should keep track of *which* die produced which number. Imagining the dice to be colored red, white, and green, he showed that, for example, the combination (3, 3, 3) could be found only one way, with all dice turning up 3, while the combination (6, 2, 1) could occur six different ways, depending on the permutation of colors. Combinations such as (5, 2, 2) could likewise show up in any one of three different configurations of dice. Thus, in the final tally there were 25 possible ways to get a 9 and 27 to get a 10 out of the 216 possible rolls, making 10 ever so slightly more probable.

So even a simple dice game could produce some confusion over the proper way to do the counting. This issue continues to confuse students of probability to the present day. Many struggle with problems like this one:

> You and two of your friends are in a group of 10 people. The group is randomly split up into two groups of 5 people each. What is the probability you and both of your friends are in the same group?

Solutions may involve distinctions like whether the groups are "labeled" or "unlabeled" and so on.

The implicit understanding present in the classical definition of probability was that the enumeration of outcomes must be done in such a way that they are *equiprobable*, or equally likely to occur. These are the tiny atoms from which all probabilities are then built. But, of course, this couldn't be an *explicit* requirement because then the definition would be circular: probability depends on probability. Most current authors attempt to dodge this by saying the outcomes must be equally "likely," but what is likelihood if not probability?

For most simple problems, the enumeration of equally likely cases was self-evident because of some apparent symmetry—the six faces of a die all seem roughly the same, for example. But what about lopsided or misshapen dice?

Unless chance was measurable some other way, there would always be some doubt the counting had been done properly. Furthermore, some problems to which we'd like to apply probability, such as forecasting tomorrow's weather or determining the guilt of a murder suspect, seem resistant to this kind of description at all. If we claim a 5 percent chance there will be a major earthquake in the next year, what are the equiprobable outcomes out of which 5 percent involve an earthquake and 95 percent don't? Attempts to imagine some theoretical population of equally likely worlds out of which ours is randomly drawn seem forced, at best. Despite the success of the classical interpretation for simple games of chance, there was an apparent need for a new, more experimentally verifiable approach to defining probabilities.

THE FREQUENTIST INTERPRETATION

Advantages: *Empirically testable.*
Disadvantages: *Doesn't work for rare events, one-time events, or past events.*

It was clear from the Duke of Tuscany's analysis of his dice problem that some connection was assumed between the probabilities of events and their occurrences in repeated observation—that is, their relative *frequencies*. The grounding of probability in observable fact was even already present in Aristotle's definition of the probable as "that which happens generally." Insisting probabilities *numerically* match empirical frequencies seemed to resolve any question about whether the decomposition of outcomes into equiprobable atoms had been done correctly, at least when the frequencies were known or knowable.

Enter Bernoulli's Law of Large Numbers, which he called his "golden theorem." Bernoulli imagined a situation where the number of cases for and the number of cases against some outcome, like the number of possible dice rolls with a given sum, were unknown. He then asked what could be expected to happen over a large number of observations. What he was able to show was that the observed frequencies would necessarily *converge* to the true probability as the number of trials got larger. Consider, for example, a game where a friend is rolling three dice and telling us the total, but we don't know whether they're 6-, 8-, or 12-sided dice or whether they're numbered in the usual way. We want to know the probability of getting a 10. We could try to figure it out by asking to inspect the dice and working out all the possible sums with their associated probabilities for ourselves, or we could just allow our friend to roll the dice a large number of times and tally up how often 10 shows up. According to Bernoulli, if the number of trials is sufficiently large—say a few million or billion—these two procedures will almost certainly give approximately the same answers.

His claim was that by establishing this convergence of frequencies to probabilities, he had provided a means of determining, to within a small error and with some "moral certainty," an unknown probability: "If, however, this is attained and we thus finally obtain moral certainty . . . then we determine the number of cases *a posteriori* [after doing the sample] almost as though it was known to us *a priori* [before doing the sample]."[7]

Bernoulli didn't confine himself to games of chance, though. In the opening of chapter 4 in part IV of *Ars Conjectandi*, he gave an example of determining a person's chance of living 10 more years by tallying up the results of men of the "same age and complexion":

> It should be assumed that each phenomenon can occur and not occur in the same number of cases in which, under similar circumstances, it was previously observed to happen and not to happen. Actually, if, for example, it was formerly noted that, from among the observed three hundred men of the same age and complexion as Titius now is and has, two hundred died after ten years with the others still remaining alive, we may conclude with sufficient confidence that Titius also has twice as many cases for paying his debt to nature during the next ten years than for crossing this border.[8]

Here, "number of cases" meant the factors determining the true probability, as in the numbers of black and white pebbles in an urn, even if those weren't countable in any other way. So a sample of 300 men would give us "sufficient confidence" to conclude the real mortality rate was the same as what we had observed. To Bernoulli, this was self-evidently the correct way to determine probabilities in all manner of different settings, from simple experiments involving urn drawing to real-world problems dealing with mortality, incidence rates of disease, occurrence of weather events, and even guilt or innocence of defendants in trials.

This all may seem obvious now, but before Bernoulli, there was no mathematical guarantee that frequencies had to match probabilities; it just felt intuitively right. He saw his main contribution, the Law of Large Numbers, as justifying this intuition. Without something like Bernoulli's theorem, how could we really know our observed frequencies wouldn't converge to some value *other* than the true single-event probability? For example, what if the real probability of rolling a double-six with two dice is 1/36, but over the long run, the frequency we actually observe stabilizes at 1/35 or 1/37? Bernoulli put this concern to rest.

In the centuries following Bernoulli, as empiricism began to crescendo, some made the leap to claiming this long-run frequency of occurrence was actually what the probability was *by definition*. Bernoulli had opened the door to this way of thinking by including examples like those above, where establishing probability was not possible except by measuring frequency. How else, for example, could you come up with the "number of cases" for and against Titius living another 10 years except by observing how often similar people lived that long? If the frequency is the only available means of calculating the probability, it may as well serve as the definition. As we'll see later, this interpretation especially gained traction after probability took hold in social science; in the mid-19th century, John Stuart Mill, Robert Leslie Ellis, Antoine Augustin Cournot, Jakob Friedrich Fries, and the logician George Boole all described probabilities in terms of frequencies of occurrence.

Perhaps the greatest proponent of the frequentist view was the English logician and philosopher John Venn (of Venn diagram fame). He came from a family with a long tradition of being ordained clergy, and after studying mathematics at Cambridge, he briefly joined the family business as an Anglican priest before returning to Cambridge to teach morality, logic, and probability theory. Reflecting his time in the pulpit, his mathematical writing often reads more like a sermon than a math paper. In his groundbreaking book *The Logic of Chance* (1866), Venn lamented that probability had not received the rigorous treatment he thought it deserved and only philosophers could provide:

> Probability has been very much abandoned to mathematicians, who as mathematicians have generally been unwilling to treat it thoroughly. They have worked out its results, it is true, with wonderful acuteness, and the greatest ingenuity has been shown in solving various problems that arose, and deducing subordinate rules. And this was all that they could in fairness be expected to do. . . . But from this province the real principles of the science have generally been excluded, or so meagrely discussed that they had better been omitted altogether.[9]

He then attempted to formalize the frequency interpretation of probability and iron out some of the practical difficulties of applying this definition that had tripped up previous authors.

For example, taking the frequentist view, we may say the chance of a coin flip coming up heads *is* the relative frequency of heads in a long series of repeated coin tosses. But how long is "long"? In any finite sequence of coin flips, the proportions of heads and tails will always add up to 100 percent, but we could expect some variation around the true "long-term" answer. The whole point of Bernoulli's theorem was to show that the variation probably wouldn't be too much, but it would still be present. We wouldn't say the probability of heads was 50.1 percent just because we saw 501 heads out of 1,000. Venn argued we need to consider the *limit* as the number of trials goes to infinity. That limiting frequency is the precise probability of the event.

However, this, of course, presents a practical impossibility: we can't flip a coin an infinite number of times. Furthermore, even flipping a real coin a *very large* number of times introduces with near certainty the fact that the characteristics of the coin will change over time—it will gradually get dented, scratched up, or bent out of shape. Venn's answer was that one should *imagine* repeating the infinite series of trials with no change to the relevant underlying factors. That is, we should imagine flipping *an imaginary coin*. Yet somehow the results of this are still held to be materially meaningful and, according to Venn, the only acceptable definition of probability.

In fact, Venn went so far down the frequentist rabbit hole he even left Bernoulli's Law of Large Numbers behind. Bernoulli, drawing on the classical interpretation, had described an experiment as having a "true probability" for a single event and had derived the probability of an observed sample falling within some bounds of that true value. However, for Venn and others who were trying to extend the reach of probability beyond the classical interpretation, there was no such thing as the "true probability" in any sense other than the long-run frequency of occurrence. From Venn's point of view, the Law of Large Numbers was tautological; it was trivial to claim the long-run frequency of an event's occurrence would *converge* to the probability if the frequency was taken to be equal to the probability *by definition*. Concerning Bernoulli's theorem, Venn wrote, "With the mathematical proof of this theorem we need not trouble ourselves, as it lies outside the province of this work; but . . . the basis on which the mathematics rest is faulty, owing to there being really nothing which we can with propriety call an objective probability."[10]

The French mathematician Pierre-Simon Laplace (whom we'll be hearing more from later on) had once written that when we look at nature, we begin to detect "a striking regularity which seems to suggest a design, and which some have considered a proof of Providence. But, on reflection, it is soon perceived that this regularity is nothing but the development of the respective probabilities of simple events, which ought to occur more frequently as they are more probable."[11] Venn dismissed this as pure nonsense, calling Laplace's statement "a somewhat pretentious re-statement of the fact already asserted,"[12] since for Venn there was no meaning to the probabilities of events *other* than how often they occurred.

Venn's dogmatic view of probability caught on in the sciences. In particular, it profoundly influenced the methods developed by the statistician Ronald Fisher, who may have been directly exposed to Venn's ideas while an undergraduate at Gonville and Caius College, Cambridge, when Venn was the college president. In his book *Statistical Methods and Scientific Inference* (1956), Fisher laid out what he understood to be the correct definition of probability, and he credited Venn with "developing the concept of probability as an objective fact, verifiable by observations of frequency."[13] We'll be coming back to Fisher and frequentist statistics.

Some technical issues with the frequentist interpretation continued to be an annoyance well into the 20th century. For example, Venn's definition presupposes that the limiting frequency actually exists, which means mathematically it's important to consider exactly what kinds of sequences we are dealing with when we talk about random trials and in what manner this convergence takes place. The requirement simply that an infinite series of coin flips have limiting frequency 1/2 would be satisfied if the coin flips came out to be HTHTHTHTHTHTHT ... with the pattern repeating forever, but that's clearly not the kind of sequence we should expect from a real coin. It's not "random" enough. But how to make randomness (from the old Germanic root *rinnana* meaning "run") precise while assuming the long-run frequency is known turns out to be harder than you might think. Venn said a truly random series with a given limiting frequency should exhibit "an order, gradually emerging out of disorder"[14] but didn't try to express this in any more exact way.

Around 1919, the Austrian mathematician Richard von Mises made a valiant effort. He defined a random sequence, which he called a *Kollektiv*, as having the property that we can't *select* any of its members to produce a subset with a frequency different from the overall limit. So, in the above example, if we picked out every other coin flip, we'd have a sequence of all heads or all tails, and these would have frequency 1 or 0, not 1/2. A truly "mixed-up" sequence should be immune to having the heads or tails results selected according to any pattern or rule.

Unfortunately, in a purely mathematical sense this doesn't quite work because we can just define the selection rule to be "select an element of the sequence if and only if the outcome is heads." Even though it seems like cheating, that's an allowable function according to the set-theoretic axioms of math, so no random sequence meeting such a broad criterion could ever exist. To get around this problem, von Mises tried for a while to *restrict* the kinds of rules we can use to select sequence terms, but in the end, he wasn't able to formalize them in a satisfactory way.

In 1940, the American mathematician Alonzo Church took up the charge, applying some of the then brand-new concepts of theoretical computer science. His idea was that the selection rule should be *computable*, meaning the decision to include a particular sequence term should be the result of some computer program applied to the previous part of the sequence as the "input data." Since there are only so many possible programs, he was able to show this did, in fact, allow for the existence of truly random sequences.

One final wrinkle, though, is that it turns out there are *still* sequences satisfying even this restrictive definition yet without the properties we'd want from true randomness. For example, the French mathematician Jean Ville observed in 1939 that, with respect to any countable collection of selection rules (which Church's computer programs were), there would always exist a sequence of heads and tails results with limiting frequency 1/2 satisfying the selection-rule requirement, but with the weird property that if you *stopped* the sequence at any finite point,[15] the proportion of heads would always be greater than or equal to 1/2. So, over the lifetime of any finite number of coin flips, we'd always have at least as many heads as tails. That, again, just doesn't feel right. For real coin flips, we'd expect the running total number of heads to be sometimes more and sometimes less than the number of tails and the proportions to oscillate above and below 1/2 on the way to stabilizing. But Church's formal definition of randomness couldn't guarantee that would happen, and the problem of replacing it with a more satisfying definition is, in fact, still open.

Putting these theoretical issues aside, though, the frequentist interpretation of probability offers an appealing degree of objectivity, which likely explains its popularity among scientists. Instead of worrying forever about whether we have enumerated all the possible equally likely cases and counted up the ones corresponding to some event of interest, we can rest assured that, *in principle*, we can always verify probabilities (at least approximately) by conducting experiments. This puts probability on the same footing as other measurable quantities of physical systems like mass, density, and specific heat. Even though a single measurement may have some embedded error, over enough repeated observations we can narrow in on the true theoretical value or at least rule out values too far outside our observed range.

Combining noisy observations of a particular quantity—say the position of a planet in the night sky—had been a practical problem facing astronomers and physicists for a long time, and the technique of averaging measurements was known as a good practice since antiquity. So, if averaging noisy measurements of a physical quantity seemed to produce reliable estimates of the true value, then averaging the observed number of occurrences of an event—that is, calculating its observed frequency—seemed a natural way to estimate the true probability of the event each time. Bernoulli's theorem seemed to back this up mathematically.

But the cracks in the frequentist interpretation begin to show when even a little pressure is applied. For example, consider the frequentist answer to the question of the probability of an extremely rare event. If asked for, say, the probability of being dealt all 13 spades in a hand of bridge, any well-trained frequentist would no doubt begin with the calculation $(13/52) \cdot (12/51) \cdot (11/50) \cdot \ldots \cdot (1/40)$, reasoning there are 13 spades out of 52 cards available for the first card dealt, 12 spades out of 51 cards remaining for the second card, and so on. The frequentist would then ultimately arrive (hopefully with the aid of a computer) at the extremely small probability of 1 in 635,013,559,600.

However, there is no empirical evidence for this value as the limiting frequency of the event, nor will there ever be, since this is several orders of magnitude greater than the number of bridge hands ever dealt. Assuming we could deal one complete hand per minute and record the results, to get through one cycle of over 635 billion hands would require 1.2 million years. And, remember, in all that time our expected number of occurrences of the all-spades hand in question would be one! What if we saw two or none? Presumably, to establish the above probability as correct with any accuracy, we would need many *more* observations, taking perhaps billions or trillions of years.

Even that may not satisfy the frequentist. Imagine that after our impossibly long process of shuffling and dealing out trillions of bridge hands was complete, we arrived at a different limiting frequency than the one predicted—say we got something like twice as many perfect all-spades hands as expected. In that case, the frequentist would certainly claim something had gone *wrong* with the shuffling and dealing of the cards, so they were not truly "randomized" or had not been independent of one another! So there's an inherent circularity in the definition of probability in terms of frequency: the probability is the frequency we would get—but only if the conditions of the experiment were managed correctly, which we could discern only by the frequency coming out to be nearly equal to the probability.

Evidently even the adherents of this definition do not hold to it strictly in practice, since even in simple thought experiments like this one, they are able to compute probabilities with absolutely no basis in empirical frequencies, relying on rules of calculation from *somewhere* that are so strong they would trump the frequencies even if they *had* been observed. Where do these principles come from, and how are they justified? We'll return to this point later on when we discuss Edwin Jaynes's view of probability and logic.

Where the frequentist interpretation struggles most, though, is in handling probabilities of one-time events. For example, if we want to say a certain politician has a 29 percent probability of winning the next election, how can this statement even be given a meaning in terms of frequencies? The election will happen only once, and even if we *could* imagine repeating it, à la Venn's imaginary coin flips, what conditions would have to change each time in order to produce different outcomes at all? If the election were held again, wouldn't all the voters vote the same way?

The question of what exactly counts as an independent trial and what characteristics these trials need to share becomes especially complicated when applying the frequency model to social science. It's easy to recognize a legitimate trial of a coin toss when we see one, but what if we wanted to estimate, say, the probability that a given person will live to be 90, as an actuary would need to do for a mortality table? In principle, the frequentist interpretation would tell us to estimate this probability by finding a large collection of similar people—as in Bernoulli's "same age and complexion" requirement—and counting the proportion among them who live past 90. But similar in exactly what ways? The more identifying characteristics we include—male, age 40, nonsmoker, overweight, family history of diabetes, etc.—the smaller the reference class becomes until we're necessarily left with only the one person with whom we started. So are these probabilities only ever 0 or 1?

Too large a reference class would also pose a problem because it might give us accurate estimates of frequency that are practically useless. For example, here is a way to give very accurate weather forecasts for, say, the chance of rain in your local area without any meteorological training or equipment: simply look up the long-run historical frequency of rain for that area, and give this number as your predicted chance of rain *every day*. After a few years or so have gone by, the observed frequency will closely match your prediction, assuming the climate on the whole continues roughly as it was. But from day to day, your predictions won't incorporate any *new* information that would actually be useful for people trying to decide whether they need to bring an umbrella that day.

Venn had actually raised this issue of what should count as a trial for estimating a given probability, which later became known as the *reference class problem*, in *The Logic of Chance*: "Every individual thing or event has an indefinite number of properties or attributes observable in it, and might therefore be considered as belonging to an indefinite number of different classes of things."[16] But he failed to provide a satisfying answer, as has every other frequentist since.

Similarly, we can't apply a frequentist interpretation to events in the past because the frequency model allows only for contingent future outcomes we *would* observe if we conducted a series of experiments. So there is no (nontrivial) frequentist answer to questions like these: What is the probability that the mass extinction at the end of the Cretaceous period was caused by a comet? What is the probability that Shakespeare's plays were written by Francis Bacon? Such questions require a version of probability suitable for *inference*, which the frequentist interpretation simply can't easily handle.

THE SUBJECTIVE INTERPRETATION

Advantages: *Flexible, applies everywhere.*
Disadvantages: *Flexible, applies nowhere.*

During the later years of his life in the 1740s and '50s, Thomas Bayes, an English Presbyterian minister and amateur mathematician who had studied logic and philosophy at the University of Edinburgh, became greatly interested in the subject of probability. One possible explanation for his interest was that, around the same time, the Scottish philosopher David Hume argued in *An Enquiry Concerning Human Understanding* that valid conclusions about the world could not be arrived at through experience. Hume's general point, later referred to as the *problem of induction*, was that we have no way of *knowing* experience is a guide for valid conclusions about the future because if we did, that claim could be based only on past experience. In other words, we only think the future will be like the past because as a rule "The future will be like the past" has always worked in the past. This means the argument from experience is necessarily circular. In section 10, titled "Of Miracles," he directed particular criticism toward the idea of empirical evidence for miracles, arguing one should not rely on supposed eyewitness testimony of their existence, such as the accounts of the Resurrection given in the Gospels.

Bayes may have been particularly motivated for religious reasons to prove Hume wrong by providing an explicit counterexample of valid induction, which he attempted to do by addressing a problem raised by de Moivre some 30 years earlier.

The problem was one of probabilistic inference. To take a slightly simpler version of the example Bayes considered, imagine the following dice game:

> Your friend rolls a six-sided die and secretly records the outcome; this number becomes the target *T*. You then put on a blindfold and roll the same six-sided die over and over. You're unable to see how it lands so, each time, your friend (under the watchful eye of a judge, to prevent any cheating) tells you *only* whether the number you just rolled was greater than, equal to, or less than *T*. After some number of rolls, say 10, you must guess what the target was. What would be your strategy for guessing, and how confident would you be?

Probability textbooks like de Moivre's could give some of the probabilities needed for the answer but couldn't solve the whole thing. For example, suppose in one round of the game we had this sequence of outcomes, with G representing a greater roll, L a lesser roll, and E an equal roll:

$$G, G, L, E, L, L, L, E, G, L$$

If the target number is 3, for instance, then we could say the probability of any given roll being greater than *T* is 3/6, or 1/2 (corresponding to rolls of 4, 5, and 6), and the probability of being less than *T* is 2/6, or 1/3 (corresponding to rolls of 1 and 2). The remaining probability of rolling an equal number is always 1/6. Applying the rule of multiplying probabilities for independent trials, we could say the probability of getting that particular sequence is

$$\left(\frac{1}{2}\right)\left(\frac{1}{2}\right)\left(\frac{1}{3}\right)\left(\frac{1}{6}\right)\left(\frac{1}{3}\right)\left(\frac{1}{3}\right)\left(\frac{1}{3}\right)\left(\frac{1}{6}\right)\left(\frac{1}{2}\right)\left(\frac{1}{3}\right), \text{ or } \frac{1}{69{,}984}$$

But all that assumes the target value is 3 to begin with. Similar calculations could apply to other choices of *T*, but then what? The probability rules of Bernoulli, de Moivre, and company couldn't tell us how to assemble these numbers into an answer to the *inverse* question: What is the probability that *T* = 3 given the observation?

To accomplish this, Bayes needed to prove an identity involving the *conditional probabilities* for any two events *A* and *B*, an equation we now call *Bayes' theorem*:

$$P[A \mid B] = P[A]\frac{P[B \mid A]}{P[B]}$$

The probabilities with the vertical bars refer to conditional probabilities—that is, *the probability of event* A *assuming event* B *has happened*, and vice versa. So A could represent a hypothesis, like our guess about the target number, and B could represent our observation, the sequence of greater, lesser, and equal results we're given. With elegant simplicity, the rule tells us how the probability of A given B, the quantity we want, is related to that of B given A, the probability we already calculated. Thus, assuming we have the other necessary ingredients, we can use the equation to switch from one logical direction to the other. Amazingly, this little equation—and what it implies about the nature of scientific inference—has been the source of an immense amount of controversy spanning four centuries now. The remainder of this book will consist mostly of exploring that controversy, but it all starts here.

The identity was likely known before Bayes and becomes apparent if we think of probabilities as relative proportions. Suppose, for example, we're interested in two characteristics among a group of individuals—say whether someone is a clown (A) and whether they wear big shoes (B). Maybe we are given that 20 percent of the people in this group are clowns (P[A]), 25 percent of the people wear big shoes (P[B]), and 80 percent of the clowns wear big shoes (P[B|A]). Say we see someone wearing big shoes, and we'd like to know the chance that they're a clown (P[A|B])—that is, we want the proportion of B who are also A. The natural way to compute the ratio is to find out how many big-shoed clowns there are (A and B) and then divide by the total number of people with big shoes. Imagine the total population consists of 100 people. Then we are assuming 20 are clowns, of whom 80 percent—that is, 16—are big-shoed clowns. As a proportion of the 25 people with big shoes, this intersection group makes up 16/25, or 64 percent, so this is the *conditional* proportion of those with big shoes who are clowns. Bayes' theorem just shows these quantities relate the way we have reasoned here.

The main innovation of Bayes's solution, though, presented and clarified somewhat after his death by his friend and fellow minister Richard Price, was that this conditional probability idea would apply to *events, no matter the order in which they occur in time.* So, if we were interested in events E_1 and E_2, with E_1 necessarily having the chance to occur first, we could reason just as well about $P[E_1|E_2]$ as we could about $P[E_2|E_1]$, even though the latter seems more natural. This gave Bayes a tool to use for inverse problems, relating the "forward-looking" probabilities he could already compute to the "backward-looking" ones he wanted to compute.

In our example problem involving the six-sided die, we would say *condition-ally* on the observed data D recording our sequence of greater, lesser, and equal outcomes, the probability of the target being 3 could be found by

$$P[T = 3 \mid D] = P[T = 3] \frac{P[D \mid T = 3]}{P[D]}$$

We already handled the term $P[D \mid T = 3]$ when we multiplied the probability fractions together and got 1/69,984. This is, as it usually will be, the easiest part, since it's the most "ordinary" probability calculation in the bunch; it represents the probability of getting a certain sequence of outcomes assuming a given setup. To finish the calculation, we have two missing pieces: $P[T = 3]$ and $P[D]$, unconditional on any assumed target value. The first term, $P[T = 3]$, represents our initial probability for the proposition $T = 3$ before we make any observation. In this case, since the target was the result of the initial roll of the die, we would say $P[T = 3] = 1/6$. Similarly, we would give the same prior probability to $T = 1$, $T = 2$, etc.

The denominator term requires the most attention, but Bayes recommended a general strategy: break the probability down into the possible cases and sum them. That is, we know the target number must be one of the numbers $1, \ldots, 6$. For each possibility, we can carry out calculations similar to the previous one to get the conditional probabilities of the data assuming $T = 1$, $T = 2$, up to $T = 6$. The rules of (ordinary, forward-looking) probability tell us the total probability for D is then the sum

$$P[D] = P[T = 1] \cdot P[D \mid T = 1] + \ldots + P[T = 6] \cdot P[D \mid T = 6]$$

That is, T could equal 1 (with the given probability) and we'd observe D with the conditional probability $P[D \mid T = 1]$, or T could be 2 and we'd observe D with conditional probability $P[D \mid T = 2]$, etc. Since these are mutually exclusive cases forming "pathways" to the data D, we add them up to get the total probability of observing D.

Our prior probabilities for the various target values were uniform with probability 1/6 for each, as shown in figure 1.1.

But after applying Bayes's process for the various target values for the given observation D, we find the *posterior* probability assignments $P[T = t \mid D]$ are concentrated around $T = 4$, as shown in figure 1.2.

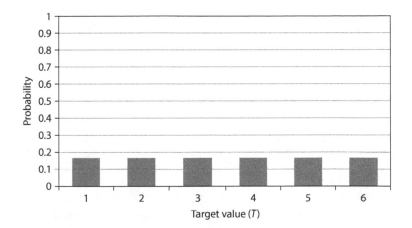

FIGURE 1.1

Prior probabilities for the target guessing game.

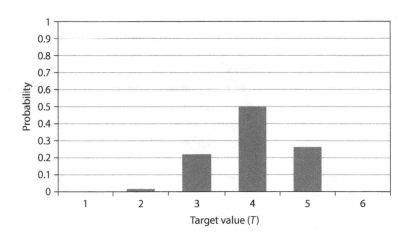

FIGURE 1.2

Posterior probabilities for the target guessing game using Bayes' theorem.

In fact, we are 50 percent confident $T = 4$, with some possibility remaining that $T = 3$ or $T = 5$ and a very slim chance $T = 2$. Note we have ruled out $T = 1$ and $T = 6$ completely. We could have accomplished this by deductive reasoning, since any observation of a "greater" roll rules out the target being 6 and any "lesser" roll rules out the target being 1 and we observed both. However, there's no need for us to make a special case out of these, since Bayes' theorem works perfectly well when the probability is 0, meaning the data is impossible given a particular hypothesis. (We'll return to this point when we connect Bayes' theorem with deductive logic.)

This same process will apply to any problem of inference among multiple hypotheses. In fact, this will be the only procedure we'll ever need to use to do probabilistic inference—for the rest of this book and for the rest of our lives. To help keep the pieces organized, we'll arrange them in a table, which I'll refer to from now on as an *inference table*. The following are the steps for any problem, and the corresponding inference table is shown in table 1.1.

1. Enumerate all the possible hypotheses, $H_1, \ldots, H_n$ and consider their probabilities not including any observation, $P[H_1], \ldots, P[H_n]$. These are the *prior probabilities*, or *priors* for short.
2. For a given observation of data, D, compute the probability of that observation assuming each hypothesis is true, in turn. These are the *sampling probabilities* for the data given the hypotheses.
3. Compute the probability of arriving at the observation D by means of any one of the hypotheses by multiplying the prior by the sampling probability: for example, $P[H_1] \cdot P[D|H_1]$ and so on. We'll call these the *pathway probabilities*. Summing them gives the total probability of the data:

$$P[D] = P[H_1] \cdot P[D \mid H_1] + \ldots + P[H_n] \cdot P[D \mid H_n]$$

TABLE 1.1 **All-purpose inference table**

Hypothesis	Prior probability	Sampling probability	Pathway probability
H_1	$P[H_1]$	$P[D|H_1]$	$P[H_1] \cdot P[D|H_1]$
H_2	$P[H_2]$	$P[D|H_2]$	$P[H_2] \cdot P[D|H_2]$
. . .	. . .	. . .	. . .
H_n	$P[H_n]$	$P[D|H_n]$	$P[H_n] \cdot P[D|H_n]$

4. Once this calculation is accomplished, the inferential probability for each hypothesis is easy to find, since it is just the *relative proportion* of that term in the preceding sum. These are the *posterior probabilities*, which, according to Bayes' theorem, are given by

$$P[H_i \mid D] = P[H_i] \, \frac{P[D \mid H_i]}{P[D]}$$

That is, the posterior probability for a hypothesis given data is the proportion of the total probability for the data represented by the pathway through that hypothesis. What this also means, practically, is that once we've computed the priors and the sampling probabilities, all the real work is done. The rest is just accounting.

With the aid of his theorem, Bayes was able to give a complete answer to the question he considered. A consequence of this inference was that future events could, under some assumptions, be logically predicted to have roughly the same chance of occurring as they had been seen to occur in the past. This showed, somewhat contrary to Hume's claims, that, through rational application of probabilities, one *could* arrive at knowledge from experience or at least something very *close* to knowledge—not absolute certainty as in a logical deduction, perhaps, but probable judgment in the sense of Cicero and the ancient Greeks. It wasn't a complete answer to the problem of induction, but it was close.

Whether Bayes himself believed he had disproved Hume we have no way of knowing. Some historians such as Stephen Stigler at the University of Chicago have suggested that since Bayes did not find the counterexample sufficiently convincing because it relied on some assumptions he could not justify, he delayed publishing his results. When presenting Bayes's results to the world, Price did not shy away from emphasizing their philosophical and religious significance. Contemporary reprints of the essay show Price intended the title to be "A Method of Calculating the Exact Probability of All Conclusions founded on Induction."[17] In his publication, he added this preamble: "The purpose I mean is, to shew what reason we have for believing that there are in the constitution of things fixt laws according to which things happen, and that, therefore, the frame of the world must be the effect of the wisdom and power of an intelligent cause; and thus to confirm the argument taken from final causes for the existence of the Deity."[18] That is, somewhere in the calculation of probabilities for Bayes's rule, Price thought he saw evidence for God.

Somewhat later, Pierre-Simon Laplace, apparently unaware of Bayes's solution, published his own version of the theorem and showed its application to problems of inference. Laplace was, in our current sense, more Bayesian than Bayes himself had been, since Laplace was ready to divorce probabilities from frequencies or empirical fact altogether. The big question concerned the prior probabilities needed to do Bayesian inference: Where did these come from?

Bayes had confined himself to a problem where the prior probabilities are somewhat obviously uniform, like our target value probabilities in the earlier example. He had still broken new philosophical ground by showing probabilistic inference could go backward in time, but his treatment of probability suggested he considered these probabilities to exist independently of the person doing the analysis. Laplace allowed probabilities to depend on the observer's state of knowledge. He opened his treatise on the subject, *Théorie analytique des probabilités* (Analytical theory of probabilities), published in 1812, with this definition: "The probability for an event is the ratio of the number of cases favorable to it, to the number of all cases possible when nothing leads us to expect that any one of these cases should occur more than any other, which renders them, for us, equally possible."[19] The "for us" made Laplace's probabilities *subjective*. This freed him up to compute inferential probabilities for all manner of propositions based on all manner of observations. As he described it, "Probability relates partly to our ignorance, partly to our knowledge."[20]

Laplace included, perhaps tongue-in-cheek, an example computing the probability the sun would rise tomorrow given it had risen every day in recorded history. With a general mathematical tool he called the *rule of succession*, he arrived at the probability $(d + 1)/(d + 2)$, where d is the number of days of previous observation. Assuming, say, 5,000 years of recorded observations would give something like $d = 1,825,500$, for a probability of 0.99999945205.

The point was not so much the exact value as the fact it was less than 1. Anything less than total certainty the sun would rise was seen as a ridiculous conclusion, and Laplace was roundly criticized for this, especially by frequentists like Venn. In *The Logic of Chance*, Venn wrote of this example, "It is hard to take such a rule as this seriously"[21] because, among other reasons, the probability was manifestly *wrong* in the frequentist meaning, since it would not match the long-run frequency of occurrence.

However, his critics failed to properly understand that Laplace *also* rejected this conclusion as a misapplication of his rule! He had derived the rule of succession for inference problems about systems like Bernoulli's urn, where some unknown proportion of colored balls determines the single-event probability of a trial. Laplace included as an assumption that we begin from *complete ignorance* about the proportion, so we assign it a uniform probability of any value between 0 and 1 and assume that the resulting days, if given, are like independent "draws" from the urn. However, in real life we know the sun is *not* such a system, and we have a great deal more understanding of the sun than just whether it has risen each day. In Laplace's own words: "But this number [the probability of the sun coming up tomorrow] is far greater for him who, seeing in the totality of phenomena the principle regulating the days and seasons, realizes that nothing at present moment can arrest the course of it."[22] Laplace was not giving some objective probability the sun would rise but merely the subjective probability one would *assign*, consistent with Bayesian updating, starting from a position of total ignorance of what the sun even is.

Bayes loosened the lid on the idea of subjective probability, and Laplace removed it completely. Over the course of the next two centuries, many prominent philosophers and mathematicians took up the idea that probability *meant* the speaker's degree of belief in a given proposition. John Maynard Keynes took this position in his *Treatise on Probability* (1921),[23] and Frank Ramsey defined probability in personalist terms in his essay "Truth and Probability" (1926).[24] The British physicist Harold Jeffreys played perhaps the greatest role in preserving the Bayesian view of probability in science into the mid-20th century, applying Bayesian techniques with great success in the domain of geophysics and clarifying many of the practical considerations in his textbook *Theory of Probability* (1939). As Jeffreys wrote: "The essence of the present theory is that no probability, direct, prior, or posterior, is simply a frequency."[25]

Philosophically speaking, there were a few major problems to overcome, though. One was that Laplace's definition of probability as a ratio of cases—and all the mathematics of probabilities that followed as a result—seemed unsupported. If we think of probabilities as ratios of equiprobable events, as in the classical definition, or as limiting frequencies, as in the frequentist definition, then these properties are self-evident. But if we are willing to scrap all of that and say probability is just a "degree of belief," measured on a scale from 0 percent to 100 percent, why should these axioms hold at all? If my degree of belief in the proposition "It will rain tomorrow" is 30 percent, why should that mean my degree of belief in "It will not rain tomorrow" is 70 percent? And if the basic property $P[A] = 1 - P[\text{not } A]$ did not have to hold, what hope was there for Bayes' theorem or any other familiar mathematical result?

The Italian statistician and actuary Bruno de Finetti, working in the 1920s and '30s, attempted an answer by defining probability in terms of betting prices. Instead of a vague degree of confidence, he assumed the probability of an event could be made "operational" as the price the speaker would agree to pay for a bet of one unit on whether the event occurred. So a probability assignment of 30 percent for it raining tomorrow was equivalent to an agreed-upon price of

$0.30 for a $1 bet on whether it would rain, and so on. In a sense, this stayed faithful to the origins of probability in Pascal and Fermat's analysis of gambling games; Bayes, too, had phrased all his probabilities in terms of "expectations," meaning fair values of bets.

This concrete idea of probabilities then allowed de Finetti to impose additional structure that would constrain the probabilities a rational agent would give. The key idea was a *Dutch book*, a portfolio of bets that would earn a sure profit (in modern language, we would call this an *arbitrage strategy*). He showed that unless a person's probability assignments—that is, betting prices—followed the rules of probability, a Dutch book was always possible. For example, suppose my probability of it raining tomorrow were 30 percent, meaning I'd pay $0.30 for a chance to win $1 if it rained, but my probability of it not raining were 80 percent, so I'd pay $0.80 to win $1 if it didn't rain. Someone could make a Dutch book against me by selling me *both* bets for a total price of $1.10 with the certain profit of $0.10, since they would have to pay out only $1 in either case.

Using similar assumptions about the coherence of bets placed over multiple times, de Finetti was even able to show Bayes' theorem for conditional probabilities had to hold. So the whole of mathematical probability was intact if probabilities were thought of as betting prices of a rationally coherent agent, meaning someone who was able to avoid price combinations that would make them certain losers.

However, another problem remained, which was that de Finetti's betting prices could still be totally arbitrary as long as they held together as a system of probabilities. That is, he could not tell anyone what price they *should* pay for a given bet. This obviously ran afoul of those instances of probability that seemingly *could* be verified empirically. Nothing would constrain a person, under the subjective definition, from believing they had a 1-in-30 chance of winning at roulette, instead of the conventional answer of 1-in-38, and proceeding to bet endlessly against unfavorable odds in a casino. Because they would *probably* lose but not *certainly*, de Finetti's assumptions didn't prevent this kind of thing from happening.

Similarly, a weather forecaster might predict a 99 percent chance of rain every day for a year and, after seeing only maybe 15 percent of those days turn out to be rainy, shrug off any criticism as a matter of different degrees of belief. Degrees of belief could be as personal as movie opinions and restaurant reviews. How could a subjective degree of belief ever be wrong?

Laplace had attempted to pin down probability assignments as the ratio of cases when there was no reason to prefer one case over another, which came to be known as the *principle of insufficient reason*. Keynes rebranded it as the *principle of indifference* and emphasized that it applies only when *we have no information* suggesting unequal probabilities. But here, again, the objective versus subjective problem made itself known. Were the probabilities actually equal in fact, or do we just treat them as equal because we don't know any better? If the latter, then certainly there were examples where we'd compute the wrong answers through ignorance, contradicting empirical experience. Wasn't this a concern?

Also, an annoying technical problem came up when trying to apply the principle to continuous parameters (as even Bayes and Laplace had considered). The problem was that there seemed to be no consistent probability distribution for continuous values representing total ignorance. Suppose an unknown quantity had a continuous value—for example, the length of the side of a square sheet of paper known to be between 0 and 10 inches—but that's all the information we're given. We might say ignorance would force us to assume it is uniformly distributed, meaning it's just as likely to be in any range of a given size within the allowable limits. This would mean in particular that it had equal probabilities of being smaller or larger than 5 inches. But the same ignorance about the side length could equally well apply to the area of the square (or any other derived quantity), so total ignorance would imply the area, known to be between 0 and 100 square inches, should *also* have a 1/2 chance of being less than 50 square inches. However, since area and side length are related by the equation $A = S^2$, these two conclusions are *inconsistent*; the squares with side lengths less than 5 inches have areas less than 25 square inches, not 50. Clarification of how the principle of indifference would defend against these so-called Bertrand paradoxes—after the French mathematician Joseph Bertrand, who first introduced a similar problem in 1889—would take a long time to develop (and in some cases is still ongoing). In the meantime, these problems caused logicians like Boole and Venn to reject the principle altogether in favor of objective frequencies.

THE AXIOMATIC ANSWER

Advantages: *Mathematically rigorous.*
Disadvantages: *Doesn't answer the question.*

In 1900, the German mathematician David Hilbert gave an address to the International Congress of Mathematicians about what he considered to be the most important outstanding problems at the time. Known ever since then as the *Hilbert problems*, the 23 challenges he issued (10 during the presentation and the rest in print) set the course of mathematics for the 20th century. They attracted the attention of the greatest mathematicians of the century, and those who contributed significantly to solving any of the problems reached an even higher echelon of greatness, becoming what mathematician Hermann Weyl called "the honors class."[26]

The second of these 23 problems was to prove that the axioms of arithmetic are consistent, and the sixth was to establish an axiomatic treatment of probability, as it was required for statistical physics. Both desires were in keeping with a great push in the mathematical community around that time to examine the foundations of mathematics and shore up parts of various disciplines long taken for granted. Most notably, Bertrand Russell and Alfred North Whitehead published their famous *Principia Mathematica* in three volumes over the period 1910–1913. Contained within was an attempt to formalize the rules of symbolic logic from which all mathematical truths could be derived.

It was against this backdrop that, in 1933, Russian mathematician Andreï Kolmogorov published his *Foundations of the Theory of Probability*,[27] which was the first modern axiomatic treatment probability had received. His key idea was that probability could fit within the somewhat new mathematical domain called *measure theory*, developed over the late 19th and early 20th centuries by Émile Borel, Henri Lebesgue, and others. A *measure* in its most abstract form is a way of associating nonnegative numbers to subsets of a given master set in a way that satisfies certain constraints, such as the measure of the union of disjoint sets being the sum of the measures of each set. In Kolmogorov's translation, the master set, or *sample space*, would represent all the possible outcomes of some random situation or process, like the 216 possible outcomes of rolling three dice, and the subsets would be the possible *events*, like the dice summing to 9 or 10. The measure of a set would be the probability of that event occurring, standardized so that the measure of the whole space was 1.

Kolmogorov's axioms cleaned up a number of issues of theoretical probability, particularly questions having to do with infinite sets or infinite collections of sets. Using the powerful measure theory, Kolmogorov was able to prove several important theorems involving limits of sequences of random variables and other things that had vexed probability theorists for years, including a stronger version of Bernoulli's Law of Large Numbers.

However, the axiomatic system Kolmogorov developed, and all the rich, beautiful theory it entailed, did not answer the questions of where probabilities came from in the first place or how to understand their meaning in the real world. Any example of a probability measure space would assume the basic probabilities had been given from the start. The axiomatic system could apply equally well to objective and subjective probabilities (at least granting something like de Finetti's ideas about coherent betting prices). It was clear from his writings that Kolmogorov thought of probabilities as frequencies, which made these axioms seem more natural, but there was nothing mathematically to disallow other interpretations.

PROBABILITY AS LOGIC: THE SYNTHESIS

By the middle of the 20th century, philosophers, mathematicians, statisticians, and working scientists who used probability had rallied around Kolmogorov's axiomatization of the subject as a mathematical system, but there was still no general agreement on the interpretation of probability in any real problem. Instead, there seemed to be a patchwork of overlapping and somewhat related concepts, straddling the line between objective and subjective, between ontology (the study of things *as they are*) and epistemology (the study of our *knowledge* of things). Sometimes probability meant empirical frequency, as most statistical practice would have it (more on this later), and sometimes it meant subjective degree of belief, as Laplace and his Bayesian descendants would use it in problems of inference. Frequency interpretations seemed to be more suitable for games of chance or repeated randomized experiments such as sampling randomly from a population. Degrees of belief seemed more applicable to forecasting one-time events, testing hypotheses, and making inferences about the past. Most probability textbooks tried to remain as neutral as possible on the subject and move swiftly to proving results about probability measure spaces and random variables. No single definition of probability seemed adequate to cover all these cases.

Indeed, for many people this is how the story ends, with probability simply meaning different things at different times or in different applications. Any way of assigning numbers to subsets of some sample space in a way that satisfied Kolmogorov's axioms could be called a *probability*, and all the mathematical properties one could want (and then some) would come for free. So what harm is there in sometimes applying the idea to proportions of a set of equiprobable outcomes and sometimes applying it to betting prices assuming coherence and sometimes applying it to limiting frequencies of a series of experiments?

Well, one major way this causes harm, as we'll see abundantly, is through the confusion it causes when the concepts are mixed together, particularly for scientific inference. Our assessment of the probable truth of a hypothesis is clearly subjective, but it's constructed using sampling frequencies for the data, which are objective. When subjective combines with objective, which one trumps the other?

However, there were some in the late 20th century who continued to seek a synthesis, a way of interpreting frequency-based probability and belief-based probability as different instances of the same underlying phenomenon. The closest anyone had come thus far had been in 1921, when Keynes published his logical theory of probability, including the principle of indifference. So logic was where these new authors began. But a new form of logic was clearly needed because classical logic dealt only with certain deductions, not probabilities. As the physicist James Clerk Maxwell had described in 1850, probabilistic thinking was more natural in some ways than classical logic anyway because no proposition in the real world is ever truly certain: "The actual science of logic is conversant at present only with things either certain, impossible, or entirely doubtful, none of which (fortunately) we have to reason on. Therefore the true logic for this world is the calculus of Probabilities, which takes account of the magnitude of the probability which is, or ought to be, in a reasonable man's mind."[28]

Keynes had attempted to define probability as a "degree of rational belief,"[29] and some practitioners continued to argue for this definition and to make more precise the source of that rational belief. For example, Harold Jeffreys defined probability as the support one proposition gives to the truth of another: "We introduce the idea of a relation between one proposition p and another proposition q, expressing the degree of knowledge concerning p provided by q."[30] But still there was the question of why the mathematical laws of probability should hold. As Ronald Fisher wrote in a 1934 article, part of a series of back-and-forth arguments with Jeffreys:

> Keynes establishes the laws of addition and multiplication of probabilities, by stating these laws in the form of definitions of the processes of addition and multiplication. The important step of showing that, when these probabilities have numerical values, "addition" and "multiplication," as so defined, are equivalent to the arithmetical processes ordinarily known by these names, is omitted. The omission is an interesting one, since it shows the difficulty of establishing the laws of mathematical probability, without basing the notion of probability on the concept of frequency, for which these laws are really true, and from which they were originally derived.[31]

For example, the only reason the probabilities of the event A and the event (not A) add up to 100 percent is that *proportions* behave that way.

This was the real problem at the heart of the feud between the different interpretations of probability: *there was a strong case to be made that a frequency and proportion understanding of probability was inadequate for all situations, but the agreed-upon axiomatic system was based on these elementary frequency and proportion concepts.* If probability was not somehow defined in reference to proportions, why should the rigid axioms of Kolmogorov's mathematical probability, and all the theory that depended on them, be expected to hold true? What would stop us from assigning whatever degree of belief we wanted to any proposition?

De Finetti's Dutch book argument had shown that the rational prices we would agree to pay to bet on propositions would need to obey the rules of probability, but that interpretation left two major holes. For one thing, it couldn't say if a probability was right, and without resorting to long-run frequencies, we have no other basis for coming up with these numbers. Second, and somewhat relatedly, the *probability* we associate to an event might not match the *price* we'd pay to bet on it. People are willing to play gambling games at prices significantly higher than their pure expected value based on the probability of winning; this is what keeps casinos and state lotteries in business. That doesn't mean we should change our probability assignment, just that it must come from somewhere else.

The biggest breakthrough toward uniting the two interpretations came in 1979, when the American physicist Richard Cox proved the mathematical result now called *Cox's theorem*. The main idea was this: start over from scratch, and think of probabilities not as relative measures of some subsets of a sample space, as Kolmogorov had done, but rather as *degrees of plausibility given to propositions*, more in line with the subjective interpretation. However, to keep these numbers internally consistent, Cox added the requirement that they respect the *propositional calculus* of ordinary logic and obey other principles in agreement with common sense.

Cox loosened the rigid requirement that the probability of proposition A and the probability of proposition (not A) sum to 1, saying only that if the plausibility of A was given, the plausibility of (not A) should be *determined*. So the plausibility of A could be 40 percent, say, while the plausibility of (not A) could be 80 percent—but only if anytime *another* proposition B had a plausibility 40 percent, we'd also have a plausibility of (not B) equal to 80 percent. There was just some fixed rule that would get you from the 40 percent to the 80 percent.

Furthermore, Cox required these plausibilities to behave in such a way that if *A* became more plausible, (not *A*) would become less so. This seemed like commonsense behavior for the plausibility of propositions and their negations—having greater confidence in *A* is the same as having less confidence in (not *A*)—and importantly this was not based on any idea of probabilities as proportions. Similarly he assumed that if the plausibility of *A* and the plausibility of *B* given *A* was known, then the plausibility of (*A* and *B*) would be determined, and if either ingredient became more plausible, the combination would likewise become more plausible.

From these and other similar assumptions about how plausibilities of propositions should behave (plus other technical conditions on the mathematical functions involved), Cox was able to show there always exists a way of *rescaling* his plausibilities such that the rescaled values satisfy the familiar axioms of probability. In other words, probability became a *convenient system of units* to describe plausibilities with nice mathematical properties, like using degrees Kelvin to describe temperature instead of degrees Fahrenheit. This gave an entirely new justification for Kolmogorov's axioms that had nothing whatsoever to do with frequencies or relative proportions and that was also free of the artificiality of de Finetti's coherent betting prices.

The most complete synthesis was achieved by Edwin Jaynes, another American physicist, beginning with a series of papers in the 1980s and culminating in his survey *Probability Theory: The Logic of Science*,[32] published posthumously in 2003. He was for many years a professor of physics at Washington University in St. Louis and had already made a career studying statistical mechanics and particle physics separately from his interest in theoretical probability. But his practical experience as a scientist—particularly in these domains, where probability had made great inroads—led him to believe there was something missing in the way most practitioners thought about probability.

In Jaynes's view, probability was entirely about *information*—specifically, the degree of certainty a rational person should have given incomplete information about an event or process. Anyone who attempted to make it "objective" by defining probability in terms of empirical frequencies was committing what he called the "mind projection fallacy"[33]—that is, confusing the speaker's mental state with the objective facts of the outside world. He rejected entirely the idea of verifying probabilities empirically; he said this would be "like trying to verify a boy's love for his dog by performing experiments on the dog."[34]

Instead, having been motivated by Cox's result and drawing further inspiration from George Pólya's writings about the way mathematicians reason in practice, Jaynes defined probabilities as plausibilities satisfying certain commonsense properties similar to Cox's, rescaled (per Cox's theorem) so that the familiar equations of probability would hold true. Jaynes emphasized the role of the background information these probabilities depended on by always including a "conditional on X" in the notation, where X captured the assumptions of that problem. So, for example, Jaynes's version of the rules of probability included two main calculation rules, the sum rule:

$$P[A \mid X] + P[\text{not } A \mid X] = 1$$

and the product rule:

$$P[A \text{ and } B \mid X] = P[A \mid X] \cdot P[B \mid A \text{ and } X]$$

Depending on your background, these equations may be anywhere from trivial to mystifying, but really the ideas are simple and familiar. The first rule expresses the basic fact that the probability of a proposition and its negation must sum to 1. That concept is obvious for proportional measures of probability, but here it is the result of rescaling plausibilities *so that* these equations hold true. Once you have "unlearned" the idea that probability means proportion, it takes a substantial amount of work to get back to this "obvious" equation.

We already encountered the second equation, without the "conditional on X" decoration, when we discussed Bayes' theorem. The proposition (A and B) could refer, say, to a particular person having two characteristics simultaneously; the ingredients on the right-hand side are then the probability of having one characteristic and the conditional probability of having the second one assuming the first one. The notation expressing all these statements as "conditional on X" is Jaynes's innovation, but it captures a simple yet powerful idea: that all three probabilities are dependent on some underlying state of knowledge making us uncertain about this particular person in some way.

But Jaynes did not just allow for arbitrary assignments of probabilities, in the loosest subjectivist sense. Instead, he required probabilities to satisfy constraints that depended on the information assumed in any given problem. He did this by considering the possible *transformations* of the background information and the implications these might have for probability assignments. If, for instance, a problem could be transformed into another problem in such a way that the person doing the probability assignment was in the *same state of knowledge*, the probabilities they assigned must be the same.

A simple example would be drawing a ball from an urn containing two balls—say one black and one white. Letting A be the proposition "We draw the black ball" and B be the proposition "We draw the white ball," we might imagine a state of information that consists *only* of proposition X: "Exactly one of A or B is true." However, if we *interchange* the labels A and B, proposition X would be transformed into X' = "Exactly one of B or A is true." Since this is logically an *equivalent* statement to our original X, we are in the same state of information. In summary, switching the proposition we called A with the one we called B had no effect on our state of knowledge. According to Jaynes, this means that, conditionally only on X, propositions A and B must have the same probability. And since they're mutually exclusive and exhaust all the possibilities in this problem, their probabilities must also sum to 1, meaning both A and B have probability 1/2 given X. That is, in Jaynes's notation, $P[A \mid X] = P[B \mid X]$ = 1/2. Different states of information might give rise to different probabilities. For example, we could have had background knowledge that the black ball was on top and we would draw from the top; this would result in A being certain, while B is impossible.

Jaynes even imagined this kind of reasoning being programmed into a robot, so probabilities would become nothing more than a dispassionate numerical *processing* of the available information loaded up as assumptions of a given problem. This mirrored similar initiatives in the world of computer science to represent logical reasoning, meaning deductive proof, as Boolean algebra, the arithmetic of ones and zeros native to digital computers.

In fact, Jaynes's version of probability as logic showed that probability calculations are simply extensions of logical deduction from true-or-false arithmetic to computation with other numbers. For example, a basic step of deductive logic called the *rule of contraposition* is that the implication "A implies B" is logically equivalent to the implication "(not B) implies (not A)." If we know that whenever it rains (A), Mr. Smith always takes the bus to work (B) and we see one day that he did not take the bus (not B), then we are logically justified in concluding it must not be raining (not A).

We can describe this deduction as a special case of Bayes' theorem. Thinking of the proposition (not A) as our hypothesis and the proposition (not B) as our data, according to the usual procedure, to see what consequence the data has for our hypothesis, we need to consider all the pathways to the data to calculate the posterior probability for the hypothesis given the data. But the only pathway other than the one through (not A) goes through the proposition A, which we know gives a sampling probability for (not B) equal to 0, since $P[B \mid A]$ is 1. Therefore, when we express the pathways as relative proportions, the probability of (not A) given (not B) is 1. In other words, given (not B), (not A) is certain, the rule of contraposition.

Other basic deductive rules like the *rule of modus ponens*—if we are given the premise "*P* implies *Q*" and that *P* is true, we conclude *Q* is true—follow by similar reasoning. In this way, logical deduction is just a special case of reasoning with probabilities, in which all the probability values are zeros or ones.

THE DIFFERENCE BETWEEN VALIDITY AND TRUTH

Jaynes's criticism of the frequentists was extremely harsh. First, he not so politely asked them to stop using the word *probability* if they were just going to mean *frequency* anyway because this was nowhere near faithful to all the ways probability had been used in making judgments (as in Cicero's "guide to a wise person's life") since before mathematical probability even existed. He suggested their misuse was only confusing everyone:

> It seems to us that, if Mr A wishes to study properties of frequencies in random experiments and publish the results for all to see and teach them to the next generation, he has every right to do so, and we wish him every success. But in turn Mr B has an equal right to study problems of logical inference that have no necessary connection with frequencies or random experiments, and to publish his conclusions and teach them. The world has ample room for both. Then why should there be such unending conflict, unresolved after over a century of bitter debate? Why cannot both coexist in peace? What we have never been able to comprehend is this: If Mr A wants to talk about frequencies, then why can't he just use the word "frequency"? Why does he insist on appropriating the word "probability" and using it in a sense that flies in the face of both historical precedent and the common colloquial meaning of that word? By this practice he guarantees that his meaning will be misunderstood by almost every reader who does not belong to his inner circle clique.[35]

But Jaynes would certainly admit some repeated experiments like flipping coins or rolling dice *did* result in stable frequencies. His point was that these frequencies actually *are* properties of those physical systems (including, most importantly, the variability in the initial conditions—the position, velocity, angular momentum, etc. of all the objects in play) that exist independently of the person doing the probability calculation. The probability could not be measured, but the frequency could be.

Why should probability and frequency ever give the same number? Because, Jaynes argued, we assign probabilities for these physical systems based on a lot of detailed knowledge of them, including their previously observed frequencies! That is, for an example such as flipping a coin, we know a lot about coins, including that they are (roughly) physically symmetric, and we know a lot about what it means to flip them, including that the initial conditions when thumb is applied to coin are somewhat uncontrollable and that the outcome depends very sensitively on what happens in that instant. Therefore, we correctly process our *ignorance* of the initial conditions of a particular coin flip, along with our other knowledge of the symmetry of the coin, into a 50/50 probability for that flip coming up heads.

This also gave Jaynes a way to address examples of extremely rare events, like our all-spades bridge hand. We know, from practical experience handling cards, that they are roughly identical apart from the number and suit printed on each one, and we know that those differences don't affect how any particular card moves around when shuffled in a deck. Furthermore, we know that tiny differences in the exact mechanical details of each individual shuffle accumulate to large differences in the final order of cards.

So our knowledge of those physical aspects of the cards, combined with our understanding of the chaotic nature of the shuffling process and our ignorance of all the fine details of a particular shuffle, leads us to assign the probability 13/52 to the first card dealt being a spade, the conditional probability 12/51 to the next, and so on until we arrive at the astronomically low chance of being dealt all 13 spades we computed previously. That realm of "symmetric ignorance" is the place these calculations originate from, and our agreement on the nature of that ignorance is what makes us so confident in them that even a billion years of observed card dealing could never persuade us otherwise.

In particular, the number we came up with for the probability would be our answer whether we understood that the process was going to be repeated or whether we expected that it would happen only one time, after which the deck would be thrown into the fireplace and destroyed. In this way of thinking, we don't have to imagine an impossible series of card shuffles to confirm for ourselves that our probability calculation is correct.

However, we might, for various reasons, be led to ask what we expect the long-run frequency of this bridge hand to be if, hypothetically, a long series of similar shuffles were possible. The same argument we used to deduce the one-time probability could be extended with the extra observation that we don't have any information to connect the results of one shuffle of the deck with the results of any other. That is, in our understanding of cards and how they work, we would assign successive shuffles probabilities as though they are independent. Putting this together with our previous answer and a Bernoulli-esque Law of Large Numbers argument, we would expect within some small margin (and with high but not perfect confidence) that, for a very long series of trials, the frequency we'd observe would match our probability answer for the bridge hand.

It might seem that we've just argued in a circle. We can't simply say the probability of getting an all-spades hand is equal to the long-run frequency (as the frequentist interpretation would suggest we must) because it's impossible to actually do such an experiment and as a result we don't know the frequency. But we do know certain things about the physical properties of cards that allow us to give a probability answer anyway, which, as it turns out, is equal to what we think the frequency probably is. So we do sort of know the frequency.

But here's the difference: we based our probability answer not on an assertion that it's the frequency we would obtain but rather on the tangible facts about cards that we know are true from actual experience. The logical dependency goes like this:

Known facts about cards ⇒ One-time probability of a perfect bridge hand ⇒
    Predicted frequency of this hand occurring over the theoretical long run

whereas the frequentist answer is something more like this:

"Known" frequency of this hand occurring over the theoretical long run (?) ⇒
    One-time probability answer (by definition)

Furthermore, and most importantly, we could hold those assumptions of symmetry *provisionally* as a hypothesis and allow practical experience to change our minds as evidence accumulates. So we might say, for example, that we're pretty sure the results of a shuffle don't influence the results of the next one, but if we checked our theoretical predictions against observation, we might become less sure. The frequentist interpretation has no such wiggle room.

According to Jaynes, probabilities are a consequence of the assumptions present in the background. And *the assumptions we make might be incorrect.* For example, say a person with no knowledge of magicians or sleight of hand were asked for the probability that someone could blindly draw a particular card from an apparently well-shuffled deck. They would be perfectly justified in giving the answer 1/52, *consistent with their background knowledge*. If a magician then exhibited this ability over and over, it wouldn't make the person's probability calculations wrong, just founded on ignorance. Similarly, for any of our urn-drawing examples, we might reach in and find the urn is empty! This doesn't mean our calculation was incorrect. Our probability robot processed the inputs given to it in the correct way; it's just that we may have given it garbage and should expect garbage in return, as is the case with every algorithm.

Exactly the same phenomenon occurs in ordinary logic: a deduction may be *valid*, meaning it appropriately follows the rules of deduction, yet lead to a *false conclusion* because one or more of the premises are false. For example,

If it's Wednesday, there was once life on Mars.
It's Wednesday.

──────

∴ There was once life on Mars.

is a valid deduction even though the premise is faulty and it may not even be Wednesday.

In the probability-as-logic interpretation, our probability assignments do not need to be *right*, in the sense of being verifiable externally by some experiment; they need only to be a *valid* expression of the information we are given and the assumptions we hold to be true.

Trying to force probability answers to be right or wrong by claiming probabilities were the frequencies one "would" obtain as the limit in an infinite number of trials especially annoyed Jaynes, who had years of practical experience as an experimental physicist. He chastised mathematicians sharply for fabricating these claims out of pure thought—imagining all experiments as drawing from some "infinitely large urn" if need be—and then trying to claim they had done so objectively:

In other cases, such as flipping a coin, making repeated measurements of the temperature and wind velocity, the position of a planet, the weight of a baby, or the price of a commodity, the urn analogy seems so farfetched as to be dangerously misleading. Yet in much of the literature one still uses urn distributions to represent the data probabilities, and tries to justify that choice by visualizing the experiment as drawing from some "hypothetical infinite population" which is entirely a figment of our imagination. Functionally, the main consequence of this is strict independence of successive draws, regardless of all other circumstances. Obviously, this is not sound reasoning, and a price must be paid eventually in erroneous conclusions. This kind of conceptualizing often leads one to suppose that these distributions represent not just our prior state of knowledge about the data, but the actual long-run variability of the data in such experiments. Clearly, such a belief cannot be justified; anyone who claims to know in advance the long-run results in an experiment that has not been performed is drawing on a vivid imagination, not on any fund of actual knowledge of the phenomenon. Indeed, if that infinite population is only imagined, then it seems that we are free to imagine any population we please.[36]

As Jaynes suggested, no probabilist would ever hold themselves to such a standard anyway because no limiting frequencies had ever really been observed or ever would be:

Unfortunately, to maintain [the frequentist] view strictly and consistently would reduce the legitimate applications of probability theory almost to zero; for one can (and most of us do) work in this field for a lifetime without ever encountering a real problem in which one actually has knowledge of the "true" limiting frequencies for an infinite number of trials; how could one ever acquire such knowledge? Indeed, quite apart from probability theory, no scientist ever has sure knowledge of what is "really true"; the only thing we can ever know with certainty is: what is our state of knowledge?[37]

So frequentist probability was, in Jaynes's view, complete nonsense. It existed primarily as a way for scientists to maintain a self-satisfied notion of objectivity and sidestep the question of what background knowledge and preconceptions they were bringing to any given problem.

Jaynes's notion of probability as logic made the importance of these assumptions explicit while simultaneously showing his theory could reproduce all the frequentist results if loaded up with background assumptions consistent with the real physical behavior of the apparatus of the random experiment. In this way, he achieved a synthesis of subjective and objective probability.

The point was that if our background assumptions had a kind of *symmetry*, our probabilities would also have to have this symmetry and thus be *forced* to have particular numerical values. Jaynes formalized the principle of indifference Keynes had postulated (for Jaynes, it was a theorem), and in doing so, he *split the difference* between the subjective and the objective. Jaynes's probabilities were subjective in the sense of being dependent on the assumptions brought to a problem by a particular person, but they were objective in the sense that any two people reasoning rationally (following Cox's rules for plausible reasoning) from the same starting assumptions would necessarily arrive at the same answer.

The real power of this interpretation of probability first starts to become apparent when we apply it to problems of inference, as in the problems that motivated Bayes and Laplace. Remember, one of the questions gestured at but left unanswered in those problems was where the prior probabilities came from. Bayes had constructed his example so the priors were manifestly uniform, and Laplace had assumed them mostly as a mathematical convenience.

In the Jaynesian view, *there is no difference between prior probabilities and any other kind of probability*. We are justified in assigning uniform probabilities any time our background knowledge makes us indifferent between the various propositions—that is, if we have no information to lead us to prefer one or the other. This is exactly the same justification that leads us to assign sampling probabilities, like saying our chance of rolling a 6 with a fair die is 1/6. As Jaynes pointed out, the latter is usually easier to do only because the information about the sampling process has been specified more clearly than the background assumptions have:

> From the start it has seemed clear how one determines numerical values of sampling probabilities, but not what determines the prior probabilities. In the present work we shall see that this was only an artifact of an unsymmetrical way of formulating problems, which left them ill-posed. One could see clearly how to assign sampling probabilities because the hypothesis H was stated very specifically; had the prior information X been specified equally well, it would have been equally clear how to assign prior probabilities.[38]

But according to Jaynes, specifying the information that determined these prior probabilities was crucial to any problem of inference. We could no more hope to make an inference without priors than we could hope to make an inference without data.

As an exercise in this way of thinking, and for a bit of fun, let's see how probability as logic handles a couple of the most famous paradoxes of probability that usually give headaches to students (and teachers!) in traditional probability courses. In each example, we'll see that the hard work we need to do to clarify and make explicit the background information we have when making the probability assignments is all front-loaded. The actual conclusions will follow almost immediately.

THE BOY OR GIRL PARADOX

This problem has been part of the lore of probability theory since at least 1959, when Martin Gardner published it in *Scientific American*:[39]

> Mr. Smith says, "I have two children and at least one of them is a boy." What is the probability that the other child is a boy?

Although it may not appear so at first, this has the structure of a hypothesis test. We are interested in the hypothesis "Both children are boys," and we have some observed data: at least one of the two children is a boy. So we can apply Bayes' theorem (with Jaynes's notation) as with any hypothesis and data propositions:

$$P[H \mid D \text{ and } X] = P[H \mid X]\frac{P[D \mid H \text{ and } X]}{P[D \mid X]}$$

As Jaynes instructs us, we therefore must first consider our prior probability assignments. Knowing nothing other than the fact that someone has two children, what probability do we assign to our hypothesis? It is almost certainly the intent of the problem that we take this probability to be 1/4. However, it may be worth inspecting that assumption a bit. There are four possible cases we can imagine,[40] but do we necessarily consider them equally likely? For ease of reference, imagine we think of the older of the two as child 1 and the younger as child 2; we can then label the propositions as follows:

B_1 = "The older child is a boy"; G_1 = "The older child is a girl"
B_2 = "The younger child is a boy"; G_2 = "The younger child is a girl"

And we have the four cases:

B_1 and B_2
B_1 and G_2
G_1 and B_2
G_1 and G_2

If we know *only* that exactly one of these four cases holds true, then a Jaynesian transformation argument (say switching the labels B_1 and G_1 or B_1 and B_2) tells us our state of knowledge makes us indifferent to the four. Therefore, we must assign them each probability 1/4.

However, in reality we probably know much more than that! We could, if we were so inclined, find actual rates of occurrence of (boy, boy) sibling pairs and so on, and we may find not all four combinations appear equally often. They would, at minimum, be somewhat confounded by the incidence of identical twins sharing a gender. At any rate, since it's not supposed to be the point of the problem, let's just assume that whatever state of information we have initially can at least be well approximated by one in which we assign each case probability 1/4. These are our prior probabilities for these various propositions.

In true Jaynesian fashion, though, once we have done this work of assigning prior probabilities, the rest of the problem is pretty trivial. Suppose we represent the data proposition as:

$$D = \text{"At least one child is a boy"}$$

Our hypothesis, "Both children are boys," is just the case $(B_1$ and $B_2)$. Using our general template of Bayesian inference, we can build table 1.2. Here, all the sampling probabilities are 1 or 0, since whether at least one of the children is a boy is determined deductively in each case. As a relative proportion, the pathway we care about, corresponding to the hypothesis $(B_1$ and $B_2)$, represents $(1/4)/(3/4) = 1/3$ of the sum. So, as before, our posterior inferential probability for the hypothesis $(B_1$ and $B_2)$ given the data is 1/3.

TABLE 1.2 Boy or girl paradox: at least one child is a boy

Hypothesis H	Prior probability $P[H \mid X]$	Sampling probability $P[D \mid H \text{ and } X]$	Pathway probability $P[H \mid X]P[D \mid H \text{ and } X]$	Relative proportion $P[H \mid D \text{ and } X]$
B_1 and B_2	1/4	1	1/4	1/3
B_1 and G_2	1/4	1	1/4	1/3
G_1 and B_2	1/4	1	1/4	1/3
G_1 and G_2	1/4	0	0	0

Most people are surprised at this result at first because we are so thoroughly conditioned to think the gender of someone unknown to us—in this case, the "missing" child—should be a 50/50 proposition. But the point here is the information "at least one child is a boy" is a weird kind of observation to have, and it puts us in a funny position where we can rule out only one of four possible cases, leaving the other three equally likely. If it had been phrased as "It is not the case that both children are girls," the inference would be more clear.[41]

To see a more "normal" situation, imagine we are given instead that Mr. Smith's *older* child is a boy—that is, $D = B_1$.

Then our inference table would look like table 1.3. And our posterior probability for $(B_1$ and $B_2)$ would be $(1/4)/(1/2) = 1/2$. That is, given the gender of a *particular* child, we assign equal probabilities for the gender of the *other* child, in keeping with ordinary common sense.

We can even imagine different states of background knowledge or different kinds of information that could shift the probabilities around. For example, maybe instead of knowing directly that Mr. Smith has at least one boy, we could have the data proposition $D =$ "Mr. Smith *tells us* he has at least one boy." Now the problem becomes much more complex because our given observation involves the actions of a *person*, so we need to consider what we know about this person and his behavior. Our background assumptions could incorporate knowledge about Mr. Smith, such as an assumption that he had always wanted a daughter and would surely tell us at every opportunity if he had a girl. Under those assumptions, we have sampling probabilities for hypothesis $(B_1$ and $G_2)$ and hypothesis $(G_1$ and $B_2)$ equal to 0. And our inferential probability for $(B_1$ and $B_2)$ equals 1! That is, deductively, we could reason that Mr. Smith passed

TABLE 1.3 **Boy or girl paradox: the older child is a boy**

Hypothesis H	Prior probability $P[H \mid X]$	Sampling probability $P[D \mid H$ and $X]$	Pathway probability $P[H \mid X]P[D \mid H$ and $X]$	Relative proportion $P[H \mid D$ and $X]$
B_1 and B_2	1/4	1	1/4	1/2
B_1 and G_2	1/4	1	1/4	1/2
G_1 and B_2	1/4	0	0	0
G_1 and G_2	1/4	0	0	0

up an opportunity to mention a daughter, so he must not have one, per our assumptions about him.

Likewise, we could imagine all kinds of fanciful scenarios about how we got this information, such as D = "We rang the doorbell of Mr. Smith's house and a boy answered the door." Then we need to consider the impact of different assumptions about Mr. Smith's household door-answering policies, such as "If Mr. Smith had a son, his son would always answer the door" [posterior probability for (B_1 and B_2) = 1/3] versus "Mr. Smith's children would answer the door equally often" [posterior probability for (B_1 and B_2) = 1/2].

Another variant comes about if we assume something particular about the boy in question, such as D = "At least one child is a boy born on a Tuesday." Following similar reasoning as previously (and with the same caveats), we could argue our way into probability 1/7 under either "one boy" hypothesis and into something like probability $1 - (6/7)^2$ assuming two boys, since the chance in a two-boy family of having at least one born on a Tuesday is the chance they're not both born on some other day. Our posterior probability for (B_1 and B_2) then turns out to be the very surprising number 13/27, or about 48 percent. This suggests the general pattern that the more specific information we know about this mystery "at least one" boy, the closer we are to pinning him down to being a particular one of the children, leaving us in the familiar situation of knowing one child's gender and assigning a 50/50 chance for the other's.

THE MONTY HALL PROBLEM

This problem is likewise now legendary in probability circles. First posed in a letter to *American Statistician* in 1975 and made famous by Marilyn vos Savant in her "Ask Marilyn" column in *Parade* magazine in 1990, it's a slightly abstracted version of a situation faced by contestants on the game show *Let's Make a Deal*, hosted by Monty Hall from 1963 to 1976 (and occasionally in various revivals thereafter):

> Suppose you're on a game show, and you're given the choice of three doors: Behind one door is a car; behind the others, goats. You pick a door, say door 1, and the host, who knows what's behind the doors, opens another door—say door 3, which has a goat. He then says to you, "Do you want to pick door 2?" Is it to your advantage to switch your choice?[42]

In her column, vos Savant said you should switch, and by doing so, you would increase your chance of winning from 1/3 to 2/3.

Intuition may suggest it doesn't matter whether you switch because there are two remaining doors that can be hiding the car and it seems you have no reason to prefer one or the other. Many, many people saw it that way at the time, and they expressed no small amount of contempt toward vos Savant for claiming otherwise. She received over 10,000 letters about the problem and devoted three subsequent columns to defending her answer. Among those attacking here were numerous people with advanced degrees in math and statistics. For example, one math professor wrote, "As a professional mathematician, I'm very concerned with the general public's lack of mathematical skills. Please help by confessing your error and, in the future, being more careful."[43] Someone with a PhD wrote, "You blew it, and you blew it big! Since you seem to have difficulty grasping the basic principle at work here, I'll explain. After the host reveals a goat, you now have a one-in-two chance of being correct. Whether you change your selection or not, the odds are the same. There is enough mathematical illiteracy in this country, and we don't need the world's highest IQ propagating more. Shame!"[44] Another professor wrote, "Our math department had a good, self-righteous laugh at your expense."[45] Another PhD wrote, "May I suggest that you obtain and refer to a standard textbook on probability before you try to answer a question of this type again?"[46] And yet another wrote, "You made a mistake, but look at the positive side. If all those Ph.D.'s were wrong, the country would be in some very serious trouble."[47] Even the great mathematician Paul Erdős was thoroughly convinced switching didn't matter. However, all the PhD's were wrong, and vos Savant was right. Switching doors does improve your chance of winning from 1/3 to 2/3, as we can now show using Bayes' theorem.

As in the Boy or Girl Paradox, though, before computing any probabilities, we must faithfully account for our assumptions about what we have just seen and the circumstances in which we imagine ourselves. We can structure our analysis, again, as an inference about a hypothesis: in this case, the hypothesis that our chosen door was correct, conditional on some observations—namely, everything else given in the problem. However, since the content of our observations is more complex than in the last problem, some subtle considerations emerge when we start trying to write things out.

Let H_1 be the hypothesis that the car is behind door 1, with H_2 and H_3 similar. For starters, we'd like to say our prior probability assignment for H_1 is 1/3. Why is this? One argument might be indifference; we were given only that the car was behind one of the three doors, so it feels like our ignorance is symmetric.

However, in a real-life situation we could also have come to this probability through repeated observation of the show. Maybe we know the car is more often behind door 1 because the producers prefer it that way. Or maybe we know the producers do a poor job of randomizing the car's location and never repeat the same door from week to week; this might lead us to higher levels of confidence in the car's location. Since none of this was specified in the problem, though, we'll assume indifference and assign the probability 1/3.

Similarly, we need to think a little carefully about *what it is we have actually observed*. A faithful record of all of the observations given in the problem would be something like this:

$$D = \text{"The host opened door 3" and "Door 3 revealed a goat" and}$$
$$\text{"The host offered a switch to door 2"}$$

We can eliminate H_3 immediately, since that would be inconsistent with our observation—that is, the probability of this observation conditional on the car being behind door 3 is 0. The subtlety emerges in assigning the sampling probabilities for D given H_1 and H_2, and this is where the usual statement of the problem and the usual solutions tend to leave out some important details.

Under assumption H_1, the car is behind our chosen door 1, so we know the host has a choice of which other door to open, door 2 or door 3. Do we have any reason to suspect he prefers one or the other? Nothing is stated in the problem, so we'll go ahead and assume indifference here as well, but, again, we *could* imagine having information that might guide us. Similarly, and more problematically, we will assume the host will *surely offer the switch* no matter what—maybe we know this to be a rule of the show. But nothing in the problem as stated explicitly gives us this information. We'll return to this assumption in a moment.

For now, though, we will make these assumptions, which are the standard ones given in most solutions. With these in place, we have all the necessary ingredients for our inference, since we can say our state of knowledge X is such that

$$P[H_1 \mid X] = P[H_2 \mid X] = P[H_3 \mid X] = 1/3$$

and

$$P[D \mid H_1 \text{ and } X] = 1/2; \; P[D \mid H_2 \text{ and } X] = 1; \; P[D \mid H_3 \text{ and } X] = 0$$

TABLE 1.4 The Monty Hall problem: standard assumptions

Hypothesis H	Prior probability P[H | X]	Sampling probability P[D | H and X]	Pathway probability P[H | X]P[D | H and X]	Relative proportion P[H | and X]
H_1 (door 1)	1/3	1/2	1/6	1/3
H_2 (door 2)	1/3	1	1/3	2/3
H_3 (door 3)	1/3	0	0	0

Putting these into an inference table results in table 1.4. Our posterior probability for H_1 is the relative proportion (1/6)/(1/6 + 1/3), or 1/3, with the remaining probability of 2/3 going to H_2. Thus, switching doors is definitely to our advantage, doubling our chance of winning.

But let's see what difference a slight change to some of the standard assumptions could make. For example, we could imagine knowing that the host has an extreme hatred for door 3 and will open it only if absolutely necessary not to reveal the prize. Under that assumption, we have the probability for D under H_1 equal to 0, since in that case the host could have opened door 2 instead. And we see the only possibility remaining is door 2, so we should definitely switch!

On the other hand, maybe the host loves door 3 and will always open it if given the chance. Under that assumption, we have the probability for D under H_1 equal to 1, and the resulting inference table is table 1.5. This time our posterior probability for H_1 is (1/3)/(2/3) = 1/2, so switching doesn't matter!

TABLE 1.5 The Monty Hall problem: the host loves door 3

Hypothesis H	Prior probability P[H | X]	Sampling probability P[D | H and X]	Pathway probability P[H | X]P[D | H and X]	Relative proportion P[H | D and X]
H_1 (door 1)	1/3	1	1/3	1/2
H_2 (door 2)	1/3	1	1/3	1/2
H_3 (door 3)	1/3	0	0	0

It all hinges on what we assume about the host's feelings about door 3, which we could dial anywhere from absolute hatred to absolute fondness and get a range of answers from "You must switch to have any chance of winning" to "It doesn't matter at all whether you switch."

Returning to the assumption about the host necessarily offering the choice to switch, we could, of course, imagine this is not the case. Suppose the host is malicious and wants us to lose. Since he knows where the car is, he could decide to offer the switch *only* if we chose correctly in the first place. If we are wrong, he simply opens our choice of door and moves on to the next contestant. Under those assumptions, our observation is now incompatible with assumption H_2, so the posterior probability for H_1 is 1 (it's the only case consistent with what we've seen). As a result, switching reduces our chance to 0!

In fact, this turns out to be pretty close to the behavior of Monty Hall (who understood the problem and its solution). In an interview in the *New York Times* years later, he revealed that the rules of the show actually did not force him to offer a switch at all, and he was free to do whatever he thought would most heighten the drama of the show. As he put it, "If the host is required to open a door all the time and offer you a switch, then you should take the switch. But if he has the choice whether to allow a switch or not, beware. Caveat emptor. It all depends on his mood."[48]

We have one final variant to consider: What if the host doesn't know where the car is? (In this version, we're explicitly told otherwise, but this part is often left out.) It may even be the case that he *usually* knows but happened to forget this one time. Maybe we see him give a heavy sigh of relief when door 3 is opened and the car isn't there. Under any of these assumptions, we give the probability assignments

$$P[D\,|\,H_1 \text{ and } X] = P[D\,|\,H_2 \text{ and } X] = 1/2$$

because our observation includes the fact door 3 was opened. Even assuming H_2, it could just as well have been door 2 that got opened; by luck, it happened not to be. The resulting inference table is table 1.6. And our posterior probability assignment is 1/2. So, once again, switching adds nothing! In the original problem, it is the host's knowledge of what to do if the car is behind door 2 that allows him to add information and shift the balance of probabilities. Confusing our ignorance for the host's is what led to the intuitive—but wrong—answer to the problem.

But it all depends on what we assume.

TABLE 1.6 The Monty Hall problem: the ignorant or lucky host

Hypothesis H	Prior probability $P[H \mid X]$	Sampling probability $P[D \mid H \text{ and } X]$	Pathway probability $P[H \mid X]P[D \mid H \text{ and } X]$	Relative proportion $P[H \mid D \text{ and } X]$
H_1 (door 1)	1/3	1/2	1/6	1/2
H_2 (door 2)	1/3	1/2	1/6	1/2
H_3 (door 3)	1/3	0	0	0

WHAT PROBABILITY ISN'T

Note at no point while solving either the Boy or Girl Paradox or the Monty Hall Problem did we make reference to frequencies or proportions of some hypothetical population. Most standard solutions for the Boy or Girl Paradox will involve imagining "sampling a random family from among those with two children" or an even more grotesque idea like "selecting from an urn full of children." Some, like Gardner himself, have argued that the question as stated is ambiguous because of its "failure to specify the randomizing procedure,"[49] whatever that means. There's no need for any such specification. Where in the problem does it say anything about something being chosen at random?

Similarly, solutions to the Monty Hall Problem often involve simulating results, using a random number generator, to show that, over the long run, a strategy of switching leads to winning the car about two-thirds of the time. (This was how Erdős eventually became convinced.) These simulations will produce the "right" frequencies, though, only if the behavior of the virtual host is coded up to match the assumptions we've made—in particular, choosing an available door not hiding the car at random and always offering the contestant the choice to switch. So, if those assumptions are inconsistent with our actual state of knowledge, the simulation proves nothing.

Probability as logic has no need for such artificialities. Using Bayesian logical analysis, we can analyze each problem as though it is a single occurrence, with no need to think about some hypothetical random sampling procedure delivering us this information or some long-run sequence of events. We simply condition on what we know or are told is true, and we process the available information into a probabilistic inference. That inference will necessarily involve some assumptions, sometimes complex ones about the people involved—especially if our observation includes someone *doing* something or *telling* us something rather than just some propositional fact. But first we need to *slow down our thinking* and carefully inspect all the information we're given and the assumptions we're willing to stipulate to finish the problem. Those assumptions might be wrong. Maybe Monty Hall has more fondness for door 3 than we assumed, and maybe Mr. Smith is equally proud of his son and his daughter, and maybe there never were two children to begin with, and maybe the game show is rigged against us.

Jaynes's essential point bears repeating: *probability is about information.* It comes from what we observe, what we are given, and what we hold to be true. We should want our assumptions to be true, and, indeed, if we are able to conceive of alternatives, probability as logic tells us how we should go about testing those alternatives based on observed fact, how surprised we should be to see one kind of observation or another, and how much evidence this provides in support of one hypothesis or another. But the only valid tools for processing that information, if we hold to the general rules for consistent reasoning given by Cox and Jaynes, are the rules of probability—specifically, Bayes' theorem. Cox's theorem shows that these rules depend on something more fundamental than ideas of frequency and proportion. In fact, they are derivable just from a weak assumption that our probabilistic reasoning process is logically consistent. Any form of reasoning that is "non-Bayesian" violates these conditions and is necessarily illogical.

These example brainteasers reveal the power of the probability-as-logic interpretation and show the steps required to assign probabilities in any problem of inference. The consequence for skipping any of the steps in these problems is harmless (other than some mild embarrassment), but in the real world, it often is not. In the next chapter, we will see several very real instances of people's lives being ruined because of a misuse of probabilistic inference.

2

THE TITULAR FALLACY

Mathematics, a veritable sorcerer in our computerized society, while assisting the trier of fact in the search for truth, must not cast a spell over him.

~Justice Raymond L. Sullivan, majority opinion in *People v. Collins* (1968)

I f you ever wondered how choosing a ball from an urn became a standard example of a probability question, this is how.

For over a thousand years, from the Middle Ages to modernity, the Republic of Venice was ruled by a leader called the *doge*, who, like European monarchs of the time, was the chief civil, military, judicial, and religious officer of the state. The key difference was the doge did not inherit his position. Instead, he was elected for life by the noblemen of Venice. The process of selecting a doge was unbelievably complex, involving numerous committees and randomly chosen subsets of those committees. The procedure established in 1268 and followed with only minor modification until the last doge was elected in 1789 went like this:

- First, 30 members were randomly chosen from among those serving on the aristocratic Great Council, of whom there were something like 480.
- From those 30, a committee of 9 was randomly chosen by lot.
- Those 9 voted to form a committee of 40, with each member requiring 7 votes of approval.
- The committee of 40 was then reduced by lot to 12, who elected a committee of 25, each requiring 9 votes.

- That committee of 25 was reduced by lot to 9, who elected a committee of 45, each needing 7 votes.
- Those 45 were reduced by lot to 11, who elected a committee of 41, each needing 9 votes.
- Finally, those 41 elected the doge by simple majority vote.[1]

Although it may seem overly complicated, the system had certain advantages. A recent mathematical analysis showed the protocol created a substantial probability of choosing a winner without a majority faction of support, which would have protected minority interests while still ensuring the more popular candidates were more likely to win.[2] Due to its multilayered probabilistic nature, the process may have also discouraged corruption, since anyone seeking to sell their vote in exchange for a favor would have little guarantee their preferred candidate would be in a position to deliver. Even the senior advisers of the previous doge couldn't exert much influence over the outcome, which limited their ability to appoint a successor and establish a dynasty.

The election process also involved an appealing amount of ceremonial pomp. Like their Byzantine neighbors, the Venetians were not exactly known for a love of simplicity. Each random selection was accomplished by picking a ball, or *balota*, from a clay urn. (Our English *ballot* comes from this same Venetian word.) The drawing was managed by a boy called the *balotino*, himself semirandomly selected, as he was the first boy a particular council member saw after finishing prayers at St. Mark's Basilica on election day.

Now imagine you're a Venetian noble in 1457. The last doge, Francesco Foscari, was forced to abdicate after his son was convicted of bribery and corruption, and Francesco died shortly thereafter. The council is ready to elect a new doge. You don't like the looks of the balotino, though. He's got a shiftiness about him and just so happens to be best friends with the grandnephew of Pasquale Malipiero, a leading candidate to be the new doge. Your family and Malipieros have been feuding for generations, and you desperately want to stop Pasquale's ascent to power. Out of those chosen in the first drawing of 30, about half are Malipiero supporters, a bad start for you but not a guarantee of his victory thanks to the complexity of the electoral system. Next comes the second draw, to reduce these 30 to a committee of 9. This stage is critical. If Malipiero's faction gets 7 or more members, they can completely dictate the rest of the process and elect him the doge by forming subsequent committees entirely from their own ranks.

The drawing begins, and it's a disaster. The first 7 people chosen are all Malipiero supporters. How can this be?! You want to cry foul and accuse the balotino of corruption, but you don't have a solid basis for your argument. Maybe he pulled some trick and rigged the drawing process, or maybe it was simply good fortune. Falsely accusing the balotino—and, by extension, the new doge-elect—could lead to your exile or worse. What do you do?

This was the kind of problem that motivated Jacob Bernoulli to spend the last portion of his life studying probability and developing many of its most important ideas. What makes it especially difficult is that it's a question of inference. We have a sample of balls from an urn, and we need to reach some conclusion about the urn's contents: Was it probably fair or probably rigged? The probability of getting 7 Malipieros in a row out of an urn that was supposedly only 50 percent Malipiero is straightforward enough to compute. The hard direction is the other way. How do we use what we observed to make probability statements about what we don't know?

At this point in the book, if things have gone according to plan, you should be convinced of two inescapable propositions: (1) probability is best understood as the plausibility of a statement given some assumed information, not just the frequency of occurrence of events, and (2) the correct process of probabilistic inference involves a faithful accounting of background information and assumptions, which is conceptually tricky and enormously difficult.

However, as we noted briefly in the last chapter, not everyone sees it that way. There is a competing school of thought—in fact, now the dominant one—which holds that all probability questions are answerable using frequencies. Even though Bernoulli predated this school, he was largely responsible for inspiring it. In a way, frequentism all started with Bernoulli's suggestion that one could measure the probability of an event—like pulling a particular ball from an urn—by repeating the process many times and recording the frequency with which it happened. Replace "measure the probability" with "define probability" in that construction, and you have the frequentist interpretation in a nutshell.

But Bernoulli also said something a bit more subtle. Not only did he establish that as the number of trials grew to infinity, the observed frequency would ultimately converge to the true probability, but also he put bounds on how accurate the estimates would be for any given finite sample size. That is, he calculated how likely it was that an actually realized estimate of the probability would fall within some margin of the truth. As we'll see, this emboldened him to say with some degree of confidence that an unknown probability was likely close to what was observed. Bernoulli's argument amounted to a method of inference using only sampling probabilities (the probability of obtaining some sample given a possible urn mixture), the kinds of things that are measurable with frequencies.

In other words, he could make an inferential statement—he could accuse the hypothetical balotino of corruption—and the only ingredients to his argument were the basic probabilities on which everyone agreed.

The appeal of such a procedure, were it possible, is obvious, since it suggests we can bypass the hard work of precisely describing our state of knowledge, enumerating all the background assumptions we bring to the table, etc., when making a probabilistic inference. Instead of all that, we can simply compute a sampling probability or measure it empirically, turn the handle on an inference machine, and accept whatever answer the machine gives us.

In this chapter, we're going to take a closer look at the inner workings of that machine. Because of the examples we have considered, we already know the answers it produces must sometimes be illogical. We can learn a lot, though, by watching the process in slow motion to see exactly where it goes haywire. What's even more interesting is how incredibly tempting this method of inference can seem when described in the right way. In particular, we'll start with Bernoulli's urn-drawing problem, where in the right light it will appear for all the world that the sampling probabilities—that is, the frequencies—*are* actually enough information from which to make inferences. Putting aside all the abstract mathematical gobbledygook for a second, it does just seem intuitively obvious. You want to figure out what's inside an urn? Take a large sample and see.

But even in this simple and contrived example, there's a lot going on. We can start to get a sense of how our intuition about sampling and learning could be misleading us if we step, very slowly, through a similar problem and compare the Bayesian/logical conclusions with what Bernoulli's approach would give us. By the end of that process, we'll have revealed both what Bernoulli was missing and why his answer seemed intuitive in the first place. To go ahead and spoil the ending, Bernoulli's approach ignores the role of prior information and alternative hypotheses, and it feels right only because it gives the same answers as the Bayesian approach when the alternatives are clear and there is no important prior information, which is generally the case when we're talking about an urn.

The real problems start to show up, then, when we do have strong prior information, as is often the case in real-life experience. When that happens, we can expect any inference that ignores that information to lead us to some pretty awful places. As we'll see, the fallacy at the heart of Bernoulli's argument—that sampling probabilities are sufficient to make inferences—has found its way into all manner of high-profile situations where it's caused serious damage. One of those, unfortunately, is modern statistics.

BERNOULLI'S BARGAIN

Let's take the story back to circa 1700, to Bernoulli's *Ars Conjectandi* and the Law of Large Numbers, his "golden theorem." The headline of the theorem is that the true probability of any event will be borne out in the frequency with which the event occurs over a number of trials as that number goes to infinity.

In the finite meantime, though, some fluctuation in observed frequencies is always possible just as a matter of chance. In 100 flips of a fair coin, we might not see exactly 50 heads and 50 tails. Could we see 55 and 45, 60 and 40, or even 100 and 0? There are several moving parts to the problem: the number of observations could be changed, as could the size of the observed fluctuation and how often a fluctuation of that size was seen. To prove that the desired convergence would happen, Bernoulli had to understand the exact relationships among these parts. How much variation could we expect in a given number of observations, and how often could it be expected?

Before Bernoulli devised his theorem, it was already commonly understood that more observations should tend to produce less overall variation in the frequency. As he wrote, "Even the most stupid person, all by himself and without any preliminary instruction, being guided by some natural instinct (which is extremely miraculous) feels sure that the more such observations are taken into account, the less is the danger of straying from the goal."[3] The nagging question was whether enough observations could ultimately produce a *sufficient* amount of certainty or whether it would cap out at some level. That is, for any given interpretation of *close*, could there be some uncrossable upper limit of certainty—say 95 percent or 99 percent—we could claim as the chance that our observed ratio would be close to its true probability? As Bernoulli put it, "It certainly remains to inquire whether, when the number of observations thus increases, the probability of attaining the real ratio between the number of cases, in which some event can occur or not, continually augments so that it finally exceeds any given degree of certitude."[4]

To answer the question, he had to do some hard calculations with probabilities. For starters, he referred to the ratio of the number of ways a particular event could happen (the "fertile" cases) to the number of ways it could fail to happen (the "sterile" cases) as $r{:}s$, thus making the true probability of the event $r/(r + s)$. Specifying the lower limit of variation as $(r - 1)/(r + s)$ and the upper limit as $(r + 1)/(r + s)$, he calculated the chance of getting an observed ratio between these limits for a given sample size, n. Observing how this varied with n, he was then able to prove the following: "It is possible to take such a number of experiments that it will be in any number of times (for example, in *c* times) more likely that the number of fertile observations will occur between these limits rather than beyond them."[5]

To take a numerical example, suppose an event—such as a particular type of lightbulb burning out sometime in the next year—has a true probability of 40 percent, or 2/5. This probability can be thought of equally well as any multiple of 2 divided by any multiple of 5, like 20/50, 200/500, 2,000/5,000, and so on. Fixing one of these in mind—say 20/50—we can add and subtract 1 to the numerator to get a small interval around the true probability—in this case, the interval from 19/50 to 21/50, or from 0.38 to 0.42. Bernoulli showed that, with a large enough sample of observations, we can be very sure the observed ratio of fertile:total will fall within this interval. That is, with enough lightbulbs, we can be very sure that between 38 percent and 42 percent of them will burn out in the next year. How big a sample we need depends on how sure we want to be, which Bernoulli expressed as the odds of the observation being inside the interval versus being outside, the number he called *c*. For example, we may require odds of 99 to 1, meaning a chance of 99/100 the ratio we observe is within our tolerance and a chance of 1/100 it falls outside our tolerance.[6]

Bernoulli's proof of the Law of Large Numbers therefore not only justified the idea that observed frequencies would approximately match true probabilities over the long run but also gave a concrete number of observations one would need to achieve a particular accuracy of measurement with a particular likelihood. By examining how this sample size depended on the true probability value, he found that the worst-case probability (requiring the largest sample) was always 0.5. So whatever sample size was necessary for that case would actually work simultaneously for all probabilities.

This is where things started to go off the rails. Because his sample size would work for any probability, Bernoulli made the leap to saying we could *learn* the probability, to within some margin and with some certainty, from a *particular* observation of that size. His theorem seemed to concern a trade-off like the one in project management: "Fast, good, or cheap. Pick two." Here, Bernoulli offered this bargain: "Precise estimates, high certainty, or small samples. Pick two." If we're willing to pay the price of a high sample, he could promise we'd obtain whatever combination of accuracy and certainty we desired.

To see how this would work in practice, imagine an urn has a true fraction of white:total pebbles, but the numbers are now *unknown* to us. We agree that the ratio determines the probability of drawing a white ball. We'd like to learn about that fraction by observing the ratio white:total in a large sample of pebbles taken from the urn.

Suppose we set a large enough sample size, whatever Bernoulli quoted us before, so that the sample ratio has a 99 percent probability of being within 2 percent of the true urn fraction.[7] This means that if the urn fraction is actually 40 percent, we have a 99 percent chance of observing a sample ratio between 38 percent and 42 percent. The same will be true for any hypothetical value of the urn fraction. In general, if the urn fraction is assumed to be x percent, then with 99 percent probability, we expect to see a sample ratio between $(x - 2)$ percent and $(x + 2)$ percent. Now suppose that, when we actually do the experiment, we get a sample ratio of 35 percent. Bernoulli's inferential claim—supposedly a consequence of his proof—was that we are now justified in concluding with 99 percent probability that the true urn fraction is between 33 percent and 37 percent. The Law of Large Numbers was Bernoulli's guarantee that we would get what we bargained for.

Could it be that simple?

Putting aside the technical details, Bernoulli's theorem can be paraphrased as

> The observed ratio in the sample will be close to the true urn fraction with high probability.

The idea at the heart of Bernoulli's guarantee, then, is that the concept of *closeness* in that statement is symmetric: if X is close to Y, then Y is close to X. So the theorem could apparently be rephrased as

> The true urn fraction will be close to the observed ratio in the sample with high probability.

With the sampling accomplished, couldn't we conclude the sample ratio is still likely to be close to the true value? That is, if we know it is highly likely the observation will be within 2 percent of the truth, couldn't we just as well claim with the same certainty that the truth is within 2 percent of whatever we actually observe?

Well, not quite. Mathematically speaking, the difficulty lies in what each probability is *conditioned* on. In short, it's the difference between saying "*Whatever* sample we get will probably be close to the truth" and saying "This *particular* sample is probably close to the truth." To understand that difference, we need to be explicit about what Bernoulli's theorem actually says and doesn't say. This is slippery territory, so it's important that we tread slowly and carefully.

First, we need some notation. Let F stand for the true fraction of white:total pebbles in the urn. This fraction is unknown to us, so we may use probabilities to express our state of uncertainty about its true value. Even though F is a fixed constant, our uncertainty makes it, in the usual confusing mathematical terminology, a *random variable*, something to which we can associate probabilities. We're able to make statements of the form "F is probably between 20 percent and 30 percent" or "The probability that F is 74 percent is 0.1" and so on based on whatever information we currently have.

We may also, in the course of making other probability statements, *condition* on F having a particular value, meaning we assume it temporarily for the purposes of a calculation. We may refer to such probabilities, for example, as "the probability of drawing a white pebble, conditional on the urn fraction F being 60 percent."

Another quantity we care about is the ratio of white:total pebbles we get in the sample. Before we take the sample, this is also unknown to us, and we'll denote the true sample ratio we ultimately get by S. We're allowed therefore to talk about the probability that $S = 0.4$ or the probability that S is between 0.12 and 0.15 and so on. It may seem more natural to think of probabilities for S in terms of repeated samples, but really it's just as well if we assume the sampling is going to happen only once and we just don't know the outcome yet. We can also form probability statements that are conditional on S having a particular value, just as we did with F. It may be useful to do this when we have actually done the experiment or cracked open the urn to count all the pebbles (so, say, we now do know the true value of S or F) or when we want to make a provisional "what-if" assumption (to explore the consequence if, say, we observe a possible sample S or if F is equal to 0.5). The conditioning process looks identical whether we know the information to be true or we are just assuming it hypothetically for that calculation.

Finally, we can make probability statements involving both unknowns, typically by conditioning on one having a certain value and seeing how this affects our uncertainty about the other. Bernoulli's theorem takes this form. As in our previous example, if, say, the urn fraction F is assumed to be 0.40 for a suitably large sample, the probability that the sample ratio S is within 0.02 of 0.40 is more than 0.99. In fact, Bernoulli's theorem produces an infinite class of such probability statements simultaneously for different hypothetical values of F other than just 0.40. The same sample size, it turns out, also guarantees that if F is 0.62, S will be within 0.02 of 0.62 with the same certainty and similarly for 0.417 and so on. To avoid having to rewrite all infinity of the statements every time, we'll let f stand in as a placeholder for every possible value we might want to sub in. It will happen that the probability statements in Bernoulli's theorem are true regardless of what value we assume, or, to put it more "mathily," they are true for all f.

Holding those conventions in place and continuing to ignore the other technical details, we can state the actual conclusion of Bernoulli's theorem in this way:

For all f, $P[S$ is close to $f \,|\, F = f]$ is high.
(The probability that our to-be-observed sample ratio will be close to the true urn fraction is high for any particular hypothetical true urn fraction.)

This follows as a result of tallying up the probabilities for getting different sample data under the assumed hypothesis. We've been referring to those probabilities as the *sampling probabilities*. The problem is that the preceding claim is an altogether *different* statement from the *inferential probability* statement,[8] which says that, conditional on any given value of sample ratio s, the urn fraction is probably close to s:

For all s, $P[F$ is close to $s \,|\, S = s]$ is high.
(The probability that the true urn fraction is close to any particular already observed sample is high no matter the particular observed sample ratio.)

The first, the one guaranteed by Bernoulli's theorem, is a statement about probabilities for S assuming F; the second, which is the one we actually want in order to make inferences, involves probabilities for F assuming S. Bernoulli's lack of a clear idea of conditional probability, and a notation that could express it concisely, led him to mistake one for the other.

In fact, the notational issue largely persists to this day. Most probability and statistics textbooks do not include the "conditional on" part in any probability statement except those that are explicit examples of conditional probability, usually treated as a different kind of probability from the ordinary, unconditional kind. For us (as we established in the previous chapter), *all probability is conditional probability*. Without that habit of thinking, it's an understandable mistake to describe the probabilities in both preceding statements as "the probability of S being close to F," forgetting that they come from different conditional assumptions.[9]

For anyone without a mathematician's love of precise notation, though, this will probably feel like a very minor distinction at this point. So what if the sampling probability statement is not perfectly identical to the inferential probability statement? Surely the fact that the sample ratio and the urn fraction are very likely to be very close to each other must give us *some* way to infer the latter from the former, even if Bernoulli's theorem technically points only in the other direction.

You and Bernoulli would not be alone in thinking that. The confusion of sampling and inferential probability arguments seems to have come up essentially everywhere people have tried to use probabilistic thinking to learn from data. For example, a few years ago I sat in on a computer science lecture at, let's just say, a very prestigious technical university. The topic was the theory of machine learning—in particular, the fundamental question of whether there is a way to prove that an algorithm, given enough data, will eventually converge on a correct understanding of the truth. The professor constructed a model for learning an unknown constant he called μ in terms of a quantity derived from the data called v, and he showed (with a slightly fancier version of Bernoulli's argument called Hoeffding's inequality) that it was highly probable that v would be close to a given μ. So, in the next line, he simply reversed the two and claimed it was therefore highly probable that μ would be close to a given v . . .

Beginning with his 1930 paper "Inverse Probability," the statistician Ronald Fisher referred to this method of inference as "fiducial."[10] The term *fiducial*, meaning "faithful," has origins in astronomy and land surveying, describing a fixed point of reference for establishing position or distance. Fisher's idea, just like Bernoulli's, was that, when trying to learn about an unknown quantity from a sample, either the unknown quantity could be taken as fixed and the sample shown to be close to it, or the sample value could be taken as fixed with the underlying quantity therefore necessarily being close to *it*. From either perspective, close was close.

In Fisher's case, this was no innocent confusion. As we'll see in more detail later on, Fisher was highly motivated to argue that inference could be done using only sampling probabilities for reasons that involve the very meaning of probability and its place in science.

And the problem we identified—that the inferential probability statement is not the same as the sampling probability statement—is not minor at all.

If we cast the problem in terms of Bayes' theorem, we can begin to grasp the real magnitude of what's going on here. For any hypothesis H and observed data D, the theorem tells us

$$P[H \mid D] = P[H]\frac{P[D \mid H]}{P[D]}$$

The inferential probability is the one on the left-hand side (How probable is the hypothesis given the data?). The sampling probability that Bernoulli focused on exclusively is only the numerator of the second term on the right-hand side (How probable is the data given the assumed hypothesis?). So how the sampling probability affects the inferential one will depend on the other terms in the equation. In particular, we need to know $P[H]$ (How probable do we consider the hypothesis without knowing the data?) and $P[D]$ (How probable is the data, considering all possible alternative hypotheses together?).

Both terms involve other background information we may have about the likelihood of the hypothesis and its alternatives. So to begin to open a crack between the two preceding probability statements, all we need to do is imagine different states of information that might lead us to different prior probability assignments. As in the problems we considered in the last chapter, the prior probabilities, together with the sampling probabilities, then determine the pathway probabilities to the observed data, which immediately give us the posterior probabilities for the different values of the unknown.

Once the crack between sampling and inference is open, we will widen it into a gaping abyss into which we will throw most of modern statistics along with a disturbingly large amount of published scientific results. For now, though, let's make Bernoulli break his promise.

A MISHAP AT THE CANDY FACTORY

For the sake of a more vivid example, instead of counting pebbles in an urn, imagine we've been hired to do quality control at a candy factory, and we're sampling colored candies from an enormous bin. Being specially manufactured for the Christmas season, these candies are all either green or red. Suppose the possible fraction F of green:total candies in the bin is limited to two possibilities: either 1/3 or 2/3. This could happen if, say, the machine at the factory is supposed to produce one red candy for every two green candies, but there's some possibility the colors got mixed up. We think it's likely that the green fraction in the bin as a whole is 2/3, but we're not totally sure. We take a scoop of candies from the bin and count the red and green candies in our sample.

Under these assumptions, we are still justified in making Bernoulli's Law of Large Numbers claim: if $F = 1/3$, then we should expect to see something close to 1/3 of the candies in our sample being green, and if $F = 2/3$, we should expect to see something close to 2/3 being green.

To give some numbers, suppose we grab a sample of $n = 30$ candies. According to Bernoulli's calculations, the number of green candies, G, we could get in the sample then has what's called a *binomial distribution*, meaning the probability of getting a particular value of $G = g$, conditional on any assumed value of $F = f$, is given by the formula

$$\binom{30}{g} f^g \left(1 - f\right)^{30-g}$$

The binomial coefficient at the beginning of the formula expresses the fact that we may count the candies in any order; it is the number of different sequences of g green candies and $(30 - g)$ red ones. The remaining terms express the probabilities of each individual candy being one of the two possible colors, and our assumption that the draws are independent means we multiply these probabilities together.

Graphically the binomial distribution has a bell curve shape with a center around the value corresponding to the exact fraction F. Assuming $F = 1/3$, we'd have the probability distribution shown in figure 2.1, with the height of each bar representing the probability of observing that many green candies in our sample of 30. The peak of the bell curve—that is, the most likely value—is at

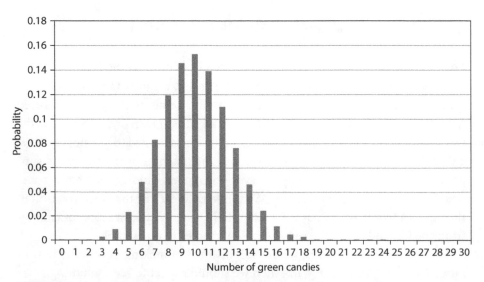

FIGURE 2.1

Probability of observing each number of green candies in the sample assuming $F = 1/3$.

$G = 30 \cdot F = 10$, which would make the sample exactly 1/3 green, in perfect agreement with the fraction in the bin. For $F = 2/3$, we'd have a similar graph centered around $G = 30 \cdot F = 20$.

We could then follow Bernoulli's idea and find an interval around this peak that constitutes almost all of the probability—that is, it is morally certain. Suppose our threshold for moral certainty is about 97 percent. We'd find that the range from $G = 5$ to 15 (inclusive) does the trick. That is, assuming the bin fraction is 1/3, we can say we're morally certain the sample will have between 5 and 15 green candies in it. We'd have a similar result for $F = 2/3$, with a range of highly probable values for G between 15 and 25. This means that, either way, our sample ratio $G/30$ is morally certain to be within 5/30, or about 17 percent, of the true bin fraction F.

None of this is controversial; every probability theorist from Blaise Pascal to Edwin Jaynes would agree with our calculations thus far. The problem comes about only when we start using these probabilities to do the inference.

Imagine we get something close to a 1/3 sample—say 8 green candies out of 30. What would Bernoulli say about our inference? Well, Bernoulli's idea is to invert the relationship given earlier, so that it's morally certain our observed ratio $G/30$ is within 5/30 of F. If we grant the symmetry of the idea of closeness—that is, if we think of that statement as giving a probability for F instead of G—we are led to say we're about 97 percent sure the bin fraction is within 5/30 of 8/30, or within a range from 10 percent to 43 percent, which seems an awfully vague conclusion under the circumstances.

But let's see how the probabilities play out in our Bayesian analysis. The problem is, thus far, that we haven't accounted anywhere for the fact that we initially expected the fraction to be 2/3. Maybe the condition $F = 1/3$ corresponds to an unlikely manufacturing error—say someone mixing up the supplies of red and green food coloring—to which we assign a very low prior probability, something like 0.01 percent, or 1 in 10,000.

Putting these probabilities into an inference table, as we saw in the last chapter, would yield table 2.1. So, after pulling a sample of only 8 green out of 30, we'd assign the mix-up hypothesis a probability of around 62 percent because the extremely small likelihood of this error occurring roughly balances the extremely small chance of getting this sample had the error *not* occurred. But we'd still allow for a 38 percent chance of the fraction being 2/3, way outside Bernoulli's range of moral certainty from 10 percent to 43 percent.

Or, if we are initially indifferent between the possibilities—we don't know whether it is supposed to be two reds for every green, or vice versa—and

TABLE 2.1 **Candy factory inferences: the unlikely mix-up**

Hypothesis H	Prior probability $P[H\mid X]$	Sampling probability $P[D\mid H \text{ and } X]$	Pathway probability $P[H\mid X]P[D\mid H \text{ and } X]$	Relative proportion $P[H\mid D \text{ and } X]$
$F = 1/3$ "Mix-up"	0.0001	$\binom{30}{8}\left(\frac{1}{3}\right)^8\left(\frac{2}{3}\right)^{22} = 0.1192$	$1.192e^{-5}$	$\approx 62\%$
$F = 2/3$	0.9999	$\binom{30}{8}\left(\frac{2}{3}\right)^8\left(\frac{1}{3}\right)^{22} = 0.000007$	$7.277e^{-6}$	$\approx 38\%$

TABLE 2.2 **Candy factory inferences: initial indifference**

Hypothesis H	Prior probability $P[H\mid X]$	Sampling probability $P[D\mid H \text{ and } X]$	Pathway probability $P[H\mid X]P[D\mid H \text{ and } X]$	Relative proportion $P[H\mid D \text{ and } X]$
$F = 1/3$	0.5	0.1192	0.0596	$\approx 99.99\%$
$F = 2/3$	0.5	0.000007	0.000003	$\approx 0.01\%$

therefore assign equal probabilities to the bin fraction being 1/3 and 2/3, we would build the inference table shown as table 2.2 with the same sampling probabilities but new priors. This gives a posterior probability for $F = 1/3$ of around **99.99 percent**. That is, instead of giving odds of about 32:1 that F is "somewhere" between 10 percent and 43 percent, we are confident in the exact value $F = 1/3$ with odds of about 10,000:1! Here, again, we see the effects of our strong prior information. Bernoulli's conclusion included a range of values we know to be impossible—namely, anything not exactly equal to 1/3 or 2/3—so the probability quoted was far too conservative.

The Bayesian inference gives us the ability to sort this out, whereas Bernoulli's analysis is stuck reporting the same conclusion and the same accuracy with no way of incorporating this additional information. These examples show Bernoulli's bargain may lead us to a somewhat reasonable inference, or it may fail spectacularly, depending on what background information we bring to the problem. Bernoulli's approach ignores such information altogether.

This is what I will refer to from now on as *Bernoulli's Fallacy*: the mistaken idea that sampling probabilities are sufficient to determine inferential probabilities. The Bayesian analysis reveals that this way of thinking misses out on two other essential ingredients: (1) what available hypotheses we may have to explain the data some other way, with their own associated sampling probabilities, and (2) what prior probabilities we assign to the various hypotheses in play. Bernoulli's mistake was not just confusing the sampling statement in his theorem and the inferential statement he wanted to make, although that implies Bernoulli's Fallacy as a consequence. His real mistake was thinking he had all the necessary information in the first place. No amount of clever manipulation can ever produce inferential probabilities about a hypothesis purely from its sampling probabilities. How likely the hypothesis makes the data is simply not enough information from which to draw an inference.

Before we move on to plumbing the depths of the awful conclusions Bernoulli's Fallacy can lead us to, let's pause for a moment to consider why it ever seemed reasonable in the first place. We have seen how strong prior information can create a disparity between Bernoulli's answer and the Bayesian one. The situations Bernoulli was interested in, though, were ones in which it may have been reasonable to say *no such prior information existed.*

That is, imagine our initial state of knowledge of the true bin fraction has *uniform* probabilities covering all possible fractions from 0 to 100 percent. This could reasonably be the case if all we know about the bin is that it contains some mix of red and green candies but we have absolutely no idea what the ratio is likely to be. Then the Bayesian inference process would involve multiplying each sampling probability for the different fractions by the same prior probability. This would produce pathway probabilities that are just proportional to Bernoulli's sampling probabilities for that given sample ratio. But after expressing these as relative proportions, those prior probabilities cancel out!

The result of all of this—say for our preceding example, assuming an observation of 8 green candies out of 30—is that we would represent our posterior probability assignments for the bin fraction using the graph shown in figure 2.2 (with the x-axis scaled to be the number of greens out of 30). And we would have drawn so nearly the same picture as Bernoulli's distribution (moved over to be centered at 8 rather than 10, consistent with his symmetric closeness idea) that we would reach all the same conclusions almost exactly! In this case, we are, indeed, about 97 percent confident that the urn fraction is between 3/30 and 13/30, just as Bernoulli's analysis concluded.[11]

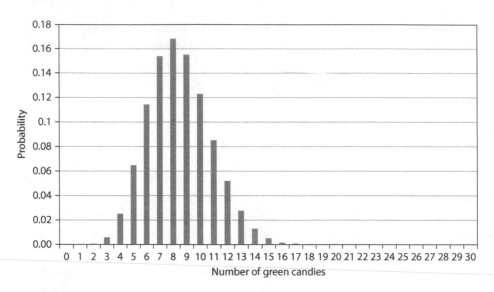

FIGURE 2.2

Posterior probabilities for the true number of green candies out of 30 assuming uniform priors.

So, in fairness to Bernoulli, whose name we are presently about to malign, given the problems he was considering, the approach he recommended was actually not at all unreasonable. We could say it was secretly the Bayesian reasoning in disguise, assuming a state of uniform ignorance about the true single-event probabilities. This likely explains why it seemed (and still seems!) to be a generally correct procedure that would produce good results for the toy examples he had in mind, things like sampling from an urn with unknown ratios that could equally plausibly be anything from 0 to 1. Working as he did over half a century before Thomas Bayes provided the crucial insight about inference and conditional probabilities (and another two decades before Pierre-Simon Laplace did so in such a way that anyone noticed), Bernoulli didn't yet have the conceptual tools to interrogate the latent assumptions he had been making. Statisticians of the 20th century who rejected Bayesian thinking and doubled down on Bernoulli's mistakes, on the other hand, had no excuse. We'll come back to them.

First, it will be illustrative to open up even more distance between the kinds of inferences Bernoulli's "sampling probabilities only" approach and the Bayesian approach produce. We'll do so by considering some real-life instances of probabilistic reasoning, from courtrooms to doctor's offices and elsewhere, in which the same essential fallacy as Bernoulli's has led to disastrous results.

THE CERTAIN OCCURRENCE OF UNLIKELY EVENTS

A common trope among "news of the weird" human interest stories is the "incredible coincidence": a couple has three children with the same birthday in separate years, a man and his brother are both hit by trains while bicycling in different countries on the same day, a woman strikes up a conversation with a stranger in an airport who turns out to have just left her an introductory voice mail, etc. Often these stories are accompanied by a quote from a statistics professor giving the odds against such a thing happening by chance—a million to one, a billion to one, and so on—or sometimes comparing the probability to that of other rare events—like winning the Powerball lottery and then being struck by lightning. Much has been written about the general phenomenon that these coincidences are actually much more common than we tend to think, owing mostly to the number of possible chances each coincidence is given to happen (the large number of three-child families in the world, say) and the number of such events that we would recognize if given the chance (we would care just as much if the brothers had been sisters or if they had been hit by cars instead of trains, etc.). Persi Diaconis and Frederick Mosteller called this the "law of truly large numbers": given a large enough sample size, any outrageous thing is bound to happen.[12]

Often left unstated, though, is why the occurrence of something highly improbable should be inherently noteworthy *at all*. For example, here is a way to produce an outcome almost certainly never seen before in human history and nearly certain never to be repeated: shuffle a deck of cards. The resulting permutation, if the cards are shuffled correctly, should occur only about once every 52 factorial shuffles—that is, $52 \cdot 51 \cdot 50 \cdot \ldots \cdot 2 \cdot 1$—because this is the number of possible shuffles, all of which should be equally likely. This number is on the order of 10^{68}, or one hundred million trillion trillion trillion trillion trillion. Every person on earth could shuffle cards once every nanosecond for the expected lifetime of the universe and not even put a dent in that number. So why is there not a news story written every time a deck is shuffled about the astronomically unlikely event that just took place? Why aren't statisticians brought in to comment that this was like winning the Powerball eight times in a row?

The reason is that the vast, vast majority of those possible shuffles do not contain any recognizable *pattern* that is suggestive of any alternative hypothesis other than "pure chance." These alternatives are usually secretly lurking in the background as unlikely suspects until some particularly improbable data reveals them. For example, under the assumption that a coin is fair, the sequences of 20 flips

HHTHTTTHTHHHTTTTHTHHT and HHHHHHHHHHHHHHHHHHHH

have exactly the same probability: $(1/2)^{20}$, or about 1 in 1 million, which is a pretty small chance. But only the latter is suggestive of an alternative hypothesis: that the coin is biased or even has heads on both sides, which would make the observed outcome *certain*. The same property is what makes "pseudorandom" number generators succeed or fail. A good generator is one that produces results (like those on the left) that don't trigger any of our available pattern-recognizing hypotheses (like those on the right).

Bayesian reasoning can help us detect these patterns—but only if we load up our inference table with the possible alternatives and their prior probabilities. For example, if we initially assume only that the coin is *very probably* fair but has, say, a 1-in-10,000 chance of being double-headed (maybe we've heard of the existence of such coins but consider them extremely rare), then our Bayesian inference table for the sequence of 20 heads in a row would look like table 2.3. And our posterior probability assignment for the double-headed coin theory would be about **99 percent**. So the data being *so* unlikely under the assumption of chance has revived our alternative hypothesis and has, in fact, almost completely convinced us that it's true.

Of course, not all unlikely data has this effect on any given hypothesis. Whether an observation counts against a hypothesis depends entirely on the alternatives. It may be that an unlikely event even makes the hypothesis *certain*. For example, if H was the proposition "Mr. Smith bought a lottery ticket" and D was the event "Mr. Smith won the lottery," then even though the probability of D given H is vanishingly small (it's very unlikely someone would win the

TABLE 2.3 **Coin flipping inferences after 20 consecutive heads**

Hypothesis H	Prior probability $P[H\|X]$	Sampling probability $P[D\|H \text{ and } X]$	Pathway probability $P[H\|X]P[D\|H \text{ and } X]$	Relative proportion $P[H\|D \text{ and } X]$
Fair coin	0.9999	$(1/2)^{20} \approx 9e^{-7}$	$9e^{-7}$	$\approx 1\%$
Double-headed coin	0.0001	1	0.0001	$\approx 99\%$

lottery, assuming only that they bought a ticket), we would not say D makes H unlikely at all. Quite the opposite; you need a ticket to win!

This is why particular *forms* of unlikely occurrences are so noteworthy. They carry an enormous potential energy like a coiled-up spring that could be released to launch an unlikely alternative hypothesis, such as the idea that something other than chance is at work, into the heights of near certainty. We humans are optimistic beings who are generally very skilled at finding patterns in things, so when we see some evidence that the world is not totally random and some underlying order is guiding events, it's natural that we get a frisson of excitement. That feeling is an unlikely alternative hypothesis being elevated. But the data being unlikely is not, in itself, always enough to give us that feeling.

Even the *possibility* of observing such unlikely patterned data, and the effect it *would* have on our inferences, can tell us something about our present state of mind. In the previous coin-flipping example, we know that a similar sequence of 20 *tails* in a row is suggestive of some other kind of bias toward tails. But unless this is included in our available universe of hypotheses with some non-zero prior probability, we could never reach this inference no matter what we saw. So instead of just the two hypotheses given earlier, we should include a third one, involving, say, a double-tailed coin. And once we start thinking about it, we should admit bizarre possibilities like coins that always alternate results between heads and tails (or are made to do so by some trickery on the part of the person doing the flipping), so that sequences like

HTHTHTHTHTHTHTHTHTHT

could have some inferential meaning for us, other than our just deeming them unlikely. Disturbing though it may be, we must have some prior probability, maybe close to 0, that a coin like that could exist.

PROBABILITY AND THE PARANORMAL

Suppose someone, say Ms. Stewart, claims to have ESP abilities, by which she means that if you think of a whole number between 1 and 10, she can tell you the number with certainty. What probability would you assign to her claim being true?

If at first it seems like the answer is surely zero—you don't believe in ESP, period!—perhaps the following thought experiment can elicit a nonzero

TABLE 2.4 **ESP inferences: from initial skepticism to uncertainty**

Hypothesis H	Prior probability P[H | X]	Sampling probability P[D | H and X]	Pathway probability P[H | X]P[D | H and X]	Relative proportion P[H | D and X]
E = Ms. Stewart has ESP	10^{-10}	1	10^{-10}	≈ 50%
C = She's just guessing	$1 - 10^{-10}$	10^{-10}	≈10^{-10}	≈ 50%

probability from you. Imagine conducting an experiment over which you have complete control. You may set the conditions however you want, but the basic setup should include you thinking of a number and Ms. Stewart guessing it (with the results honestly recorded), and the controls of the experiment should allow only two possibilities, meaning the only hypotheses allowed by your background assumptions are E: "Ms. Stewart has the ability she claims to have" and C: "Ms. Stewart is guessing numbers by chance." Under these conditions, how many correct guesses in a row would it take just to make you *uncertain* between E and C?

If the answer is any finite number, then it can be used to work backward and deduce your prior probability assignment for the ESP hypothesis E. For example, if you said it would take 10 guesses in a row—a pretty skeptical answer—then your prior probability for E must be about 10^{-10}, or 1 in 10 billion, because this is the prior probability that, when put into the inferential machine, would result in about a 50:50 posterior chance for E and C, as shown in table 2.4. The pathways are close to equal, so the posterior probabilities are each about 1/2, meaning you're on the fence between belief and doubt.

It may be, of course, that, in any *real* experiment, we would always admit the possibility of there being some trick, so the previous conditions would never actually be satisfied.

This exact situation came up during a famous series of experiments conducted in the 1940s by "parapsychologists" Samuel Soal and Kathleen Goldney to demonstrate the existence of ESP. Over several long sessions, one of their subjects, Gloria Stewart, was able to guess which symbol (a picture of an animal) out of a group of five was being thought of by an assistant in the adjacent room at a rate somewhat higher than pure chance would dictate. She correctly guessed 9,410 out of 37,100, about 25.3 percent, whereas the chance expectation was 20 percent.

How unlikely was this result? Letting C again be the hypothesis that she was purely guessing, Bernoulli's formula gives us the following probability of her getting this many correct:

$$P[9,410 \text{ correct out of } 37,100 \mid C] = \binom{37,100}{9,410}(0.2)^{9,410} (0.8)^{27,690} \approx 10^{-139}$$

That is, the probability is a number beginning with 138 zeros after the decimal point. This puts even our deck-shuffling probability to shame!

But with that previous lesson in mind, we know we ought to *compare* this with the probability under the alternative hypothesis. Supposing for simplicity that the meaning of ESP here is an increased chance of guessing cards correctly to exactly this degree, we can compute the probability of getting this result using Bernoulli's formula again. Thinking of the subject getting an answer correct as drawing a white pebble from an urn, it would be as though she were drawing from an urn with 25.3 percent white instead of one with 20 percent white like the rest of us. Letting $E_{0.253}$ denote the hypothesis that she has ESP in exactly this way, we'd find

$$P[9,410 \text{ correct out of } 37,100 \mid E_{0.253}] = \binom{37,100}{9,410}(0.253)^{9,410} (0.747)^{27,690} \approx 0.005$$

So the probability under this alternative hypothesis, relative to that under the chance hypothesis, is about *136 orders of magnitude* greater, or a difference about 100 billion times greater than the ratio in volume of the observable universe to a single proton. Why, then, does this not lead us to conclude with near certainty that it's true?

That was certainly the conclusion Samuel Soal wanted people to draw, and many did. For quite some time, the Soal-Goldney experiments were cited as positive proof that ESP was real. And these were no cheap back-alley illusions. Soal was a respected mathematician who had lectured at Oxford and Queen Mary College in London, where he received his doctorate. He was a Fulbright scholar and held an honorary fellowship in psychology from Birkbeck College at the University of London. He had first become interested in the occult after one of his brothers was killed in World War I. Like many people in England at the time who had lost loved ones in the war, he sought out ways of communicating with the departed. The difference was that Soal conducted his observations with a scientific approach. He had used rigorous statistical methods to analyze many previous experiments in telepathy and had concluded every time that no telepathic effect was seen. His positive results were published in *Nature*.[13]

Soal had gone to elaborate lengths to design the protocols of his experiment in such a way that cheating, or even assistance given to the subject *unintentionally*, seemed impossible. For each trial, his assistant doing the "transmission" was given a card telling him which animal to think about. Cards were chosen according to a random series of digits from 1 to 5, with the card that corresponded to any particular digit *also* randomized every so often. The final card was seen only by the assistant, who was physically separated from Stewart by a wall. The experiments were monitored by a panel of observers. As the philosopher C. D. Broad wrote, "Dr. Soal's results are outstanding. The precautions taken to prevent deliberate fraud or the unwitting conveyance of information by normal means are described in great detail, and seem to be absolutely watertight."[14] A biology professor at Yale, G. Evelyn Hutchinson, wrote, "Soal's work was conducted with every precaution that it was possible to devise."[15] If any experiment was going to scientifically prove the existence of ESP, this would be the one, and Samuel Soal would be the person to do it.

In order *not* to believe him, then, we would need some *other* hypothesis that could also explain the data. Suppose we admit a small residual possibility that the results were due to fakery of some kind. Let F denote this hypothesis, and let's suppose it explains the given data about as well as the hypothesis that Stewart really has ESP, so it has a sampling probability somewhere on the order of 0.005. The problem for Soal is that no matter how much scrutiny his experiments were placed under, we'd likely assign a *prior* probability for F that was several orders of magnitude greater than our prior probability for the ESP hypothesis. Let's say that all his precautions led us to a conservative $P[F|X]$ of 1 in 10,000, or 10^{-5}. Our Bayesian inference table would then look something like table 2.5.

So even though the observational data has completely killed the "chance" hypothesis, it has elevated the fakery hypothesis just as much as the hypothesis that the subject really has ESP, with the result that there is *still* a difference of five orders of magnitude in the posterior probabilities we'd give the one over the other. Simply put, there's nothing Soal could ever do to shake us of our suspicion that the results were due to some kind of deception rather than ESP because we had a deep skepticism of the ESP claim from the start, as reflected in our prior probabilities, and deception explains the data just as well as his theory.

TABLE 2.5 ESP inferences with the possibility of deception

Hypothesis H	Prior probability $P[H \mid X]$	Sampling probability $P[D \mid H \text{ and } X]$	Pathway probability $P[H \mid X]P[D \mid H \text{ and } X]$	Relative proportion $P[H \mid D \text{ and } X]$
$E_{0.253}$ Ms. Stewart has partial ESP	10^{-10}	0.005	$5e^{-13}$	$\approx 0.001\%$
C She's just guessing	$1 - 10^{-10} - 10^{-5}$	10^{-139}	$\approx 10^{-139}$ [effectively 0]	$\approx 0\%$
F Fakery	10^{-5}	0.005	$5e^{-8}$	$\approx$ **99.999%**

As David Hume wrote in *An Enquiry Concerning Human Understanding* (with the argument that may have inspired Bayes to prove the theorem we have just applied): "No testimony is sufficient to establish a miracle, unless the testimony be of such a kind, that its falsehood would be more miraculous, than the fact, which it endeavours to establish."[16] Soal even seemed to recognize as much. In his book *Modern Experiments in Telepathy*, he wrote that if the experimenters "are not to be trusted, then there is no point whatever in their doing experiments."[17]

By way of epilogue, years later it was discovered that Soal had doctored his data to incrementally improve Stewart's rate of successful guessing. He had strategically added extra 1s to his lists of random digits to generate the card sequences; then, if the guesses for those cards had corresponded to a 4 or a 5, he simply wrote over the 1 to make it a 4 or 5 to match the guess.[18] So the F hypothesis turned out to be correct after all.

REDUCTIO AD SO WHAT?

The larger lesson of these examples is that, in order to draw any conclusions from data, even very unlikely data, we need to consider what hypothesis we are testing and against what alternatives. Those alternatives may not be the intended ones and can often be revealed if we think about what conclusions we *would* draw based on some possibly suggestive observations. With only one hypothesis specified, though, we have no way of making any inferences at all. We're left in the position of poor Guildenstern from *Rosencrantz and Guildenstern Are Dead*, losing 90 coin flips in a row and not knowing whether to interpret his bad luck as divine intervention, a suspension of natural laws, or "a spectacular vindication of the principle that each individual coin spun individually is as likely to come down heads as tails and therefore should cause no surprise each individual time it does."[19]

Part of Bernoulli's Fallacy, then, is that the proposed method of inference makes no explicit mention of alternative hypotheses at all. His statement that, for example, under the assumption that the proportion of white:black pebbles in the urn is 2:3, the ratio in any large sample is "morally certain" to be close to 2:3 is *meaningless* on its own. It's morally certain that if we shuffle a deck of cards and deal a five-card poker hand, it will *not* be the ace of clubs, jack of spades, 8 of hearts, and 5 and 3 of diamonds, but that doesn't mean anything noteworthy would necessarily have happened if it were. Unlikely events happen all the time, as in the story Richard Feynman sardonically recounted in a lecture about the scientific method: "You know, the most amazing thing happened to me tonight . . . I saw a car with the license plate ARW 357. Can you imagine? Of all the millions of license plates in the state, what was the chance that I would see that particular one tonight? Amazing!"[20]

Fisher, in his final book, *Statistical Methods and Scientific Inference*, described what he considered to be the fundamental logic of using statistical tests to reject chance as an explanation: "The force with which such a conclusion is supported is logically that of the simple disjunction: Either an exceptionally rare chance has occurred, or the theory of random distribution is not true."[21] Fisher's statement is true—in the most pedantic sense possible. Another way of phrasing it for the situation he considered would be to say, "If the chance hypothesis is true, then what we observed is very unlikely." Since by rules of logic "A implies B" is equivalent to "B or (not A)," we can grant Fisher's disjunction as correct. But the logical truth of this either-or statement was never the point.

Fisher intended our distaste for accepting an unlikely proposition to drive us to the other conclusion: if we take it as given that something very unlikely has probably not occurred, then we should conclude the chance hypothesis is very probably false. But what he and his disciples misunderstood was that something being "exceptionally rare" doesn't necessarily mean it's *interesting*, nor should we necessarily count it as strong evidence of anything. Exceptionally rare chances occur constantly, and the mere fact of their rarity may have absolutely no bearing on what conclusions we draw from having witnessed them. In particular, Fisher's "or" statement is certainly not an exclusive *or*; even after rejecting the chance hypothesis in favor of some other theory, we may find our given observation is still extremely unlikely. Unfortunately, this mistaken argument is the basis for much of what now passes for statistics, as we'll see later.

It is extremely tempting to think that if a hypothesis makes something very unlikely and we have nevertheless observed that thing, this must mean something about the hypothesis is wrong. A similar but valid line of reasoning exists in ordinary deductive logic. If we have as a premise that if P is true, then Q is *impossible*, but we also know that Q is true, then we can conclude P is false. For example, if the Earth is flat, it would be impossible to sail around it; however, we can sail around it, so we know the Earth is not flat. As a technique of proof, this form of argument is known as *reductio ad absurdum* ("reduction to the absurd") and has been used since the time of the pre-Socratic philosophers in ancient Greece. Formally, it could be written as the following deduction:

If P is true, then Q is impossible.
Q is true.

_____

$\therefore$ P is false.

It might seem we can change "impossible" to "very unlikely" in that formula and not have lost much. Maybe instead of concluding that P is false, we can at least make a valid inference that P is very unlikely. (Jordan Ellenberg, for example, in *How Not to Be Wrong*, calls this the "*reductio ad* unlikely" argument.[22]) However, that small change makes all the difference in the world! As we have seen, not only are unlikely events sometimes *not* cause for concern, but also they may even be certain to occur. So what? Everything depends on what other hypotheses are available and whether they might better explain the assumed observation.

In Bernoulli's example, the alternatives we are *implicitly* meant to consider are those corresponding to different proportions of pebbles in the urn. Restricting our attention to those hypotheses, we could say the proportions in a sample have some inferential content. But we will still need to have prior probabilities for these hypotheses to draw any conclusions, a point we will now take up in more detail.

BASE RATE NEGLECT

Suppose that, after a routine checkup, your doctor informs you that you have tested positive for a rare disease and the test for this disease is, sadly for you, highly accurate. When pressed for details, they tell you that the test will always come back positive when the disease is present and that 99 percent of people without the disease will test negative. What do you estimate your chances are of having the disease? If you answered 99 percent (or any number at all, really, given the available information), you have fallen into a common fallacy, and you are not alone. When presented with this question, many people, including many doctors, will reach the same conclusion. An informal survey once revealed that, of 100 doctors, 95 of them reasoned incorrectly in similar situations, getting the answer wrong by a factor of 10 or more.[23]

The news for you, in fact, could be very good. Depending on the *overall incidence rate* of the disease in the population, your likelihood of having it even though you tested positive could be very small. To see how this plays out in Bayesian reasoning, consider the possible hypotheses:

Y: "You have the disease in question" and N: "You don't have the disease"

Then we can express the given accuracy rates of the test in terms of probabilities (with X representing all the assumptions leading us to these assignments):

$$P[\text{"You test positive"} \mid Y \text{ and } X] = 1 \text{ and}$$
$$P[\text{"You test negative"} \mid N \text{ and } X] = 0.99,$$
$$\text{so } P[\text{"You test positive"} \mid N \text{ and } X] = 0.01$$

The key ingredient most often overlooked, however, is the prior probability of the hypotheses Y and N—that is, the overall incidence rate of the disease in the population. In this hypothetical, we assume the disease is very rare, so maybe something like a rate of 1 in 10,000 people is a reasonable assessment. These give the following prior probabilities:

$$P[Y \mid X] = 0.0001; P[N \mid X] = 0.9999$$

By now, we are experienced pros at turning all these givens into a Bayesian inference for the probability we actually want, $P[Y \mid \text{"You test positive" and } X]$. That is, we are conditioning on the thing we have observed, the proposition D: "You test positive." (See table 2.6.)

As a relative proportion, the first pathway to the data represents a little under 1 percent probability. That is, even after having tested positive for the disease with a test that is very accurate (as these things go), you still have less than a 1 percent chance of really having it. We see very simply that the possibility of

TABLE 2.6 **Inference after testing positive for a rare disease**

Hypothesis H	Prior probability $P[H \mid X]$	Sampling probability $P[D \mid H \text{ and } X]$	Pathway probability $P[H \mid X]P[D \mid H \text{ and } X]$	Relative proportion $P[H \mid D \text{ and } X]$
Y You have the disease	0.0001	1	0.0001	$\approx 1\%$
N You don't have the disease	0.9999	0.01	$0.9999 \cdot 0.01 = 0.009999$	$\approx 99\%$

any false positives at all, combined with the overwhelming prior likelihood of not having the disease, leads to a very high chance of any given positive test result being a false positive.

Now all of this could be affected by *other* tests or by other observations such as a particular set of symptoms you've experienced that led you to have the test in the first place. Our posterior probability assignment here could become the new prior probability for a second round of tests, and the ultimate conclusion might be that you very likely do have the disease. But the general lesson, as always, is that unlikely hypotheses should require a lot of evidence to raise them up into the realms of plausibility. Any particular observation may not do the trick, especially if it could be better explained by a more plausible alternative. But many people, including practitioners, do not appreciate that fact.

In the world of medicine, this fallacy is called *base rate neglect*, since our intuitive answer failed to take into consideration the base rate of occurrence in the population. Real-life examples are found all the time in screening tests for conditions like dementia and various cancers. Suppose an otherwise healthy woman in her 40s goes for an annual mammogram, and the scan finds an anomaly consistent with breast cancer. How worried should she be? Well, assuming some realistic test accuracy numbers—say the scan will correctly identify 90 percent of cancers and will turn up a false positive in 7 percent of women without cancer—and, critically, a *low incidence rate* for breast cancer for women similar to this patient—say 0.5 percent—we can process the chances in a Bayesian inference. And we see in table 2.7 that the posterior probability for having cancer is only about 6 percent. So this may be cause for concern, and perhaps enough to warrant some further tests, but it is *far* from probable that the cancer diagnosis is real, even given the test results.

TABLE 2.7 The effect of low base rates on cancer screening tests

Hypothesis H	Prior probability $P[H \mid X]$	Sampling probability $P[D \mid H \text{ and } X]$	Pathway probability $P[H \mid X]P[D \mid H \text{ and } X]$	Relative proportion $P[H \mid D \text{ and } X]$
Cancer	0.005	0.9	0.0045	≈ 6%
No cancer	0.995	0.07	0.06965	≈ 94%

TABLE 2.8 Screening tests for drunk driving: a low base rate

Hypothesis H	Prior probability $P[H \mid X]$	Sampling probability $P[D \mid H \text{ and } X]$	Pathway probability $P[H \mid X]P[D \mid H \text{ and } X]$	Relative proportion $P[H \mid D \text{ and } X]$
Drunk driver	0.001	1	0.001	≈ 2%
Not a drunk driver	0.999	0.05	0.04995	≈ 98%

Other examples abound, occurring basically any time we try to detect something rare by means of an imperfect test. For example, imagine a checkpoint is set up on the highway to identify drunk drivers, and the screening test (for example, taking a breathalyzer test or having to say the alphabet without singing) has some rate of false positives. Assuming a low proportion (say 0.1 percent) of those people being checked are actually driving drunk, we might end up with an inference table like that in table 2.8, with posterior chances of only about 2 percent that someone is a drunk driver given a positive test result.

Or suppose some screening process is implemented at airport security checkpoints to try to identify terrorists based on travel patterns or other personal information. Since the population of terrorists is extremely small, the low base rate will effectively distort the results of any imperfect screening criterion (as all such tests will be). Numbers on the order of those in table 2.9 could seem reasonable.

The posterior chance of someone being a terrorist, given that they match all the criteria, is around 0.01 percent, or 1 in 10,000. How many such screenings are already taking place in airports around the world? Do you suppose the security agents treat the suspects they've identified as though they are almost

TABLE 2.9 **Airport screening for terrorists: an extremely low base rate**

Hypothesis H	Prior probability P[H | X]	Sampling probability P[D | H and X]	Pathway probability P[H | X]P[D | H and X]	Relative proportion P[H | D and X]
Terrorist	0.000001	1	0.000001	≈ 0.01%
Not a terrorist	0.999999	0.01	0.00999999	≈ 99.99%

certainly innocent and wrongfully caught up in an imperfect net? What about similar screenings for illegal immigration or voter fraud?

The pattern that unfortunately compounds these problems is this: the more rare something is, generally the more *severe* it is as well, so a risk assessment biases us toward a cautious approach of allowing more false positives in order to avoid false negatives. That is, in statistical terms we should want our tests for serious problems like cancer or drunk driving or terrorism to have a high *sensitivity*, which usually carries with it a low *specificity*. The Bayesian analysis shows right away the by-product of that cautious approach: as a relative proportion, we end up with a lot of false diagnoses or falsely accused suspects. We could even possibly accept all that if we come to terms as a society with what these numbers mean, but first we need the inferential tools to assemble the pieces.

THE PROSECUTOR'S FALLACY

Other unfortunate examples of the same type of logical error are found in courtroom arguments that a suspect under trial must almost certainly be guilty because the circumstances of the case are so *unlikely*.

Sally Clark was a fairly affluent woman who had worked in banking in London before training as a solicitor and joining a law firm in Manchester, England, in 1994. In September 1996, she gave birth to an apparently healthy baby boy who died suddenly less than three months later. Clark had been alone with the child at the time, and her claim was that he had fallen unconscious and stopped breathing shortly after she put him to bed. Following the incident, she fell into a deep depression, sought out counseling, and was in recovery when she had another baby boy, this one three weeks premature, in November 1997. Tragically he also died within eight weeks of being born under circumstances similar to those of her first child. Notably the second infant showed some signs of trauma, which Clark explained were likely caused by her attempts to resuscitate him before paramedics arrived or by the attempts of the paramedics themselves. She and her husband, Steve, were both arrested in February 1998, and Sally was charged with two counts of murder (the charges against Steve having been dropped).

During the trial, the fact that it was extremely unlikely for a pair of infant deaths to happen to such a family as a result of SIDS—that is, to happen by chance—was presented as a key piece of evidence. The pediatrician Roy Meadow, formerly a professor at the University of Leeds and inventor of the term *Munchausen syndrome by proxy*, gave testimony that the chance of two children from an affluent English family dying from SIDS was something like 1 in 73 million. He colorfully compared this to the chance that an 80-to-1 long-shot at the Grand National horse race would win four years in a row.[24] As he opined in his book on the subject, *ABC of Child Abuse* (in what came to be known as Meadow's law): "One sudden infant death in a family is a tragedy, two is suspicious and three is murder unless proven otherwise."[25] Based largely on this testimony and the idea that "lightning doesn't strike twice," Clark was convicted and sentenced to life in prison. The press coverage at the time reviled her as a child murderer.

Her husband, also a solicitor, quit his job to focus on her appeal. By combing through the prosecution's records, they found that the pathologist who testified about the results of medical exams on the second child had withheld key evidence from the jury—specifically, that tests for a bacterial infection of the cerebrospinal fluid had come back positive. On the basis of these findings, her conviction was overturned in January 2003, after she had spent more than three years in prison.

Meadow's statistical testimony was also widely criticized. His figure of 1 in 73 million was based on an estimate that the chance of a single child dying from SIDS in any given family similar to the Clarks was 1 in 8,543; from there, he reasoned that the chance of two such deaths in a given family would be 1 in $8,543^2$, or 72,982,849. This line of reasoning assumes the two events to be *independent*, though, so the probability of a second child dying is unaffected by the conditional assumption of a first child having died. This assumption would be negated by the possible presence of any common cause within the family, such as a genetic condition or an environmental health issue. In October 2001, the Royal Statistical Society issued a statement criticizing Meadow's independence assumption: "There are very strong reasons for supposing that the assumption is false. There may well be unknown genetic or environmental factors that pre-dispose families to SIDS, so that a second case within the family becomes much more likely than would be a case in another, apparently similar, family."[26]

Also, Meadow's bizarrely precise initial figure of 1 in 8,543 came from a study commissioned by the British Department of Health and was the result of adjust-ments applied to the overall incidence rate of SIDS at the time—about 1 in 1,300—based on certain factors that were known about the Clark family: they were an affluent couple in a stable relationship, Sally was over 26 years old, and the Clarks were nonsmokers, all of which were known to decrease the likelihood of SIDS. Critics such as mathematics professor Ray Hill at Salford University pointed out Meadow had overlooked factors that would *increase* the likelihood of SIDS for the Clark family, including that both children were boys.[27]

All of these points of criticism were important and well founded, but the single greatest problem with Meadow's testimony, and what *should* have been presented vociferously in Sally Clark's defense, was that he had been computing the *wrong probability*. That is, in our language Meadow had been focused only on the *sampling* probability of a given event—the event of two apparently oth-erwise healthy children in the same family dying suddenly in infancy—when he should have been considering the *inferential* probability for his hypothesis that the two children had been murdered. He argued, under the alternative hypothesis that they were well taken care of, that this care would have made their deaths incredibly unlikely, and he used this as evidence that the hypothesis itself was unlikely.

But this is just like the base rate neglect examples given earlier. Two children dying in infancy by *whatever* means is already an extremely unlikely event, but that is the data observation that we must condition on. The whole landscape of our probability assignments needs to change to reflect the fact that, by necessity, we are dealing with an extremely rare circumstance. And the *prior* probabil-ity we should reasonably assign to the proposition "Sally Clark murdered her two children," determined before considering the evidence, is itself *extremely* low because double homicide within a family is also incredibly rare! Included among the inferences, we should also note that some of the factors that make a couple like the Clarks less likely to have a child die from SIDS also lower our assignment of the probability that they are murderers.

Carrying through a Bayesian inference (that also corrected for the flawed reasoning in Meadow's sampling probability), Hill estimated in an article for the journal *Paediatric and Perinatal Epidemiology* that the posterior probability for the SIDS hypothesis was somewhere between 70 and 75 percent.[28] That is, a low sampling probability did not make the SIDS hypothesis an unlikely explanation for the deaths of the two children; it actually made it the significantly *more* likely explanation.

The judges of the appellate court noted that Meadow's calculations had been predicated on a number of questionable assumptions, none of which had been made clear to the jury. Furthermore, they observed that "we rather suspect that with the graphic reference by Professor Meadow to the chances of backing long odds winners of the Grand National year after year it may have had a major effect on [the jury's] thinking notwithstanding the efforts of the trial judge to down play it."[29]

Following Sally Clark's successful appeal in 2003, the attorney general ordered a review of all similar cases, and two other women convicted of murdering more than one of their own children, Donna Anthony and Angela Cannings, had their convictions overturned. A third, Trupti Patel, whose trial for the murder of *three* children was ongoing at the time, was acquitted. In all three cases, Meadow had testified as an expert witness that the chances of a family suffering multiple deaths from SIDS was vanishingly small.

After a hearing in 2005, the British General Medical Council struck Meadow from the British Medical Register for professional misconduct, though he was reinstated the following year after he appealed the decision to the country's High Court. His comeuppance, such as it was, had come too late for Sally Clark, though. People close to her said she never recovered from the traumatic experience of being wrongfully blamed for her children's deaths, and she was found dead of alcohol poisoning in her home in 2007.

In legal circles, the argument Meadow presented in these cases—that, under an assumption the suspect is innocent, the facts of the case would be incredibly unlikely, and, therefore, the suspect is unlikely to be innocent—is known as the *prosecutor's fallacy*. A famous and oft-cited example is the 1968 case of *People v. Collins*, which involved a pair of suspects, Malcolm and Janet Collins, who were arrested in Los Angeles for robbery based on matching certain characteristics given by eyewitnesses to the crime: that he was an African American man who may recently have had a beard and mustache, that she was a blonde woman who normally wore her hair in a ponytail, and that they drove a partly yellow car. A mathematics instructor at the nearby state college testified at the trial that the probability of a randomly chosen couple matching all the given characteristics was 1 in 12 million, based on multiplying the estimated probabilities supplied by the prosecution (see table 2.10).

As with Meadow's testimony in the Sally Clark case, there are obvious problems with this probability calculation, primarily because the factors in question

TABLE 2.10 **People v. Collins: claimed probability of a match for each characteristic**

Observation	Probability
Man with mustache	1/4
Woman with blonde hair	1/3
Woman with ponytail	1/10
African American man with beard	1/10
Interracial couple in a car	1/1,000
Partly yellow car	1/10

are treated as independent despite clear dependencies between them. Being an African American man with a beard should carry with it higher conditional probabilities of having a mustache, of being part of an interracial couple in a car, etc.

But *even setting those issues aside*, the reasoning of the prosecution that this low likelihood made the Collinses' innocence extremely unlikely is still flawed—for the same reason Meadow's was flawed. The probability is the *wrong way around*. If we assume, for example, that all we knew about Malcolm and Janet Collins is that they were a couple in the Los Angeles area, then our prior probability assignment for their guilt (not conditioning on their matching the given description) should be 1 divided by the number of all couples in the area at that time, maybe something like 5 million.

Then we can even grant the prosecution's claim that, assuming the Collinses were not guilty of the crime, the chance of a match with all the characteristics described by the witnesses is 1 in 12 million. If they were guilty, the chance of a match would be 1. The resulting inference table is shown in table 2.11.

Our posterior assessment of the probability of innocence would come out to be about 30 percent—strong evidence, maybe, but certainly not proof beyond a reasonable doubt on its own and a far cry from the 1-in-12-million figure the prosecution clearly hoped would cement itself in the jurors' minds. Here, again, we see that the incredibly small possibility of something happening "by chance" cannot completely overcome the prior probability against the alternative.

The California Supreme Court agreed and overturned the Collinses' convictions on appeal in a 5–1 decision. Writing for the majority, Justice Raymond Sullivan commented: "Undoubtedly the jurors were unduly impressed by the

TABLE 2.11 **People v. Collins: posterior inference with a low prior probability**

Hypothesis H	Prior probability $P[H \mid X]$	Sampling probability $P[D \mid H$ and $X]$	Pathway probability $P[H \mid X]P[D \mid H$ and $X]$	Relative proportion $P[H \mid D$ and $X]$
Guilty	1/5,000,000	1	1/5,000,000	≈ 70%
Innocent	4,999,999/5,000,000	1/12,000,000	4,999,999/(5,000,000 · 1,200,0000) ≈ 1/12,000,000	≈ 30%

mystique of the mathematical demonstration but were unable to assess its relevancy or value."[30]

Another, more recent example is the case of Lucia de Berk, a Dutch pediatric nurse who was convicted in 2003 of the murder and attempted murder of several patients under her care based on circumstantial evidence. Among the prosecution's arguments was the calculation, done by a law *psychologist*, Henk Elfers, that the probability of so many deaths or near deaths happening by chance during the same nurse's shifts was 1 in 342 million. According to professor of criminal law Theo de Roos, "In the Lucia de B. case statistical evidence has been of enormous importance. I do not see how one could have come to a conviction without it."[31]

Ultimately, de Berk's case was reopened, and the medical details of each patient's death were examined more closely with better diagnostic tools. During an appeal hearing in 2010, witnesses testified that the patients' deaths had been due to a combination of natural causes and hospital mismanagement, and de Berk was formally exonerated. But the fallacious statistical argument—that this series of deaths is improbable by chance, and, therefore, the suspect is unlikely to be innocent—should never have been allowed in the first place. As Mark Buchanan described the concern in a 2007 article in *Nature*:

> The court needs to weigh up two different explanations: murder or coincidence. The argument that the deaths were unlikely to have occurred by chance (whether 1 in 48 or 1 in 342 million) is not that meaningful on its own—for instance, the probability that ten murders would occur in the same hospital might be even more unlikely. What matters is the relative likelihood of the two explanations. However, the court was given an estimate for only the first scenario.[32]

As evidence based on the probabilistic tools of data science becomes more and more common (think social media activity, browsing history, location services, etc.), how many other juries will be dazzled into false inferences by similar arguments involving tiny probabilities?

ONE MISTAKE WITH MANY NAMES

Base rate neglect and the prosecutor's fallacy are the same thing, and both are examples of Bernoulli's Fallacy. When diagnosing a medical condition, it may seem the accuracy rates of the test by themselves are sufficient, but all the probabilities are stated in the wrong direction: P[Test result|Patient's condition] instead of P[Patient's condition|Test result]. This is exactly the same mistake committed in the prosecutor's fallacy, which argues that the sampling probabilities, P[Facts of the case|The suspect is innocent], are somehow all we need to assign a probability to the hypothesis of innocence. The same mistake shows up in Bernoulli's urn-drawing problem, where the probabilities P[Sample ratio|Urn fraction] are thought to determine the probabilities P[Urn fraction|Sample ratio].

We can make these fallacies look even more tempting if we rephrase them in Bernoulli's language of morally certain closeness. Taking the medical diagnosis example, we can make this true claim:

> The probability of the test result (positive or negative) matching the disease status (positive or negative) is high, no matter whether a person is truly positive.

In other words, given either case—that we have the disease or that we don't—the chances are extremely high, morally certain even, that we'll receive the test result that matches our underlying condition. According to the apparent symmetry of that statement, upon receiving a positive test result, we might be led to this claim:

> The probability of our disease status (positive or negative) matching our given test result is high.

That is, given a positive test result, we would think we're morally certain to have the disease.

But we have shown that this thought process leaves out critical information. The probability assignments in the two statements are conditioned on different assumptions: one is made assuming a given hypothesis and the other assuming a given observation, and these are simply not the same. The failure to appreciate (1) the need for alternative hypotheses and (2) the difference between sampling probabilities and inferential probabilities is what makes Bernoulli's argument incorrect and what has led us in the present day to numerous instances of mistaken medical advice, wrongful assessment of risk, and miscarriages of justice.

It has also, under a different guise, formed the basis of most of what we now think of as standard statistical practice. The problem, which we'll begin to explore in the following chapters, is that the dominant mode of statistical analysis these days isn't Bayesian. Since the 1920s, the standard approach to judging scientific theories has been significance testing, made popular by Fisher. His methods and their latter-day spin-offs are now without question the lingua franca of scientific data analysis. He claimed significance testing was a universal tool for scientific inference, "common to all experimentation,"[33] a claim that seems justified by its widespread use across all disciplines.

Fisher hated Bayesian inference with a passion and considered it a great historical error, "the only mistake to which the mathematical world has so deeply committed itself."[34] As a result, his methods don't have any place for prior probabilities, which he argued weren't necessary to make inferences. Significance testing uses only the probability of the data assuming a hypothesis is true—that is, only the sampling probability part of Bayes's rule. If the observed data (or more extreme data) would be very unlikely under a hypothesis, usually the *null hypothesis* of no effect, the data is deemed *significant* and considered sufficient evidence to reject the hypothesis.

Defending the logic of this approach, Fisher wrote, "A man who 'rejects' a hypothesis provisionally, as a matter of habitual practice, when the significance is at the 1 percent level or higher [that is, when data this extreme could be expected only 1 percent of the time] will certainly be mistaken in not more than 1 percent of such decisions. For when the hypothesis is correct he will be mistaken in just 1 percent of these cases, and when it is incorrect he will never be mistaken in rejection."[35]

However, that argument obscures a key point. To understand what's wrong, consider the following completely true summary of the facts in the disease-testing example (no false negatives, 1 percent false positive rate):

Suppose we test one million people for the disease, and we tell every person who tests positive that they have it. Then, among those who actually have it, we will be correct every single time. And among those who don't have it, we will be incorrect only 1 percent of the time. So overall our procedure will be incorrect less than 1 percent of the time.

Sounds persuasive, right? But here's another equally true summary of the facts, *including the base rate* of 1 in 10,000:

> Suppose we test one million people for the disease, and we tell every person who tests positive that they have it. Then we will have correctly told all 100 people who have the disease that they have it. Of the remaining 999,900 people without the disease, we will incorrectly tell 9,999 people that they have it. Therefore, *of the people we identify as having the disease*, about 99 percent will have been incorrectly diagnosed.

Imagine you or a loved one received a positive test result. Which summary would you find more relevant to you?

By ignoring the prior probability of the hypothesis, significance testing does the equivalent of diagnosing a medical condition based only on how often a patient will test positive if the condition is absent or of reaching a legal verdict based only on how unlikely the facts of the case are if the suspect is innocent. In short, significance testing would have told our hypothetical patient they probably have the disease and would have wrongfully convicted Sally Clark.

Finally, suppose the people who received positive test results and a presumptive diagnosis in our example were tested again by some other means. We would see the majority of the initial results fail to repeat, a "crisis of replication" in diagnoses. That's exactly what's happening in science today. Virtually every area of experimental science that uses statistics is now being forced to confront the fact that many of their established results are not reproducible.

Once we explore the historical origins of modern statistical methods, we'll turn our attention to the damage they have caused—and are continuing to cause—in the disciplines of science that have been built on them for decades.

The big question before we get there, though, is how the twisted logic of frequentist statistics ever became the standard in the first place. In part, this was due to how tempting Bernoulli's method of inference can sound when the argument is presented a certain way, as we've witnessed here. As we'll see, though, this was also in large part because of the extreme reactions probability provoked as it went from being a tool of gamblers and astronomers to handling questions about people and society. It is, in fact, no accident that Bernoulli's Fallacy persists to this day in the serious applications we have considered here: medical diagnoses, criminal justice, law enforcement, etc. When the chips are down, people reach for frequency-backed probabilities to provide a sense of stability. To understand that dynamic, and how frequentism and social science matured together, we'll first need to go back to the 19th century and meet probability's chief bridge builder, Adolphe Quetelet.

3

ADOLPHE QUETELET'S BELL CURVE BRIDGE

The application of this calculus [of probabilities] to matters of morality is repugnant to the soul. It amounts, for example, to representing the truth of a verdict by a number, to thus treat men as if they were dice.

~Louis Poinsot (1836)

As we saw in chapter 1, in its early days probability was mostly used to answer questions about games of chance. By the time Pierre-Simon Laplace and others had finished their work in the early 1800s, people had started applying probabilistic methods in the "hard" sciences like astronomy and geodesy, the study of the shape of the earth. A hundred years later it would become a staple of the "softer" disciplines of psychology, sociology, economics, law, population biology, etc., where it still resides today, but the trip was not an easy one. The person who perhaps most helped make the transition possible was a Belgian scientist named Adolphe Quetelet.

The move was foreshadowed by Jacob Bernoulli and by Laplace himself. In his earlier writings on probability in the 1770s, one of the first examples Laplace considered was something outside of games of chance: analyzing the birthrates of boys and girls in Paris. The problem already had a long history by that point. In 1710, the witty Scottish writer John Arbuthnot (who also dabbled in mathematics when he wasn't hobnobbing with the likes of Jonathan Swift and Alexander Pope) had considered the question of whether the number of male and female births could be seen as purely the product of chance. He looked up the records of christenings in London over the 82-year period from 1629 to 1710 and found that, in every single year, there were more boys born than girls.

Reasoning that if chance alone was at work, the results should behave like a fair coin flip, he came up with the probability $(1/2)^{82}$, or about 1 in 4.8 trillion trillion, for what had happened—as though nature had somehow flipped 82 heads in a row. He took this incredibly small probability as enough evidence to reject chance as an explanation and wrote up his results in a note titled, unsubtly, "An Argument for Divine Providence."

People with a bit more nuanced view of probability, such as Nicolaus Bernoulli, pointed out that *chance* could mean more than just a 50/50 coin flip. For example, chance could still be at work even if the results were like drawing a ball from an urn with an unequal mixture, the kind of problem his uncle had thought about in depth. Over the following years, the fact that records for things like births and deaths had an apparent regularity to them was commented on by several authors, such as William Derham and Johann Süssmilch, both clergymen who also interpreted the regularity as evidence of the divine, and by Willem 's Gravesande, whose work was praised by none other than Johann Bernoulli, Jacob's brother.

It's not entirely clear what motivated Laplace in particular to think about this question except possibly that he didn't ultimately consider dice and urn games to be all that important. Before he devoted his life to mathematics, he had intended to study theology and become ordained in the Catholic Church, so perhaps he had some lingering religious distaste for gambling. Or maybe he, like Jacob Bernoulli, saw the potential for probability to do more than resolve gambling disputes. He would later write in the introduction to his *Théorie Analytique des Probabilités*, "Life's most important questions are, for the most part, nothing but probability problems."[1] And in his 1780 memoir on probability, he introduced the birth-ratio example by saying "As this matter is one of the most interesting in which we are able to apply the Calculus of probabilities, I manage so to treat with all care owing to its importance."[2] It also happened to be an opportunity for Laplace to try his techniques on a real-world problem with a fairly simple setup and an enormous amount of available data.

Consulting the records of the Academy of Sciences for the years 1745–1770, Laplace determined that 251,527 boys and 241,945 girls had been born in Paris during that time, which seemed to indicate a slight bias, about 51 percent, toward a baby being a boy. The question was whether this could be due to chance variation even if the true probability was 50 percent. With the same procedure he had used to interpret urn-drawing problems, he found a probability of approximately 10^{-42} that the true rate of male births was less than or equal to one-half. In other words, with very, very near certainty, he concluded that a newborn baby in Paris was more likely to be a boy than a girl. Even though the observed bias was small, the huge sample size indicated that it was almost certainly a real effect. He drew similar conclusions about the birthrates in London, with data covering a longer time span, and he was even able to calculate a high probability that the London rate was higher than the Paris rate. He theorized that there must be some characteristic of London that "facilitates the birth of boys: it may depend on climate, food, or customs."[3]

While it didn't quite rise to the level of proving the existence of God, as some might have wanted, this kind of extension of probability from urn drawing to real questions of relevance to society was entirely in keeping with Bernoulli's original ambitions, and it nicely coincided with a new general trend of collecting data in service to the social good. John Graunt, haberdasher by day and demographer by night, had made a breakthrough in London in 1662 when he used weekly mortality records to design an early warning system to detect outbreaks of bubonic plague in the city. Even though the system was never actually deployed, it opened people's eyes to the rich possibilities of data gathering and its usefulness to the state. By the 1740s, prominent thinkers such as the German philosopher Gottfried Achenwall had taken to calling this kind of data *statistics* (*statistik* in German), the root of which is the Latin word *statisticum* meaning "of the state."

Statistical practice in the early days was often purely descriptive, but as soon as any inference from the data was required, the usual questions of uncertainty, measurement error, and chance variation would show up too. So statistics and probability were destined to be together from the start. Probability seemed to offer some useful strategies for addressing these issues, but the application of probability to statistics questions involved something of a logical leap.

Simple examples like guessing the contents of an urn had been quite amenable to probability models, and applying the same idea to error-prone physical measurements like the recorded position of a planet had seemed reasonable enough, but could one really use the same inferential process to handle questions about people? Were the multitude of factors that determined whether a baby was a boy or girl really the same, in some sense, as black and white pebbles in an urn? One could make definitive and falsifiable statements about the contents of an urn (just look inside if you want to know the real truth), but how could anyone ever definitively establish any proposition about human tendencies?

This is where Quetelet came in. It happened that he was uniquely qualified, by both his personality and his place in history, to build a bridge from one domain to the other, usher the ideas of probability across, and in the process essentially create the discipline of social science. The construction of that bridge and the fruitful marriage of probability and statistics were made possible only thanks to a mathematical function called the *normal distribution*. If Quetelet's bridge had a shape, it would be a bell curve. As we'll see, without the support it provided, the bridge between hard and soft sciences would surely have collapsed because there were too many people who opposed the union of probability and statistics and were working hard to tear that bridge down.

But the version of probability that made it across the bridge was not the same as the one that started the journey. Primarily because of the opposition just mentioned, probability had to disguise its Bayesian side and appear to be mostly frequentist to be allowed into its new territory. For Laplace and his scientific contemporaries, a probability was sometimes a quantity to be assigned based on ignorance and other times something to be measured by observing frequencies. The dual-nature interpretation of probability was what allowed Laplace to simultaneously do what we'd now understand as pure Bayesian inference about the sun coming up while also musing about nature's ability to make events "occur more frequently as they are more probable," both observations that got him mocked by John Venn, as we saw in chapter 1.

Where Laplace and the others on the hard side had enjoyed all aspects of probability, freely switching between Bayesian and frequentist ways of thinking as the situation demanded, scientists on the softer side, for reasons having to do with their own struggles for legitimacy, could accept probability only if it meant frequency. Therefore, the first encounter between probability and statistics also marked the birth of strict frequentism, a philosophy of probability that would reach adulthood sometime in the mid-20th century, as we'll witness in the next chapter. Sparked by Quetelet's work in the 1840s, frequentism began attracting serious adherents for the first time and only became more dominant the more probability was applied to questions of social importance, where people's lives and well-being were at stake.

THE FAR-FROM-AVERAGE MAN

Quetelet was born in 1796 in Ghent, then part of the new French Republic. He studied at the Ghent Lycée and excelled at mathematics. In 1819, he both completed a dissertation on conic sections, receiving the first doctorate of science from the new University of Ghent, and moved to Brussels, where he began teaching at the Athenaeum. Over the course of his career, he was enormously prolific, completing some 300 works in fields ranging from pure mathematics to astronomy to sociology, and founded numerous statistical organizations and a journal, *Correspondance mathématique et physique*. He was, without a doubt, the most influential Belgian scientist of the 19th century. In his free time, he also wrote the libretto to an opera, several volumes of poetry, and a historical study of romance.

Around 1820, Quetelet decided that what Brussels needed, to cement its status as an intellectual center, was an astronomical observatory. So he began fund-raising and garnering support among government officials to build one, with himself as the director, naturally. The one problem was that he had, at that point, no practical knowledge of astronomy, but he did not let that stand in his way. In 1823, he left for a three-month-long trip to Paris (which he managed to get the state to pay for) so that he could learn from the eminent scientists and masters of astronomy there, such as François Arago and Alexis Bouvard, both students of Laplace. These astronomers, in turn, introduced Quetelet to their friends in the community of intellectuals in the city, and it was through these friends that he discovered the Laplacean ideas of probability and its applications, which were to have a profound influence on the course of his career.

Quetelet succeeded in building his Brussels Observatory in 1828, but he also became a kind of regional consultant on matters of statistics. By that time, Belgium was part of the United Kingdom of the Netherlands, and a census of the population was planned for 1829. Past attempts at estimating population had been confounded by the turbulent political conditions of the region and the incomplete records of population change due to migration. While in Paris, Quetelet had learned about a new method of Laplace's for estimating population without conducting a full census.

Laplace's idea was this: start with the number of births in a year, of which reliable records could be found (as he had seen in his analysis of the rates of male and female births), and then multiply that figure by an estimated *ratio* of population to number of annual births. This ratio could be estimated from a full census of smaller, but similar, regions or communities. Laplace had shown the method to be accurate for estimating the population of France. Quetelet saw the appeal of saving the cost of doing a full census with only a minor loss in accuracy and proposed applying the method to his own country.

Ultimately Quetelet was talked out of this approach by the Baron de Keverberg, who was serving as an adviser on matters of state. Keverberg's argument (preserved by notes in Quetelet's memoir) was essentially that different communities would have different ratios of population:births owing to factors unique to each place:

> The law regulating mortality is composed of a large number of elements: it is different for towns and for the flatlands, for large opulent cities and for smaller and less rich villages, and depending on whether the locality is dense or sparsely populated. This law depends on the terrain (raised or depressed), on the soil (dry or marshy), on the distance to the sea (near or far), on the comfort or distress of the people, on their diet, dress, and general manner of life, and on a multitude of local circumstances that would elude any a priori enumeration. It is nearly the same as regards the laws which regulate births. It must therefore be extremely difficult, not to say impossible, to determine in advance with any precision, based on incomplete and speculative knowledge, the combination of all of these elements that in fact exists. . . . It is then doubtful that we will often find populous regions which, in this regard, can be assimilated the one with the other, and combined in the same category.[4]

We can see in Keverberg's concerns the exact same *reference class problem* that Bernoulli had once acknowledged in passing and that would become an issue for the frequentist interpretation whenever probability was applied to the complexities of life outside simple games of chance. To put Quetelet's problem in probability terms, if we want to assign probabilities to a country's population conditional on an observed number of births, we might try to base that inference on how often the relevant ratio of population:births has been observed for similar communities. But what counts as similar? As Keverberg articulated, a community of people, or even a single person, could be thought of as the combination of an endless number of variables, each with some bearing on the question at hand. By what right could we ignore these idiosyncrasies and group people together as though they were a homogeneous category? More than any mathematical problem, this conceptual obstacle proved to be the greatest hindrance to bringing probability to bear on problems of statistics.

Quetelet left his encounter with the census problem shaken but with a newfound interest in collecting and analyzing social data. Over the next several years, he tabulated and made graphical displays of a dizzying assortment of data sets: rates of birth and death by city, by season, by profession, and by time of day; measurements of people's heights, weights, strength, and rates of growth; crime reports and rates of drunkenness; suicides; marriage rates and the seasonal variation in rates of conception; and many more. His goal, perhaps antagonized by the Baron de Keverberg's skepticism, was to investigate analytically all the ways people *were* the same or different and to create a theory of *social physics*, a set of laws governing society that could be an equivalent of Kepler's laws of planetary motion and other immutable principles of the hard sciences. In this pursuit, Quetelet was his own Tycho Brahe, gathering massive amounts of data for himself the way Brahe had for Kepler.

In 1835, Quetelet compiled the results of his many observations in the two-volume book *Sur l'homme et le développement de ses facultés, ou Essai de physique sociale* (On man and the development of his faculties, or essays on social physics). It was, in a sense, the first work of social science as a quantitative discipline. The tantalizing subject matter and Quetelet's eloquent writing style made it an instant success. A glowing review in the *Athenaeum* that described the publication of these volumes as "forming an epoch in the literary history of civilization"[5] caught the attention of Charles Darwin and (along with Quetelet's later writings) helped spur him to develop the theory of natural selection.[6]

For the rest of his career in academics, Quetelet continued collecting social data and expanding on his analytical methods. The main idea of social science that Quetelet is known for now is that of "the average man" (*l'homme moyen*). The average man, for Quetelet, was a concept analogous to the physical center of gravity. It determined the central tendency of a group of disparate people to engage in one kind of behavior or another, have one physical characteristic or another, or experience one life event or another. It was numerically just the average of whatever data set he was considering, a technique lifted directly from the standard practice of averaging astronomical observations to achieve greater accuracy. But the apparent *stability* of these average propensities across time and place gave a sense that perhaps there was some order to human society after all. Even if an *individual* person's life was subject to chaotic ups and downs, the life of the average man was remarkably predictable. As Quetelet explained, a fundamental principle of his social physics was that "the greater the number of individuals observed, the more do individual peculiarities, whether physical or moral, become effaced, and allow the general facts to predominate, by which society exists and is preserved."[7] It was an egalitarian ideal that fit well with the political mentality of postrevolutionary France.

Quetelet occasionally allowed himself to wax poetic on the virtues of the average man, resulting in such lofty characterizations as this: "If an individual at any given epoch of society possess all the qualities of the average man, he would represent all that is great, good, or beautiful."[8] Some critics pointed out that the average man (actually about half man, half woman?) would necessarily not be of the highest morals but rather mediocre. Quetelet, to his credit, did not fixate on a single average man but allowed for many different average humans, depending on different ways of grouping people. In fact, the most useful thing about the concept was that it allowed Quetelet to compare averages from *different* groups—say the average Frenchman with the average Belgian or the average Englishman—or to relate one factor with another.

But there was still the problem that real people tended to deviate from the average, and Quetelet had not yet fully addressed Keverberg's concern about the dangers of placing dissimilar people in the same group. Could anything be said about the *regularity* with which individual differences of various magnitudes could be expected? Quetelet answered in the affirmative, drawing inspiration again from a tool of astronomy: Laplace's *law of the distribution of error*, or what we now call the *normal distribution*.

The normal distribution has a familiar bell curve shape and is described mathematically by the function

$$f(x) = Ce^{-(x-\mu)^2/(2\sigma^2)}$$

This says the probability of any given value x dies off exponentially (with the square of the distance) the farther x gets from some value called μ (the mean), which determines the location of the peak of the curve from left to right. The rate of dying off is controlled by the number σ (the standard deviation), which determines how wide the curve is, with a larger σ making for a wider bell. The constant C just ensures that the probabilities add up to 1.[9] As a result of the shape of the curve, most of the probability is concentrated near the middle: a quantity with this distribution has about a 68 percent chance of being within one standard deviation of the mean—that is, within distance σ on either side of μ.

To understand why Quetelet or Laplace would care about this function, we need to take a brief sojourn back to the early 1700s and work our way back to Parisian astronomy in 1810.

"OUR METHOD"

Abraham de Moivre gets credit for being the first to come up with the normal distribution as a way to approximate the binomial distributions required by Bernoulli's urn-drawing problems. Recall that Bernoulli had shown that, assuming a ratio of white:total pebbles in the urn equal to p, the probability of getting k white pebbles out of n would be given by the formula

$$\binom{n}{k} p^k (1-p)^{n-k}$$

This could also be thought of as adding up n random numbers, each taking the value 1 or 0 with probability p or $(1-p)$, respectively. These sums could represent the addition of one more white pebble to a sample, one more vote on a Venetian committee, or generally speaking one more "success" out of a number of trials that could each end in success or failure independently with fixed probabilities.

Plotted as a function of k, as it ranges between its minimum value of 0 and its maximum value of n, the binomial distribution also yields a bell curve–like shape, with a peak location near the mean value $p \cdot n$.

The problem is that for any decently large values of k and n, this function is extraordinarily laborious to compute without a calculator. Say, for example, we want to compute by hand the probability of getting between 5 and 15 white pebbles out of 30 from an urn with ratio 1/3, as we had in our example in the last chapter. We first need to consider the probability of getting exactly 5 out of 30:

$$\binom{30}{5}(1/3)^5 (2/3)^{25}$$

To convince yourself how difficult this is, pretend you're a mathematician in the 1700s, take out your best quill and parchment, and get to work. To begin with, you need to expand the binomial coefficient

$$\binom{30}{5} \text{ as } \frac{30!}{5! \cdot 25!}$$

You should get 142,506.

But now you still need to deal with the $(1/3)^5(2/3)^{25}$ part, which you could write as $2^{25}/3^{30}$. Expanding those powers and multiplying by the previous coefficient should give the final answer. Don't forget to reduce fractions.

Did you get $\dfrac{177,100,292,096}{7,625,597,484,987}$?

Great, now just repeat that process *10 more times* for the different values of *k* and add up all the resulting fractions (good luck with the common denominator) to get the final probability.

Practically, this was a nightmare for any concrete example, and making things even more difficult was the fact that Bernoulli and others were trying to prove general *theorems* about these kinds of sums. So instead of 30, 5, and 1/3, Bernoulli would have *n*, *k*, and *p* and then would be trying to sum the binomial coefficients *algebraically* to see that they conformed to the patterns he desired—namely, that most of the probability would be concentrated near the peak for large *n*, with precise meanings given to *most* and *near*. What made Bernoulli's proof of the Law of Large Numbers so impressive was that he actually managed to *do* all that.

De Moivre, in an act of remarkable inspiration, substantially improved on the process by showing that the normal distribution function given earlier—which he essentially just guessed—was a good approximation for the binomial distribution values, at least in the symmetric case of $p = 1/2$ and k close to the peak value $n/2$. The necessary sums for any probability calculation could then be obtained (at least approximately) using tables of logarithms or techniques from integral calculus.

Laplace expanded on de Moivre's result to show that a normal distribution would give a good approximation to the binomial for any success probability p and that, even more generally, the same law would emerge as the distribution of the sum of *any* large number of independent variables that could take arbitrary values—not just 1 and 0, as in the Bernoulli examples. This was a pretty common setup, so practically speaking it meant one should expect to find the normal distribution in all kinds of settings. In particular, if a quantity such as the error in an astronomical observation could be thought of as the sum total of a large number of small independent errors—say small imperfections in the observer's telescope—the overall probability distribution of the final error could be expected to follow a bell curve.

Around the same time Laplace was proving this general result, the great mathematician Carl Friedrich Gauss was working on an entirely separate derivation involving the normal distribution. It had been known for some time in experimental physics that averaging multiple observations of the same quantity was a generally good practice to achieve an accurate estimate. The arithmetic average was a pleasing physical analogue to the center of mass of a set of points, and it happened also to be the value that minimized the sum of the *squares* of the errors in values.

For example, suppose three observations of some unknown quantity come out to be 10, 12, and 17. If we guess the true value is x, then that implies the errors in our measurements must be $(x - 10)$, $(x - 12)$, and $(x - 17)$, respectively. The value of x that minimizes the sum of the squared error terms, $(x - 10)^2 + (x - 12)^2 + (x - 17)^2$, is the average of the three observations:[10]

$$x = (10 + 12 + 17)/3 = 13$$

Astronomers of the 18th and early 19th centuries, while trying to understand certain dynamics of the solar system such as the apparent accelerations of Jupiter and Saturn and the eccentricity of the moon's orbit about the earth, were faced with similar kinds of problems involving systems of *linear equations* in certain unknowns, like the distance from Jupiter to the sun, with coefficients determined by things they *could* observe, like the relative positions of all the planets. Typically, if many observations had been collected, there were more equations than there were unknowns. This meant the systems of equations were *overdetermined*, but each necessarily involved some errors due to the measurement error of the observations. Just like the earlier example, picking a solution for the unknowns was tantamount to assigning values to these errors.

In 1805, the French mathematician Adrien-Marie Legendre published a derivation for a general technique for this kind of problem that minimized the sum of the squared errors, called, for obvious reasons, the *method of least squares*. He advertised his approach on aesthetic grounds: "By this method, a kind of equilibrium is established among the errors, which, since it prevents the extremes from dominating, is appropriate for revealing the state of the system which most nearly approaches the truth."[11] But the minimum-sum-squared-error rule wasn't the only one that would have this "appropriate" property, and Legendre didn't (and likely couldn't) supplement the argument with any more rigorous mathematical justification of the least squares method.

Gauss, as most people know, was one of the most brilliant and influential mathematicians of all time, particularly in the fields of algebra and number theory. But he was also an accomplished empirical scientist, with particular interests in optics and astronomy, and some of his greatest contributions to science had to do with handling the practical difficulties of those disciplines, particularly measurement error. In 1801, for example, the Italian astronomer Giuseppe Piazzi had caused quite a stir by discovering the large asteroid Ceres, but he was able to record only a few observations before the combination of the asteroid's and Earth's orbits caused it to vanish behind the sun. The big question, then, was where it would reappear. Predicting where and when it would come back meant combining the sparse data that was available and coming up with an estimate of the asteroid's trajectory, and it became a kind of competition within the astronomy community. Gauss, then age 24, attacked the problem with what was essentially a messier version of the method of least squares, and later that year astronomers found Ceres almost exactly where he had predicted it would be.

A few years later he formalized his approach to this kind of problem and gave it a solid theoretical basis. In an 1809 publication (in Latin) titled *Theoria Motus Corporum Coelestium in Sectionibus Conicis Solum Ambientium* (The theory of the motion of heavenly bodies moving about the sun in conic sections), Gauss described the method of least squares, calling it simply "our method," and claimed that he had been using it for at least 10 years before Legendre's publication, a claim that Legendre, understandably, did not take particularly well.[12]

The main innovation that Gauss added, though, was the expression of his derivation in *probabilistic* terms. The general situation, as Gauss saw it, was that an astronomer needs to figure out values of several unknown parameters, such as the eccentricity and tilt of some asteroid's orbit, based solely on a set of noisy observations. But because these observations all include error, a single perfect solution that fits the data exactly is impossible. Instead, what one should try to do is assign a probability distribution to the unknowns—that is, to say how likely they are to lie in one range or another.

This made the problem ripe for Bayesian analysis, and Gauss followed the Bayesian steps to reason toward a solution. First, he had to consider what the prior distributions for the parameters should be: What probabilities should he assign to the unknown physical quantities before having made any observations? For simplicity, he took these to be uniform distributions—that is, *flat priors*, representing a state of uniform total ignorance.

According to our standard Bayesian procedure, logically the next step would be to consider the conditional probabilities of making the observations we did, assuming any given true value of the unknowns, like getting a particular sample from an urn with some assumed ratio of pebbles. This is therefore equivalent to assigning probability distributions for the measurement errors. As we've seen several times by now, Bayes' theorem would then tell us the posterior probabilities for the unknowns—that is, what inferences we could justifiably draw from the data.

However, Gauss, displaying a trademark flash of ingenuity, decided instead to *turn the problem around*. He started with the answer he wanted—the least

squares solution, which he knew had an established track record of performing well empirically—and then asked this question: What assumptions would
we need to make about the measurement errors in order for that answer to be
justified in some way? In particular, it would be especially appealing if it gave a
solution for the unknowns that was *most likely*, in the sense of having the highest posterior probability. What Gauss was able to work out was that everything
would line up perfectly, meaning the least squares solution was also the most
probable, if and only if the measurement errors could be assumed to have a
normal distribution.

News of Gauss's derivation reached Laplace in Paris sometime in 1809. To
everyone else, Gauss's result would probably have seemed clever but practically
useless. He had shown that the measurement errors would need to follow a bell
curve in order to justify the least squares solution, but there was no justification
for why the errors should follow such a distribution in the first place. But when
Laplace heard what Gauss had done, he immediately saw how to put the pieces
together—and in the right order. Laplace had shown in his own theorem that
the normal distribution could be expected to show up anytime a value could
be thought of as the aggregate of a large number of small, independent influences. Here, Gauss had shown that *if* the errors in astronomical observations
were assumed to have a normal distribution, then the practice of using the
method of least squares would be theoretically bolstered by the fact that it also
maximized posterior probabilities—that is, it produced the most probable solution indicated by the data (and, furthermore, the normal distribution was the
only error curve that would lead to this appealing property). So, by making the
sensible argument that errors in astronomical observations were, in fact, of this
nature, Laplace could justify assigning the normal distribution to the errors,
and Gauss's result would then apply! The whole practice of combining astronomical observations using the least squares approach when it was required
that the quantities being observed obey systems of linear equations (which was
almost always the case) could be supported with solid theory.

Laplace rushed an addendum to a publication of his in 1810, citing Gauss's
work, stating his own theorem more clearly, and using it to justify the assumption
of normally distributed errors.

The method of least squares—still used so regularly today that we now call
it *ordinary least squares regression*—was so successful over the ensuing decades
and Gauss's derivation of the error distribution that justified it was so ingenious that the normal distribution came to be known as the Gaussian distribution, and many still refer to it by that name. Laplace's theorem on the general

ubiquity of the normal distribution was widely celebrated as well, and in 1920, George Pólya gave it the lofty name the *central limit theorem*, which it carries (in somewhat generalized form) to this day.

QUETELISMUS

That brings the story back to Quetelet. He was likely exposed to Laplace's ideas during his 1823 visit to Paris to study astronomy. In the early 1840s, he was inspired to try applying them to his own work on social physics. Quetelet's experience with practical astronomy and interest in developing similar laws that would apply to social data put him in a unique position to transfer the ideas of one into the other. Considering the problems that had been vexing Quetelet ever since Keverberg shot down his census proposal, we can immediately see why Laplace's derivation of the bell curve as a widely applicable law of errors would hold a great appeal.

Instead of thinking about the errors in an observation made through a telescope, imagine applying the same idea to the physical characteristics of a person. If, say, a person's height could be thought of as the sum total of a large number of small, independent factors—hereditary or environmental—that could each perhaps vary from person to person but that all shared a fixed underlying *probability* of occurrence, constant across all people among those of a given category, then the overall distribution within that population could be expected to follow a bell curve. It would be as though each person's height had been determined by drawing some large number of pebbles from the same urn with a given fixed urn-ratio, with each white pebble making them taller and each black pebble making them shorter by tiny increments. Since the same probabilities applied to everyone and the net effect was a summation of small, independently drawn increments, Laplace's theorem guaranteed that the result would be a normally distributed population. In this way, everyone could be at once the *same*, in the sense of being subject to the same factors with the same underlying probabilities, while still being *unique*, in the exact way that all the factors were realized for that individual.

Quetelet began looking for normal distributions in all his data sets, and he found them pretty much everywhere he looked. In one famous example, he examined the distribution of chest measurements of 5,738 Scottish soldiers, taken from the *Edinburgh Medical and Surgical Journal* (1817), and found that a normal distribution could fit the data quite well. He developed a novel technique for calculating the values μ and σ of the distribution that would best fit a set of values and found the spread in the distribution for this particular data set was on the same order (about 33 mm) as the *measurement error* one could usually expect in a rough measurement of any one person's chest size.

Perhaps thinking back again to his days as an astronomer, he took great inspiration from this idea and compared the distribution of chest sizes in his set of 5,738 soldiers to the distribution of errors one could expect from measuring the *same* soldier 5,738 times. Both were normally distributed, so he argued that we could think of the real-life soldiers "as if the chests had been modeled on the same type, on the same individual, an ideal if you wish, but one whose proportions we can learn from sufficiently prolonged study."[13] In other words, all men, at least those of a given community, were the imperfect, error-prone renderings of an ideal: the *average man*.

The normal distribution and its relationship to the error law of astronomy were exactly what Quetelet needed to give his social physics a solid footing. It gave the average man "the character of a mathematical truth."[14] If deviations from the average in any human characteristic, from purely physical aspects like height or arm length to more socially influenced variables like age of death or marriage, could be thought of as measurement errors, then after those errors were smoothed out, what would be left over would be the true *type* for whatever population was being considered at the time. Changes in the type over time would then correspond to *movement* in those tendencies, and the overall trajectory of their movement in relation to other social variables could be described just like the orbit of an asteroid around the sun. The normal law could even apply to discrete variables, such as whether a person had been convicted of a crime, if the tallies were grouped as a per capita rate by community or subpopulation, in which case the error law could describe deviations in the local rate from the global average.

Quetelet took the presence of a normal distribution (which he called *la courbe de possibilité* meaning "the curve of possibility" or *la loi de possibilité* meaning "the law of possibility") as evidence of some kind of homogeneity within any group in which he found it. He reasoned that because the distribution would reveal itself any time a characteristic was produced as a sum of a large number of accidental deviations of a homogeneous type, this meant that *every* instance of a normally distributed data set must have been generated according to such a process. As he wrote, "This symmetry in the data only exists and can only exist in so far as the elements that concur to give the mean can be traced back to one single type."[15]

That argument is unfortunately not mathematically true. There are many other, inhomogeneous ways data can be produced such that the end product will still be normally distributed. Quetelet was fallaciously relying on something like the *converse* of Laplace's theorem. But he was so taken by the suggestion of homogeneity that showed up nearly every way he grouped his data that he was unable to tell when that homogeneity was just a phantom. He would later be harshly ridiculed for his love of the normal distribution by statisticians like Francis Edgeworth, who wrote in 1922: "The theory [of errors] is to be distinguished from the doctrine, the false doctrine, that generally, wherever there is a curve with single apex representing a group of statistics . . . that the curve must be of the 'normal' species. The doctrine has been nick-named 'Quetelismus,' on the ground that Quetelet exaggerated the prevalence of the normal law."[16]

Nevertheless, Quetelet's practice of trying to fit normal distributions to data caught on as a standard in all manner of disciplines. The distribution worked so well for so many different types of data that scientists started to become more interested in situations where it failed, since a nonnormal distribution *could* logically be interpreted as a violation of the assumptions of homogeneity. In 1863, Adolphe Bertillon found what he thought to be a two-humped (*bimodal*) distribution in the heights of French soldiers in Doubs, which his colleague Gustave Lagneau took as evidence that the population of Doubs must consist of two different races, Celts and Burgundians.[17] The desire to make exactly this kind of discriminatory analysis precise attracted the attention of statisticians, particularly Karl Pearson, at the end of the 19th century and largely inspired the technique of significance testing, as we'll see in the next chapter.

A particularly odd example of normal distribution fitting was published in 1861 by an English physicist named William Spottiswoode, who tried to fit a bell curve to the directions of orientation of 11 mountain ranges to see if mountains were generated according to a common ideal "type."[18] His attempt was not even remotely successful. The work would be otherwise unnoteworthy except that it caught the eye of a friend of his named Francis Galton, who, as we'll see, was among the people most responsible for making statistics what it is today, in every way good and bad.

THE BIRTH OF FREQUENTISM

Somewhere on Quetelet's bridge between astronomy and social physics, the meaning of probability morphed from being mostly subjective to being mostly frequentist. In particular, we can see that transition happening if we compare Quetelet's usage of his beloved normal distribution with its origins in Gauss's and Lagrange's method of least squares. Gauss's argument was this: when confronted with a system of linear equations, a normal probability distribution for the measurement errors, when *combined with a uniform prior probability distribution for the unknowns*, would make the least squares solution also the most likely in the sense of having the greatest posterior probability. By using the words *prior* and *posterior*, we've made this appear to be an obviously Bayesian idea. But considering just the assumption concerning the normal error distribution, we could interpret it in one of two ways: either that distribution was something we *assigned*, in the Bayesian sense of it being a consequence of our knowledge or lack thereof, or it could be something we *observed*, in the frequentist sense of the distribution that actually manifested in real observations.

To take a simpler example, imagine for a moment we are presented with a mysterious box with a light on top that we're told flashes either red or green once per minute. If that's all the information we have about the workings of the box, then according to Edwin Jaynes's idea of indifference, we are forced by our ignorance to assign the two outcomes equal probability. Furthermore, we have no information giving us any reason to connect the outcome of one flash with the next, so we have to treat them as independent from one another. As a result, just putting these assumptions together mathematically according to the rules of probability, we will assign the number of green flashes over some period of time a binomial distribution, which, for a large number of minutes, would be well approximated by a normal distribution.

Whether that number *does*, in fact, distribute itself in a bell curve over time, though, depends on the actual inner workings of the box. Maybe the lights are the result of a random process with some bias toward red, or maybe they're actually programmed to follow some pattern, making the successive results not independent. Or maybe every thousandth day the light just flashes green all day long, creating a possibly highly skewed and hence nonnormal distribution. Our probability assignments, and any conclusions we drew from them, could still be correct, though, based on the information we are given, even if there are underlying dynamics of which we aren't aware.

The same will be true of the cumulative error present in some measurement in an experiment if we understand that error to be the sum of a number of factors about which we know very little. Our assignment of a normal distribution could be entirely valid and also inconsistent with a long series of actual observations. And yet nowadays almost everyone who deals with statistical inference in science discusses error distributions as though their *frequencies* are what matter, not the probabilities we assign to them. For instance, one often hears scientists and statisticians talking about things like the *parametric assumptions*, including an assumption that data actually is normally distributed, presumably open to being verifiably right or wrong somehow. As we'll get to in the next chapter, this is largely the result of the dominating influence in the 20th century of the frequentist school of inference, in which probability has no *meaning* other than the observed frequency over a sequence of trials.

Looking back, though, we have to give at least some of the credit for that transition to Laplace. His theory of the distribution of errors, and the corollary that one could safely apply the method of least squares to astronomical problems, involved the claim that the errors in astronomical measurements *actually were* normally distributed, not just that they were something to which we could justifiably assign a normal distribution. This was a claim that he and others could back up empirically. Taking a large series of measurements of the same quantity, even if one didn't know the exact true value, would reveal that most of the measurements tended to cluster around some middle point, with values further out coming less frequently, and so on, in agreement with the normal curve. Nothing in Laplace's convergence theorem (the central limit theorem) required this empiricism, though; any large number of independent variables with a common probability distribution, under *whatever* interpretation of probability so long as the basic rules were satisfied, would have a sum that converged to a normal distribution.

It happened for these astronomy applications, though, that the normal distribution manifested in the real world, at least to a good approximation. Too large a deviation from the predicted normal distribution was even taken as an indication that something truly egregious had happened and that the data point should be thrown out. It was also the apparent fact that *real data* such as soldiers' heights or chest sizes seemed to conform to this distribution that gave Quetelet license to borrow the idea for his theory of social physics and to begin the confusion of probability with frequency. Using Quetelet's methods, for instance, it would be sensible to ask what the probability is of there being more than 100 murders next year, but it wouldn't make sense to ask what the probability is of the true murder rate being more than 100 per year.

Laplace would not have drawn that distinction, though, especially when it came to problems of inference. For Laplace, Gauss, and their generation of astronomers and physicists, the quantities they were most interested in were fixed unknowns, such as the true position of Jupiter, and they were comfortable

assigning "inverse" prior and posterior probabilities to these unknowns conditional on the observations they made, exactly in the way Bayes' theorem dictates.

So, in a way, it might seem like an unfortunate historical accident, just like
the situation with the classical definition and games of chance, that the assigned
probabilities came out to match the frequencies because that allowed later
authors to take frequency to be the meaning of probability. But it was also just
this correspondence that made the ideas more palatable.

Make no mistake, for many of Quetelet's intellectual contemporaries, the
idea of applying *probability*—a mathematical tool with origins in gambling,
after all—to profound questions of society was extremely unpalatable at first.
For example, one of the more controversial data sets Quetelet considered was
that of conviction rates in jury trials, broken down by year, type of crime, gender
and age of the accused, and whether the defendant was literate. (He concluded
the best situation to find yourself in, should you ever be accused of a crime in
1820s France, was to be a well-educated woman over 30 appearing voluntarily to
answer an accusation of a crime against a person.) Building on Quetelet's ideas,
the mathematician Siméon Denis Poisson (1781–1840) conducted a probabilistic analysis[19] of the same data to attempt to answer such questions as whether
the conviction rate was stable through time, what the rate of conviction suggested about the defendant's actual probability of guilt, what the effect would
be of changing the size of the jury or the rule for what majority was necessary
to convict, and so on, all perfectly in keeping with Laplace's original goals of
applying probabilistic thinking everywhere that chance seemed to operate.

The famed scientific philosopher Auguste Comte bristled at the suggestion
that probability could play any meaningful role in social science, though, calling it a "vain pretension of a large number of geometers that social studies can
be made positive by a fanciful subordination to an illusory mathematical theory
of chances."[20] Without mentioning Laplace by name, he aimed a pointed critique his way: "Is it possible to imagine a more radically irrational conception
than that which takes for its philosophical base, or for its principal method of
extension to the whole of the social sciences, a supposed mathematical theory
where . . . we strain to subject the necessarily complicated idea of numerical
probability to calculation?"[21] And one of Comte's followers, the mathematician
Louis Poinsot, publicly criticized Poisson's jury work as a "false application of
mathematical science."[22]

In these moralistic objections, and the response they elicited, we hear the continuation of a theme that began quietly with Blaise Pascal and Pierre de Fermat
grappling with the earliest questions of probability in the 1650s, became more

audible in the work of Bernoulli and Laplace, and only got louder the more probability started to encroach on the real everyday lives of people thanks to Quetelet and his successors. The theme is this: *the more important a question is to society, the more fierce the objection will be to using probability to answer that question, and the more the users of probability will retreat to frequency as a justification.* Frequentist analysis takes the observer out of the picture and says, shruggingly, that these probabilities are what we observe and this is what the data tells us, like it or not.

So it is likely no coincidence that around this same time, roughly the middle third of the 19th century, philosophers and mathematicians started to seriously consider the nature of probability and to push for the interpretation that probability was something we observe rather than something internal based on our state of knowledge. Poisson, for instance, had already distinguished between subjective probability and objective probability in the 1837 work containing his analysis of juries.

During the brief window from 1842 to 1843, these ideas were particularly *en vogue.* The great philosopher John Stuart Mill, in his 1843 book *A System of Logic, Ratiocinative and Inductive,* wrote that, in order for the doctrine of chances to apply to any given problem, "there must be numerical data, derived from the observation of a large number of instances"; otherwise, "to attempt to calculate chances is to convert mere ignorance into dangerous error by clothing it in the garb of knowledge."[23] The English mathematician and philosopher Robert Leslie Ellis likewise argued for an objective theory of probability in "On the Foundations of the Theory of Probabilities," a paper read to the Cambridge Philosophical Society in 1842 and published a year later.[24] In 1843, the French mathematician Antoine Augustin Cournot criticized some users of Laplace's methods (though not as severely as Comte had), writing "The calculus of probabilities has real importance only inasmuch as it is applied to numbers sufficiently large that one must have recourse to approximation formulas."[25] And in 1842, the German philosopher Jakob Friedrich Fries argued that probability should be thought of only in frequency terms, restricting it to questions like insurance and astronomical measurements, and that it especially should not be used for Bernoulli-esque applications like courtroom testimony because "that which cannot be calculated should not be subjected to pseudo-calculation."[26] In 1854, the logician George Boole wrote *An Investigation of the Laws of Thought,* containing a logical theory of probability that disallowed certain probability assignments not based on concrete knowledge. And in 1866, John Venn wrote what many people took to be the final word on the subject with his detailed frequency-based treatment of probability and critique of Laplace.

THE PROBLEM OF PERFECT IGNORANCE

The brigade of frequentist philosophers in the 1800s all approached the subject with their own idiosyncratic flair, but at the heart of their objections to Bayesian inference was one unifying concern: Where did the prior probabilities come from? In Bayes's original problem, which involved making a guess about the whereabouts of a ball thrown onto a table (basically a continuous version of the dice guessing game we considered in chapter 1), the correct prior probability distribution for the ball's unknown location was obviously uniform, for reasons that could be supported by a frequentist argument. If you throw a billiard ball with a sufficient amount of force onto a level table, you'll find that, over the long run, the places it comes to rest will be more or less uniformly spread out over the surface of the table. So any inferences that started with those prior probabilities would seem to be justified *empirically*. But what if you are faced with a system about which you don't have that prior knowledge? What if you know nothing?

Laplace had tried to extend the same reasoning to problems where the prior distribution was obviously not the result of frequencies, like probability assignments for the true location of a star or the true curvature of the Earth. He justified using Bayes' rule with a uniform prior probability distribution for these problems by arguing that our *ignorance* of the truth is uniform. Thus, the uniform distribution (perhaps restricted to a certain range) would play the role in probability that zero plays in counting, the natural place to start when you have nothing at all. John Maynard Keynes would later call this the *principle of indifference* in his 1921 book *A Treatise on Probability*. But in the mid-19th century, it was somewhat derisively referred to as the *principle of insufficient reason*, most notably by Boole and Venn. Ellis poetically claimed it violated a fundamental idea of *ex nihilo nil* (out of nothing, nothing).[27]

Nevertheless, it remained an accepted practice in science, mostly because there was no good alternative. As the astronomer William Fishburn Donkin noted in 1851, "A person who should dispute the propriety of dividing our belief equally amongst hypotheses about which we are equally ignorant, ought to be refuted by asking him to state which is to be preferred."[28] That is, any reason for preferring one hypothesis or another amounts to an admission that we aren't, in fact, totally ignorant about the situation in the first place.

But it turns out that perfect ignorance is not such a simple concept. Depending on how you look at it, for certain problems it might appear that being perfectly ignorant could lead to contradictory probability assignments. In chapter 1, we briefly mentioned the famous Bertrand paradoxes, which Joseph Bertrand introduced in his *Calcul des probabilités* (1889). The main thrust of these paradoxes is that a uniform probability distribution for a continuous variable isn't preserved by every possible function. So, as in our earlier example, if we are considering an unknown square, we could say we are uniformly ignorant of the square's side length or its area—but not both. Since the two measurements are related by the function $A = S^2$, they cannot both be uniformly distributed, but either could reasonably serve as a description of the size of the square.

Boole had actually raised a simpler objection that would even apply to *discrete* probability distributions.[29] We can illustrate Boole's idea with an example as simple as drawing from an urn with two balls inside. Say we know each ball can be either black or white, but apart from that, we're totally ignorant. What should be our probability assignment for the number of white balls in the urn? In one way of thinking, we could say there are three cases: the number of white balls is zero, one, or two. So perfect ignorance might dictate that we assign these equal probabilities of 1/3 each. Let's refer to this as Ignorance of Type I.

However, thinking about the individual balls *one at a time*, we could say that we give the first ball equal probabilities of being either color, so probability 1/2 for each case. Since we're supposed to be perfectly ignorant, we'd say we have no reason to draw any connection between the color of the first ball and that of the second, so we assign the second one an independent probability 1/2 of being white. It would be as though we think of the balls as binary digits (1 for white, 0 for black) and have no reason to prefer any of the four possible cases: 00, 01, 10, or 11. Call this Ignorance of Type II.

Putting the assumptions together, though, we'd see that, under Ignorance II, our probabilities for the number of white balls would look like those in table 3.1. So our probability assignments are not uniform! We're somewhat more confident that there's one black and one white ball than either of the two extreme cases, essentially because there are two ways for that situation to occur instead of just one.

TABLE 3.1 **Probability assignments under Ignorance Type II**

Number white	Probability	Reasoning
0	$1/2 \cdot 1/2 = 1/4$	Both balls 1 and 2 must be black.
1	$1/2 \cdot 1/2 + 1/2 \cdot 1/2 = 1/2$	Either ball 1 is black and ball 2 is white, or vice versa.
2	$1/2 \cdot 1/2 = 1/4$	Both balls 1 and 2 must be white.

Any inferences we make upon drawing a ball from the urn will then be different based on the two equally defensible prior probability assignments. So which one is the right type of ignorance? Boole used examples like this to dismiss the principle of insufficient reason as "an arbitrary method of procedure."[30]

It's a legitimate concern. We can push this to an extreme and see the Law of Large Numbers make yet another appearance if we let the number of balls be very large. Following the Ignorance II reasoning, we could say that, in a large urn containing N balls, we'd assign them each independently a probability 1/2 of being white, so the proportion of white balls in the urn would depend on a sum of these independent increments and would therefore follow one of Bernoulli's binomial distributions. As we know, that starts to look like a normal distribution for large N, with a peak of the bell curve around the value 1/2. As N gets larger, the width of the bell gets narrower and narrower, meaning we are more and more certain the mix of balls in the urn is close to 50/50. For example, table 3.2 shows the probabilities we would give that the proportion of white balls is between 49 percent and 51 percent as a function of N, the total number of balls. So, for values of N in the million-and-up range, we are very, very sure, with a probability starting with 88 nines after the decimal place, that the urn is within 1 percent of being evenly mixed, despite the fact that we are also supposedly totally ignorant about its contents!

These two types of ignorance are in some sense "dual" to each other: in Ignorance I, we are uniformly ignorant of the proportion of white balls in the urn, but this ignorance makes the successive ball draws *not* independent, which is essential to our ability to perform inferences by sampling. We learn something about the proportion with each successive draw, affecting our probability assignments for the next draw.

Under Ignorance II, we are seemingly as ignorant as possible because we have assumed nothing about the state of each ball or any dependence between them. Yet somehow our ignorance makes the probability assignment for the

TABLE 3.2 **Probability assignments for large urns under Ignorance Type II**

Number of balls, N	Probability of between 49% and 51% white
1,000	0.472684
10,000	0.954494336625
100,000	$1 - 10^{-9}$
1,000,000	$1 - 10^{-88}$

resulting proportions extremely sharply peaked, to the point where we are nearly certain the proportions are equal in the first place. Given this state of information, the successive ball draws *are* independent. In fact, this prior state of ignorance means we *learn nothing* by sampling from the urn, since the status of all leftover balls remains independent even after sampling. After drawing out 99,999 balls from an urn containing 100,000 and finding them all to be white, we'd still assign a 50/50 chance to the remaining ball being black. Thus, for this state of prior information, inference through sampling is impossible.

It seems therefore that we cannot be completely ignorant of everything simultaneously, and which state of information we imagine ourselves to be in at the beginning of the problem can have a dramatic effect on the kinds of inferences we can draw from data—and even whether any inferences can be drawn at all. The question, in any practical problem, would be, Which kind of ignorant are we? If we have reason to assert some logical connection between the colors of the balls—say we know they were manufactured by an assembly process with some fixed proportion, even though we don't know what it is—then we could argue for the first kind. If we have no reason to draw a connection between any one ball and any other, we could say we are ignorant in the second way; maybe this will happen if we know the balls have each been assigned colors separately by a team of monkeys without any communication among them. As we discussed in chapter 1, the key to resolving this kind of problem is always a careful consideration of what kinds of symmetry are present in our background information—and thus the kinds of transformations of the problem to which we are necessarily indifferent. But such a dissection of the exact meaning of ignorance for any given problem did not start to become clear until Jaynes finally solved one of the Bertrand paradoxes in 1973.[31]

Perfect ignorance was a fly in the Bayesian ointment for centuries. If even simple problems like these could lead to apparent contradictions in the meaning of ignorance, why should anyone trust the conclusions of any sophisticated probabilistic reasoning? It would be as though our accountant had started doing our taxes from the assumption that $0 = 1$. The whole of probability, unless it was somehow grounded in empirical observation, would seem to be on the verge of collapse at any moment. When all that was at stake was a guess about the contents of an urn, none of this may have been of much importance to anyone but a logician. But when the questions started mattering to society, serious people started demanding real answers. As we'll see next, in the 20th century probability found its way to the most significant social questions possible. These finer points about the assignment of prior probabilities and the discomfort created in the minds of those people trying to use probability on such a grand stage made the frequentist theme we heard earlier start to reach a deafening volume.

4

THE FREQUENTIST JIHAD

Social facts are capable of measurement and thus of mathematical treatment, their empire must not be usurped by talk dominating reason, by passion displacing truth, by active ignorance crushing enlightenment.

~Karl Pearson

To understand how the discipline of statistics became what it is, we now need to follow the fledgling ideas of data analysis as they moved from Adolphe Quetelet's social physics in France and Belgium to all-new applications of statistics primarily in England. This also brings us into the age of evolution, the main driver behind the development of most of the new statistical tools. In this setting, in contrast to the lyric descriptions of the average man we saw from Quetelet, the quantification of human differences started to take on a menacing undertone, infused with racism, ableism, and settler colonialism.

We'll focus on three people who had the most influence on the course of statistics in the late 19th and 20th centuries: Francis Galton, Karl Pearson, and Ronald Fisher. They were, respectively, something like the grandfather, the father, and the other-father-who-got-the-kids-in-the-divorce of modern statistics. Their collective careers spanned roughly the period from 1860 to 1960, and by the time their work was finished, all the major concepts of what we now think of as statistics were in place (including major contributions from others). Their lives were entwined in complicated ways; sometimes they were collaborators and sometimes rivals. What united them, though, was an understanding of the power of their new statistical machine to shape society according to their agenda, an agenda that, in turn, led them to shape statistics to be what they needed. In addition to being profoundly influential statisticians, all three were vocal advocates of the eugenics movement throughout their professional lives. Really, then, we have six characters: Galton, Pearson, and Fisher the eugenicists, who wanted quantitative ways to express arguments about human heredity and selective breeding, and Galton, Pearson, and Fisher the statisticians, who answered the call.

Here is also where we'll witness the growth of statistical "objectivity" and the triumph of frequentism. Up until the 20th century, Bayesian methods were still widely used, and there was vibrant debate about the relative merits of Bayesianism and frequentism. Were it not for Galton, Pearson, and Fisher, that might have remained true, but the dominant school these statisticians midwifed into the world was entirely frequentist. The interpretation of probability at the core of their methods equates the probability of an event with its frequency of occurrence over a long run of repeated trials or, more precisely in this case, repeated samples from a population. As we've seen, this shortsighted view of probability leads us directly to Bernoulli's Fallacy whenever we need to answer questions of inference. Later we'll see the logical problems this causes in present-day statistical practice and the grave consequences it has had in science through the crisis of replication. In this chapter, though, we'll try to answer the question of *why* these founders of the discipline were so dogmatic that probability had to be measurable as frequency in the first place. One possible explanation is that they were equally dogmatic about what they wanted to *do* with statistics, for which they needed to assert an authority founded on what they claimed was objective truth. In a continuation of the trend we've already observed, their methods became more cloaked in objectivity as statistics gained more political importance, until by the end the stakes were such that they couldn't allow any hint of subjectivity. Galton, Pearson, and Fisher were the mathematical equivalent of religious fundamentalists, and they claimed to follow a strictly literal reading of their holy texts.

NATURE, NURTURE, AND THE QUINCUNX

Francis Galton was born in 1822 to a wealthy family in Birmingham, England. His ancestors on his father's side had amassed a fortune in gun manufacturing and banking. His maternal grandfather, whom he shared with Charles Darwin, was Erasmus Darwin, an accomplished physician and natural philosopher in his own time, who once turned down an invitation to become personal physician to King George III. Galton, a child prodigy, was reading Shakespeare and books of poetry by the time he was six. He studied medicine at Birmingham General Hospital and King's College London Medical School for a brief time as a teenager before turning to mathematics as a student at Trinity College, Cambridge. In his early adulthood, financed by his inheritance, he traveled extensively, venturing on a solo trip across eastern Europe and then later, with accompaniment, to Egypt and the Sudan as well as the cities of Beirut and Damascus. When he was 38, he joined the Royal Geographical Society and mounted the first expedition by a European to parts of South West Africa, now Namibia (where today there is a primary school named after him). His book *The Art of Travel* (1854) was a best seller in its day and is still in print. In 1864, he and a few other notable scientists founded a weekly journal called *The Reader* with the aim of disseminating the latest scientific ideas to an educated audience; it folded after a few years but was revived under a new title, *Nature*.

Everywhere he went, Galton had a habit of counting and measuring things and people. During a particularly boring meeting, he recorded how often the attendees were fidgeting in their seats, with the results later published in *Nature*. At the 1879 Epsom Derby horse race, while other patrons were watching the race, Galton watched *them* and recorded how much the excitement caused their faces to turn pink. He kept meticulous records of women he passed on the streets, classifying each as "attractive, indifferent, or repellent" to create a "beauty map" of the British Isles (he found London to rank the highest for beauty and Aberdeen the lowest). A letter he wrote to his brother Darwin Galton during a trip to South Africa describes his process for measuring the proportions of some particularly shapely women from a distance using a sextant: "As the ladies turned themselves about, as women always do, to be admired, I surveyed them in every way and subsequently measured the distance to the spot where they stood—worked out and tabulated the results at my leisure."[1] His records are silent on what the women thought about being surveyed "in every way."

Galton was an eccentric, creative thinker whose interests ranged from meteorology to psychology, biology, and criminal forensics; his many inventions include a technique for mapping weather patterns, a method of analyzing fingerprints, and a dog whistle. He was also *incredibly* racist—in the particular ways that perhaps only a "gentleman scientist" of Victorian England could be. That is, his racism was taxonomic, driving him to study the various people of the world and arrange them in hierarchies in the service to British colonialism, and he tried to back it all up with science and statistics. His cousin's[2] publication of *On the Origin of Species* in 1859 had a profound influence on Galton's thinking on the subject, especially as it pertained to selective breeding to bring out the best traits among animals and humans.

Galton's racial theories were founded on a "survival of the fittest" reading of Darwin's evolutionary theory—with "fittest" not meaning "best suited to their environment" but rather something more like "healthiest." He believed the human races owed their relative historical merits to their triumphs over adverse conditions, and white Europeans, having endured the harshest conditions, were therefore the most "fit."

But on a prospective basis, Galton believed active human *intervention*, not just natural selection, could move the species of man toward advantageous traits. If selective breeding didn't work quickly enough, this could be accomplished by pitting various groups against each other to see who would prevail. For example, in 1873 he wrote a letter to *The Times* advocating that the coast of East Africa be given over to China to colonize, since, as he argued, their natural aptitudes at civilization building would allow them to "multiply and their descendants supplant the inferior Negro race," whom he called "lazy, palavering savages." He concluded all mankind would reap the benefits of this effort, and the lack of complaints by his contemporaries speaks volumes about the acceptability of such notions in those days. The only response letter *The Times* published to Galton's proposal was that of Gilbert Malcolm Sproat, who objected to Galton's characterization of the Chinese as "industrious" and "order-loving," saying instead that "the Chinaman, as a citizen, and also socially, is almost useless."[3]

One of Galton's enduring legacies today is the word *eugenics* (from the Greek meaning "well born"), a term he invented in 1883 in his book about families of high status called *Inquiries Into Human Faculty and Its Development*. It was a concept he had been refining over the previous 20 years of study. His theory was that most, if not all, characteristics of an accomplished family were due to nature rather than nurture (a phrase he also invented), so breeding among these families should be encouraged by the state. Conversely breeding among people of a lower quality, especially those "afflicted by lunacy, feeble-mindedness, habitual criminality, and pauperism,"[4] should be discouraged or even prohibited. His dream was a utopia in which a populace elevated by hereditary improvements in intelligence and moral character could shake off society's ills and live in paradisiacal harmony. In an early paper titled "Hereditary Talent and Character" (1865), he compared this favorably to breeding livestock: "If a twentieth part of the cost and pains were spent in measures for the improvement of the human race that is spent on the improvement of the breed of horses and cattle, what a galaxy of genius might we not create! We might introduce prophets and high priests of civilisation into the world."[5]

Galton's interest in the inheritance of abilities among prominent families, perhaps inspired by his own family, animated much of his work. His 1869 book *Hereditary Genius* gave a comprehensive account of famous people and their relatives in various walks of life—science, poetry, politics, justice, sports, etc.— in an attempt to study the degree to which *eminence*, which for Galton meant natural ability, could be expected to run in families. Clearly he thought some effects of hereditary talent were already present in what he assessed to be the superiority of white Europeans to other peoples of the world. After the chapters on different family occupations, *Hereditary Genius* also included a chapter titled "The Comparative Worth of Different Races," in which Galton concluded, based on the application of his own grading scale, that "the average intellectual standard of the negro race is some two grades below our own," meaning the Anglo-Saxons. For him, perhaps the only "race" capable of besting the Anglo-Saxons was the ancient Athenians: "The average ability of the Athenian race is, on the lowest possible estimate, very nearly two grades higher than our own—that is, about as much as our race is above that of the African negro."[6]

The grading system he used was based on the normal distribution, then called the *frequency of error*, which Galton had first encountered in William Spottiswoode's unsuccessful study of mountain ranges. As he explained the concept in the preface to the 1892 edition of *Hereditary Genius*, adding a qualification to Quetelet's theory that people would need to be "of the same race" in order for the distribution to apply:

> The method employed is based on the law commonly known to mathematicians as that of "frequency of error," because it was devised by them to discover the frequency with which various proportionate amounts of error might be expected to occur in astronomical and geodetical operations, and thereby to estimate the value that was probably nearest the truth, from a mass of slightly discordant measures of the same fact. Its application had been extended by Quetelet to the proportions of the human body, on the grounds that the differences, say in stature, between men of the same race might theoretically be treated as if they were Errors made by Nature in her attempt to mould individual men of the same race according to the same ideal pattern. Fantastic as such a notion may appear to be when it is expressed in these bare terms, without the accompaniment of a full explanation, it can be shown to rest on a perfectly just basis.[7]

It might seem at first that characteristics like worth could not possibly lend themselves to a normal distribution, but Galton claimed he was able to apply the distribution to these apparently *nonquantitative* aspects of human life using a method he created called "statistics by intercomparison."[8] The basic idea was this: Find a group of, say, 101 randomly selected people who could be deemed homogeneous according to the presence of the normal distribution in some measurable quantity such as height, à la Quetelet. Then *rank* them by whatever other quality was of interest, such as talent or genius. Since they were homogeneous in one respect, Galton reasoned all their other traits, even the difficult-to-quantify ones, would also follow a normal distribution.

This was a dubious assertion but one Galton claimed was no mere analogy. Instead, reasoning that human cognitive abilities must have a physical basis in brain chemistry or other similar factors, he argued the law of deviation should always apply:

> Now, if this be the case with stature, then it will be true as regards every other physical feature—as circumference of head, size of brain, weight of grey matter, number of brain fibres, &c.; and thence, by a step on which no physiologist will hesitate, as regards mental capacity . . . that the deviations from that average— upwards towards genius, and downwards towards stupidity—must follow the law that governs deviations from all true averages.[9]

If this distributional assumption held true, the grades could then be determined simply by the percentiles of the rank ordering. That is, the person in the group Galton judged to be 51st-best out of 101 would set the mean of the scale, the 16th-best would be one standard deviation above the mean, the person at rank 84 would be one standard deviation below the mean, and so on. This gave a reference scale for measuring the previously unmeasurable. When possible, he tried to test this predicted distribution against measurable figures such as exam results. He divided up the scale into 16 categories, labeled "a" down to "g" on the low side of average (and then a special category "x" reserved for the worst of the worst) and "A" up to "G" on the high side (with "X" for the absolute best). A table of the normal distribution showed the frequencies with which each grade could be expected in a given population. Comparing one group's "A" with another's "C" (according to Galton's judgment) and so on would allow him to establish the relative "worth" of each race.

Galton was confident in his application of the normal distribution because he knew that Laplace's theorem, the "perfectly just basis" he referred to earlier, would guarantee its presence under general conditions. In his fervent love for measuring and classifying, he was therefore a huge fan of the theorem. In his 1889 book *Natural Inheritance*, he sang its praises:

> I know of scarcely anything so apt to impress the imagination as the wonderful form of cosmic order expressed by the "Law of Frequency of Error". The law would have been personified by the Greeks and deified, if they had known of it. It reigns with serenity and in complete self-effacement, amidst the wildest confusion. The huger the mob, and the greater the apparent anarchy, the more perfect is its sway. It is the supreme law of Unreason. Whenever a large sample of chaotic elements are taken in hand and marshalled in the order of their magnitude, an unsuspected and most beautiful form of regularity proves to have been latent all along.[10]

It was, in short, the perfect tool for a mind like Galton's, seeking structure and ordered hierarchies out of the messy chaos of human variety. His interest in the distribution was, in a sense, the polar opposite of Quetelet's. Where Quetelet had used the normal error law to glorify the average man, Galton focused on the extremes. In his opinion, people who cared more about the averages were "as dull to the charm of variety as that of a native of one of our flat English counties, whose retrospect of Switzerland was that, if its mountains could be thrown into its lakes, two nuisances would be got rid of at once."[11] He wanted to know exactly how far from mediocrity people could be predicted to be and how often. Variety was a necessary component for evolution by natural selection—and therefore also for eugenics. How many more geniuses could be produced by shifting the societal distribution by a grade or two? This was the kind of question in need of an answer if his eugenics program was ever going to take off.

However, in his keen observation Galton also noticed a problem: as we saw briefly with Quetelet, there were some instances of evidently *inhomogeneous* populations that nevertheless conformed to a normal distribution. In one example, Galton considered the distribution of sizes of fruit grown on trees under clearly different conditions—mainly their exposure to sunlight. Classifying these "aspects" into different groups (small, moderate, large) made it clear that the sizes of fruit in each category would be different from those in the other categories, with each group conforming to a normal distribution with different mean sizes. Nonetheless, when all fruits were considered as *one* group, the overall distribution was *also* normal. In Galton's words, "The question is, why a mixture of series radically different, should in numerous cases give results apparently identical with those of a simple series."[12]

What Galton eventually realized was that, assuming the distribution within each group was normal, the combined distribution could still be normal if the distribution of the *aspects* was *also* normal. For example, if the aspect of exposure to sunlight was itself the sum of a large number of independent factors (directional orientation, slope of the ground, density of leaf cover, etc.), the moderate conditions would occur *more frequently* than the extremes, resulting in mean sizes of fruit that were at least approximately normally distributed. Each aspect giving rise to a normal distribution of fruit sizes would then imply the overall distribution would be an (approximately) normal *mixture* of normal distributions, which Galton showed would also be normally distributed.

Mathematically this was the kind of fact Laplace could have demonstrated easily.[13] But Galton's mathematical skills were not as strong as Laplace's. Instead, he devised a clever physical argument involving a mechanical device he called a *quincunx*, a board of equally spaced rows of pegs that could each bounce a pellet dropped in from above either left or right with equal probability.[14] At the bottom were channels to collect the falling pellets. The positions where the pellets came to rest, being the sum of some number of independent Bernoulli variables for each possible left or right bounce, were given by a binomial distribution with a peak in the middle of the board and could therefore be well approximated by a normal distribution according to Abraham de Moivre's old argument.

However, Galton's innovation was to imagine *another* set of traps being placed across the pegs about midway down so the pellets would be caught on their journey from top to bottom. He realized two things would happen: (1) the distribution at this intermediate level would *also* be binomial and therefore approximately normal, and (2) as the pellets within each individual channel from left to right were released, they would come to rest in an (approximately) normal distribution centered around *that* channel's location. In other words, each group would have a normal distribution with some mean specific to the group, and these means themselves would be normally distributed. But, of course, the result of just releasing *all* the channels was the same as if the traps had never been placed to begin with, so they would reproduce the original normal distribution. In summary, the normal mixture of normal variables was normal—QED. In this way, a normally distributed population could be made up of categorically different types as long as the types themselves could be assumed to have a normal distribution, maybe as the result of the contributions of several independent effects.

The quincunx as a thought experiment also helped Galton with another puzzle involving normally distributed data in a population: the apparent stability of certain distributions over time. Statisticians had noted it was often the case that distributions of certain characteristics of populations tended to have the same mean and standard deviation from one generation to the next, implying the distributions were the same over time. The height of the man at the 75th percentile for one generation was very close to that of the 75th percentile of his children's generation, and so on. But thinking of the previous generation's height as the aspect, like the exposure to sunlight in the model of fruit sizes, this seemed counterintuitive. If parents always had children whose heights varied according to a distribution centered around the parents' own heights, the overall distribution should *widen* over time: very tall parents should continue having very tall children, half of whom by chance would be even taller than their parents, setting the mean for the next generation, and so on. Galton's quincunx with the row of traps in the middle showed this widening behavior as well: the pellet distribution midway down the board would be more tightly concentrated toward the middle than the distribution at the bottom, and if the board was extended even further, the distributions would continue to get wider as the pellets went down.

In 1875, Galton enlisted the help of some friends to investigate this phenomenon by growing sweet peas from seeds of different sizes. Assigning the parent seeds to seven groups by size, he gave seeds from each group to several people and asked them to grow the plants to maturity, harvest the seeds, and return them to be measured. He found the seeds of the next generation grown from each group had sizes that followed a normal distribution, each with roughly the same standard deviation as the other groups but centered at different means, depending on the size of the parents. However, the key difference was that the distributions of the sizes of the progeny seeds from each group were *not* centered around the size of their parents. Instead, the mean size of each group was *closer to the whole population average size* than their parents' sizes had been. If the sizes among the parent groups were something like 40, 50, 60, 70, 80, 90, and 100, with the average being 70, then the mean size of the seeds grown from the smallest group would not be 40 but perhaps something like 45—that is, somewhat closer to the mean. Similarly, the progeny of the seeds in the largest group would have a mean size of something like 95. The further out the parent sizes were in the distribution, the greater this mean correction tended to be. In fact, Galton found the correction behaved linearly with the deviation of the parent sizes from the overall mean. He worked out an equation for what this drift had to be, in terms of the spread of the parent seeds' and progeny seeds' distributions, for the population to be stable over time.

Galton found this condition to hold true in the data of children's height relative to the heights of their parents (which he averaged to produce a single "midparent" height). The tendency was for parents at the extremes of the height distribution to have children with an average height somewhere in between their heights and the overall population mean, with a drift roughly proportional to the deviation of their parents from the mean. He imagined this being like a quincunx with the channels at the intermediate level being *angled* back toward the middle, concentrating the distribution of pellets toward the center before allowing it to widen back out.

Galton referred to this as *reversion* and later changed the name to *regression*, as in the title of the paper summarizing his results, "Regression Towards Mediocrity in Hereditary Stature" (1886). Originally he thought of the phenomenon in terms of inheritance, and at first, his theory was that this reversion was the product of previous generations expressing themselves (hence the children "reverting" to their roots): "A child inherits partly from his parents, partly from his ancestors. Speaking generally, the further his genealogy goes back, the more numerous and varied will his ancestry become, until they cease to differ from any equally numerous sample taken at haphazard from the race at large."[15]

But then he noticed the same phenomenon would manifest if the data was organized *backwards*. That is, an extreme member of the children's generation was also more likely to *have been born* from parents somewhat closer to the mean than the child was. Eventually he came to realize the same effect could be seen between any two variables, even ones having nothing at all to do with inheritance between generations. A few years after publishing his initial findings, he coined the more general term *correlation* (originally *co-relation*) to refer to the association of any two variables that tended to go together, such as a person's height and their shoe size, and he provided a formula for correlation that would reproduce his "coefficient of reversion" from the heredity studies. This is close to how the term continues to be used today. In modern language, we would say the children's and parents' heights or the two characteristics height and shoe size were *positively correlated*. The name *regression* (from the Latin meaning "go back") is still with us too in the phrase *regression to the mean* and the concept of *linear regression*, the general tool for extracting linear relationships between variables. The coefficient of this linear relationship is related by a formula to the correlation, exactly in the ways Galton uncovered.[16]

All these mathematical properties of the distributions of two variables would have been trivial for Laplace, possessed as he was of immense technical abilities. Galton's genius, on the other hand, lay in discovering these mathematical structures in his practical data and *conceptualizing* the effects by means of tangible physical examples like the quincunx or generations of pea plants, which he could use as demonstrations to help him communicate his ideas to less technical audiences. Rather than being abstract mathematical theorems, these facts, in Galton's hands, took on whole new meanings and captured the imagination of this early crop of statisticians.

THE FIRST PROFESSIONAL STATISTICIAN

The person perhaps most associated with the concept of correlation these days is Karl Pearson (1857–1936), Galton's intellectual heir. Like Galton, Pearson excelled at mathematics as a student at Cambridge but was an intellectually driven, ambitious scholar of many disciplines. After Cambridge, he studied physics at the University of Heidelberg and then, moving to Berlin, became immersed in philosophy, law, literature, history, and political science before returning to England as a professor of applied mathematics at University College London (UCL). He wrote extensively on all these topics. His bibliography lists some 100 publications, including 9 books, *before* 1893, when he first began publishing on his primary field of statistics. All told by the end of his life, Pearson was credited with a staggering 648 publications.

Pearson first encountered Galton's work while preparing a talk for a men's and women's club of UCL, in which members read and debated the social issues of the day. At first, Pearson was skeptical of Galton's (and Quetelet's) project of applying mathematics to social problems. In the manuscript he prepared for the club, he wrote:

> Personally I ought to say that there is, in my own opinion, considerable danger in applying the methods of exact science to problems in descriptive science, whether they be problems of heredity or of political economy; the grace and logical accuracy of the mathematical processes are apt to so fascinate the descriptive scientist that he seeks for sociological hypotheses which fit his mathematical reasoning and this without first ascertaining whether the basis of his hypotheses is as broad as that human life to which the theory is to be applied.[17]

Over the course of the next several years, though, as he became more acquainted with Galton's ideas and started to see their promise, Pearson eventually came around. He went on to make many important contributions to the growing body of theoretical statistics. In fact, he was so influential that he is credited by many as having created the discipline. In 1901, along with Galton and Raphael Weldon, Pearson founded *Biometrika*, which to this day is an esteemed journal of statistical methodology and theory, with Pearson serving as editor until his death. And in 1911, he founded what would become the world's first department of mathematical statistics, at UCL.

A good example of the theoretical rigor Pearson added to statistics was his famous *chi-squared test*. A major problem, going back to Quetelet and beyond, was the question of whether a set of data could reasonably be said to follow a particular distribution such as the normal distribution. Before Pearson, the best anyone could do was to eyeball it—maybe arrange results in a histogram and see whether the shape looked about right. Pearson's test provided an exact numerical measure, the χ^2 statistic, which represented a kind of "distance" between the empirical results and a theorized distribution. Too high a value and the statistician would have reason to reject the hypothesis that the data came from that distribution. It was a precursor to the enterprise of significance testing that now forms the backbone of most of statistics, as we'll see later.

Pearson was among the first to emphasize the difference between causation and correlation, the latter measured by what is now called *Pearson's rho*. For example, in a footnote in the second edition of his book *The Grammar of Science* (1900), he commented: "All causation as we have defined it is correlation, but the converse is not necessarily true, i.e., where we find correlation we cannot always predict causation. In a mixed African population of Kaffirs[18] and Europeans, the former may be more subject to smallpox, yet it would be useless to assert darkness of skin (and not absence of vaccination) as a cause."[19]

But Pearson didn't just sound this warning in the way we now think of it: a reminder not to misunderstand correlated variables as having causal relationships. Instead, he saw it as a *justification* for using Galton's ideas for data-driven inquiries in the first place! That is, with Galton's toolbox of methods, one could investigate correlated variables without any *need* for an assumption of causality, and this could still be a fruitful enterprise worthy of a theoretician. Reflecting on his career in statistics in 1934, Pearson said his great epiphany had been that "there was a category broader than causation, namely correlation, of which causation was only the limit, and that this new conception of correlation brought psychology, anthropology, medicine and sociology in large parts into the field of mathematical treatment. It was Galton who first freed me from the prejudice that sound mathematics could only be applied to natural phenomena under the category of causation."[20]

Pearson agreed with Galton on matters of race and eugenics too and didn't hesitate to ratchet up the eugenics discourse by several notches. He believed evolution was a powerful force underlying much of human society, and eugenics gave him a potent vocabulary to argue for extreme positions concerning the interests of the nation.

Leaning heavily on an idea that the laws of heredity are objective, being scientific truths "as inevitable as the law of gravity," in an address titled *National Life from the Standpoint of Science* (1901), Pearson wrote, "My view—and I think it may be called the scientific view of a nation—is that of an organized whole, kept up to a high pitch of internal efficiency by insuring that its numbers are substantially recruited from the better stocks, and kept up to a high pitch of external efficiency by contest, chiefly by way of war with inferior races."[21] He argued for "a conscious attempt to modify the percentage of [bad stock] in our own community and in the world at large."[22] Natural selection—that is, "survival of the fittest"—was, according to Pearson, the engine driving all human advancement by steadily eliminating mankind's impurities. In his view, this meant the only productive future for the "lower races of man" was a struggle between races that would drive them to achieve advances similar to those of their white superiors, but even that might not suffice:

> How many centuries, how many thousands of years, have the Kaffir and the Negro held large districts in Africa undisturbed by the white man? Yet their inter-tribal struggles have not produced a civilization in the least comparable with the Aryan. Educate and nurture them as you will, I do not believe that you will succeed in modifying the stock. History shows me one way, and one way only, in which a high state of civilization has been produced, namely the struggle of race with race, and the survival of the physically and mentally fitter race. If you want to know whether the lower races of man can evolve a higher type, I fear the only course is to leave them to fight it out among themselves, and even then the struggle for existence between individual and individual, between tribe and tribe, may not be supported by that physical selection due to a particular climate on which probably so much of the Aryan's success depended.[23]

About the treatment of Native Americans across the Atlantic, Pearson wrote: "In place of the red man, contributing practically nothing to the work and thought of the world, we have a great nation, mistress of many arts, and able, with its youthful imagination and fresh, untrammeled impulses, to contribute much to the common stock of civilized man."[24] He held up the examples of colonial genocide in America and Australia as great triumphs of human progress, as white colonists had driven out the indigenous races and made better use of the available natural resources. Perhaps with an awareness some would criticize this view as inhumane, he wrote in *The Grammar of Science*:

> It cannot be indifferent to mankind as a whole whether the occupants of a country leave its fields untilled, and its natural resources undeveloped. It is a false view of human solidarity, a weak humanitarianism, not a true humanism, which regrets that a capable and stalwart race of white men should replace a dark-skinned tribe which can neither utilize its land for the full benefit of mankind, nor contribute its quota to the common stock of human knowledge.[25]

Pearson added a footnote saying that this shouldn't be taken as a universal justification for genocide of lesser races—but only because "the anti-social effects of such a mode of accelerating the survival of the fittest may go far to destroy the preponderating fitness of the survivor."

Pearson's statistical work and his eugenicist views were thoroughly intertwined. One of the first statistical examples he considered, typical of many that would come later, was a set of skull measurements taken from graves of the Reihengräber culture of southern Germany in the 5th through 7th centuries. Using his new tests for nonnormality of data, he argued that an asymmetry in the distribution of the skulls indicated the presence of two races of people, only one of which was a close match for modern-day Germans. Examples like this one had obvious potential in eugenics. That skull measurements could indicate differences between races—and by extension, differences in intelligence or character—was almost axiomatic to eugenicist thinking. Establishing those differences in a way that appeared scientific was a powerful step toward arguing for racial superiority.

Pearson described the motivation for his statistical tests of nonnormal data by writing "In the case of certain biological, sociological, and economic measurements there is, however, a well-marked deviation from this normal shape, and it becomes important to determine the direction and amount of such deviation. The asymmetry may arise from the fact that the units grouped together in the measured material are not really homogeneous."[26] Almost at the same time, though, he was also describing his theory of nations by saying "The nation organized for the struggle must be a homogeneous whole, not a mixture of superior and inferior races."[27] So the word *homogeneous*, linking the purely statistical statement to the one from eugenics, had a particularly charged meaning, with connotations of racial purity and ethnic cleansing. Homogeneity in data and what it indicated about homogeneity of people had racial undertones from the start.

Jews posed a particularly difficult problem for Pearson. Beginning around 1910, he, like many British anti-Semites, became bothered by the influx of Jewish immigrants fleeing from pogroms in eastern Europe to Great Britain. In a series of articles titled "The Problem of Alien Immigration Into Great Britain, Illustrated by an Examination of Russian and Polish Jewish Children," which kicked off the first issue of another journal Pearson founded, *Annals of Eugenics*, he expressed his fear that these immigrants "will develop into a parasitic race."[28] The articles contained Pearson's attempts to justify that fear rigorously and quantitatively. In a large study of about 600 children attending the Jews' Free School in East London, he and coauthor Margaret Moul assessed the children on the basis of intelligence, health, cleanliness, and literacy to see if it would be appropriate for the British government to prejudicially deny entry to Jews based on race. Their judgment was unsurprisingly that it would be appropriate, since on the whole they found Jews to be inferior with only a few noteworthy exceptions:

> Taken *on the average*, and regarding both sexes, this alien Jewish population is somewhat inferior physically and mentally to the native population. . . . But we have to face the facts; we know and admit that some of the children of these alien Jews from the academic standpoint have done brilliantly, whether they have the staying powers of the native race is another question. No breeder of cattle, however, would purchase an entire herd because he anticipated finding one or two fine specimens included in it.[29]

We see here the Galton-esque comparison between people and livestock. Presumably as exceptions, Pearson may also have had in mind Jewish intellectuals like Albert Einstein and Baruch Spinoza. He studied Spinoza extensively and had in 1883 written an article for *Mind* about the influence of Maimonides on Spinoza's philosophy. We can only wonder whether Galton, with his love for people of exceptional talents, would agree with Pearson's risk assessment.

When it came to his study of Jews, Pearson was quick to reverse his fundamental position that struggle bred superior stock. He worried instead that Britain would be left with the dregs of the Jewish population that could survive persecution (a survival of the least fit?) because "such a treatment does not necessarily leave the best elements of a race surviving. It is likely indeed to weed out the mentally and physically fitter individuals, who alone may have had the courage to resist their oppressors."[30] He also abandoned traditional eugenicist thinking about what constituted desirable traits. In previous work comparing the races, eugenicists had claimed families saving a higher proportion of their income and spending less money, for example, on clothes was an indication of temperance, characteristic of superior stock. However, Pearson and Moul found on average Jewish families were spending *less* and saving *more* than non-Jewish families. So they reinterpreted this trait as a negative, arguing the Jews' lower standard of dress would allow them to economically undercut English workers in the labor market: "It is clear that the alien Jewish children are far below the average of the Gentile children, being indeed below the Gentiles of the poorer districts. . . . There seems some ground for the statement frequently made that they undersell natives in the labour market because they have a lower standard of life."[31]

Pearson and Moul subjected the Jewish children to several medical exams without consent and found they had surprisingly healthy neck glands, which could indicate they were more resistant to tuberculosis than their Gentile counterparts. This agreed with previous studies showing a lower-than-average mortality rate due to tuberculosis among Jews in London. So Pearson and Moul concocted a theory with no empirical basis that this meant Jews were quicker to visit a doctor at the first signs of illness, which saved them from dying of tuberculosis but made it a chronic illness rather than an acutely fatal one. Naturally they interpreted this as yet another strike against the Jews, since "a chronically affected population may be less efficient than one not so affected but having a higher death-rate."[32]

Taken as a whole, it becomes clear from reading these papers that Pearson, who claimed to be untainted by ordinary prejudice, had decided ahead of time what conclusion the data should support and then found a way to make it do just that.

Pearson and Galton had a close relationship for many years, although Galton was already 70 years old when they met in person. Pearson deeply admired the old man. In the opening of the *three-volume* biography he wrote of Galton, Pearson hinted that Galton, rather than Charles Darwin, would ultimately be the more renowned of the two cousins, since Darwin had merely collected facts and made hypotheses, whereas Galton provided the theory needed to *test* those hypotheses and "saw with the enthusiasm of a prophet their application in the future to the directed and self-conscious evolution of the human race."[33]

In 1904, Galton financed the creation of the Eugenics Record Office as a clearinghouse for data and research into eugenics. A few years later, to emphasize the role eugenics was supposed to play in furthering the interests of the state, it was renamed the Galton Laboratory for National Eugenics. When Galton died in 1911, he left the balance of his considerable fortune to UCL to pay for the formation of a university department of eugenics. Pearson, already serving as the director of the Galton Lab, was handpicked by Galton to be the first person to hold the chair of this department, the Galton Chair in National Eugenics. The Galton Lab still exists, now folded into the UCL Department of Biology, and the Galton Chair of Eugenics has since been (wisely) rebranded as the Galton Chair of Human Genetics.

In his dual roles as lab director and professor, Pearson had a tremendous influence on the first generation of British statisticians. The *Bibliography of Statistical Literature* by Maurice Kendall and Alison Doig lists only 26 people who were active in statistical theory in Britain between 1900 and 1914, and 12 of them were somehow directly connected to Pearson.[34] His former student, the statistician Major Greenwood, in a contribution to the *Dictionary of National Biography 1931–1940* written after Pearson died, summed up Pearson's educational career and personality:

> Pearson was among the most influential university teachers of his time . . . he had an intense and genuine belief in freedom of thought but was apt to attribute intellectual differences of opinion to stupidity or even moral obliquity. Personal relations between him and his pupils were sometimes painfully interrupted for years; but it is pleasant to record that eventually most of these broken friendships were happily resumed. . . . He was admired and feared, rather than loved, by many; in some he aroused bitter hostility.[35]

One of those people in whom Pearson aroused bitter hostility was the next to assume the Galton Chair, a statistician we have mentioned a few times already and maybe the only person who can claim to have had more influence than Pearson on the development of statistics in the 20th century: Sir Ronald Aylmer Fisher.

THE GENIUS OF SMALL SAMPLES

Fisher was born in East Finchley, London, in 1890. During his childhood years, his family was quite rich. His father, George, ran a successful art auction house on par with Sotheby's or Christie's, and the family lived in the stately Inverforth House in the Hampstead area of the city. Fisher's mother, Katie, died when he was 14, though, and his father's business collapsed the following year. They had to move to a small house in Streatham, one of the poorer parts of town, and Fisher was able to continue his education at the Harrow School only by virtue of scholarships he won.

Like Galton and Pearson, Fisher showed an early talent for math. He won a school-wide mathematical essay contest and was known among his teachers as being one of the most brilliant students they had ever met. He suffered from extremely poor eyesight since birth and was forbidden by doctors from reading under electric lights, which often required him to learn math without the use of a paper and pen. But this also helped him develop a geometric intuition about problems he could visualize in his mind's eye in ways other people couldn't. He developed a habit that was rather obnoxious, to his tutors anyway, of seeing his way to the ends of problems without needing to write down the intermediate steps.

Also like Galton and Pearson, Fisher studied at Cambridge, at Gonville and Caius College, at the time when John Venn was president. Fisher achieved first-class honors in astronomy and graduated in 1912. While still an undergraduate, he published his first paper, introducing a method for parameter estimation by maximizing likelihood, which would become one of his hallmark contributions to statistical theory (we'll return to this). He began a correspondence that year with William Gosset, a research chemist who held the job title of head experimental brewer at the Guinness distillery in Dublin. Gosset had published a new statistical test in Pearson's *Biometrika* using the pseudonym "Student" due to pressure from his employer. We still know this test as *Student's t-test*. He developed his methods to answer practical questions that came up at the brewery, such as how to select the best-yielding varieties of barley, and he had been assisted with the mathematics by Pearson, in whose lab Gosset had trained for two terms. Pearson had already developed many of the mathematical underpinnings for Gosset's ideas but failed to recognize their practical importance.

Fisher, on the other hand, was immediately impressed, later calling Gosset's work a "logical revolution" on par with the birth of deductive reasoning in the time of Euclid[36] and the inspiration for much of Fisher's own work.[37] Gosset's innovation was to develop methods that could apply to *small samples* from a population, which required an entirely new way of thinking about data from what previous statisticians had managed.

Since Gauss, people had gotten comfortable using the normal distribution to predict how often an observation would deviate from its true value, the kind of prediction that would be useful in identifying outliers among a set of astronomical records. The same formulas would dictate how often the observed average of a data set would differ by some amount from the true theoretical average, as a function of the sample size. The problem was that to do those calculations, the typical size of the error—that is, the standard deviation of the error distribution—had to be known from previous experience. This was often a nonissue for statisticians in the mold of Quetelet, Galton, and Pearson because they were working with large groups of subjects numbering in the hundreds or thousands, and they could simply use the observed standard deviation as a proxy for the theoretical one.

Gosset, working with samples of different varieties of barley, say, might have only a few data points from which to make inferences. So estimating the error size from the variability shown in the data itself was subject to error: different samples might give rise to different estimated standard deviations and therefore to different probable errors. His revolutionary idea was to calculate how variable these estimated standard deviations themselves were—that is, to find the *error of the estimated error*. Assuming the data had come from a normal distribution, Gosset was able to derive an approximate formula for the probability of observing a sample mean some distance away from the true theoretical mean, when measured on the scale of the observed variation *in the data alone*. In other words, his probability accounted for both how variable the observed average was and how variable the observed variability was. This was captured in a summary statistic now called *Student's* t.

Fisher saw the enormous potential if statistics could be applied to this kind of problem. It meant a whole range of subjects where only small samples might ever be available could be brought under the umbrella of statistics. The two found each other by means of Fisher's astronomy tutor at Cambridge, Professor F. J. M. Stratton,[38] who was also advising agriculturalists at Cambridge about how to handle statistical estimation problems with sample sizes as small as four (!). This caught the attention of Gosset, who was friends with these agriculturalists, who directed him to Stratton, who showed him Fisher's paper on the maximum likelihood method.

Interestingly, Gosset at first dismissed it as a "neat but as far as I could understand it, quite unpractical and unserviceable way of looking at things."[39] But after a back-and-forth with Fisher in which Gosset corrected a mistake in Fisher's note and then Fisher corrected a mistake in Gosset's *correction*, Fisher sent Gosset a derivation essential to Gosset's work: a proof of the exact sampling distribution for his *t*-statistic. What Gosset found most impressive about

Fisher's derivation was the facility with which he handled computations in high-dimensional spaces. Realizing he was out of his depth mathematically, Gosset forwarded the derivation to Pearson to check over:

> I am enclosing a letter which gives a proof of my formulae for the frequency distribution of z ($= x/s$), where x is the distance of the mean of n observations from the general mean and s is the [standard deviation] of the n observations. Would you mind looking at it for me; I don't feel at home in more than three dimensions even if I could understand it otherwise.
>
> The question arose because this man's tutor is a Caius man whom I have met when I visit my agricultural friends at Cambridge. . . . Well, this chap Fisher produced a paper giving "A new criterion of probability" or something of the sort. . . . Now he sends this to me. It seemed to me that if it's all right perhaps you might like to put the proof in a note. It's so nice and mathematical that it might appeal to some people. In any case I should be glad of your opinion of it.[40]

So it was through their common connection to Gosset that Fisher and Pearson first became acquainted. Pearson did not take Gosset's suggestion to publish Fisher's note in *Biometrika* at that time, but he did agree to publish a result Fisher sent him in 1914. By then, Fisher had taken a job as a statistician for the Mercantile & General Investment Company in London and was teaching math and physics— and by all accóunts hating it—in addition to doing his own research.

The note Fisher sent Pearson in 1914 was technically sophisticated. It contained his derivation for the distribution of the sample correlation coefficient for two normally distributed variables (we'll explain what all this means and why Fisher was interested in it later on). The most impressive thing was that Fisher had accomplished this derivation by considering a collection of n samples as the coordinates of a single point in n-dimensional space[41] and then using his geometric intuition to understand its properties. The language of his solution included geometric phrases like "cosine of the angle of the two radii vectors." Pearson's first response was to tell Fisher the solution looked correct and "of very great interest," and after taking some time to review the details a bit more, he wrote back to Fisher: "I have now read your paper fully and think it marks a distinct advance and is suggestive in character. I shall be very glad to publish it, as I feel sure it will lead to developments."[42] A cordial postscript even suggested Fisher reach out to Pearson's nephew who lived nearby.

Pearson suggested his group of researchers at Galton Laboratory undertake a long series of computations comparing Fisher's exact solutions with previously known approximations. Fisher's paper was published in *Biometrika* in May 1915, and the two continued exchanging letters for the next year regarding progress in the computational study and further ideas of Fisher's. In one of these letters in 1916, Pearson expressed concerns that *Biometrika* may not be able to continue publishing due to the strain of World War I. The journal had been cut off from its subscribers in continental Europe, and several of Pearson's (male) researchers had been called into military service. (Fisher was rejected from service on account of his eyesight.) He worried the journal would need to change ownership and maybe relocate to America. Fisher responded sympathetically: "It would be a most terrible loss as well as an appalling indignity, if this country cannot support such an important and valuable school of research."[43]

The interaction is worth noting only because it might have been the last kind words either Fisher or Pearson ever expressed to one another.

What happened next was a somewhat bizarre sequence of events, triggered by an editorial decision of Pearson's based mostly on a misunderstanding, and the beginning of a vicious feud between the two men. When the results of his computational study were finally ready to be published in 1917, Pearson added a section to the article, unbeknownst to Fisher, specifically criticizing Fisher's maximum likelihood method, which he had first published in 1912. The section, entitled "On the Determination of the 'Most Likely' Value of the Correlation in the Sampled Population," called out Fisher's method as being *Bayesian* and then proceeded to attack the idea of Bayesian inference on logical grounds.[44]

Pearson and Fisher (!) were committed frequentists and generally avoided inference using Bayesian methods, which they called *inverse probabilities*, since Bayes's rule worked backward from observed data to assign a probability for a hypothesis.

In the remainder of this section of his paper, Pearson critiqued the results of what he thought was Fisher's Bayesian approach, as he understood it. His argument was that Fisher's method had assumed uniform prior ignorance of any unknown quantity—in this case, the correlation between two variables in a population. But as any research scientist would know from experience, correlations tended to be mild and rarely took extreme values close to 1 or −1. Therefore, after performing a Bayesian inference for this parameter assuming a uniform prior distribution, Pearson showed the results would be counterintuitive.

That was entirely beside the point, though, since Fisher's method had not actually been Bayesian at all. What probably confused Pearson was that Fisher had proposed a new method of inference based on something called *likelihood*, which, if viewed in the right way under special circumstances, *could* be seen as

Bayesian (we'll come back to this in the next chapter). But Fisher's philosophical orientation in developing the method was frequentist, as we'll explain in more detail later.

They also quarreled over a technical point regarding Pearson's beloved chi-squared test. Fisher asked to publish his critique in *Biometrika*; Pearson refused, saying he would then need to publish his own response and space was limited. Due to Pearson's influence, Fisher had great difficulty finding anyone else willing to publish his critique. With that, their friendship came to an end. In 1919, Pearson offered Fisher the job of chief statistician at Galton Laboratory, and Fisher declined. Instead, he took a position focused on agricultural research at the Rothamsted Experimental Station in the remote town of Harpenden, north of London. Their technical arguments about the chi-squared test would go on for many years, with the consensus opinion nowadays being that Fisher had been in the right.

In the 14 years he spent at Rothamsted, Fisher gained great influence in the world of statistics by supervising numerous research assistants and visiting students. He also compiled his most important statistical work, the book *Statistical Methods for Research Workers* (1925), which contained a collection of practical methods for scientists, especially biologists, to use in problems of inference when dealing with small sample sizes—Fisher's specialty since his days at Cambridge. As he described in the preface (with a subtle jab at Pearson's large samples): "The elaborate mechanism built on the theory of infinitely large samples is not accurate enough for simple laboratory data. Only by systematically tackling small sample problems on their merits does it seem possible to apply accurate tests to practical data. Such at least has been the aim of this book."[45]

From there, the book reads much like a modern statistics textbook. It starts with a brief discussion of the meaning of statistics and its role in scientific inference and then presents a series of recipes, with no mathematical proofs, for the different tests one might use for different sorts of problems. The common refrain among these tests is the idea of *significance*. We'll see an example of a significance test worked out in detail in the next chapter, but the template is this: To test a theory—say that a new fertilizer had improved crop yields—Fisher would say to first set up a *null hypothesis*, in this case, that there had been no improvement. Then he would say to look for a *test statistic*—for example, the difference in average yields between crops given the new fertilizer versus those given the old one, standardized according to a given rule. Finally, he would advise how big the statistic had to be in order to safely reject the null hypothesis and conclude the effect was real.

The tests are arranged roughly in order of increasing complexity and supplemented by many numerical examples and tables of the relevant distributions. For an experimental scientist without advanced mathematical training, the book was a godsend. All such a person had to do was find the procedure corresponding to their problem and follow the instructions. As a result, *Statistical Methods for Research Workers* was enormously successful. It went through 14 editions between 1925 and 1970, and it became such the industry standard that anyone *not* following one of Fisher's recipes would have a hard time getting results published. Together with his later book *The Design of Experiments* (1935), it established Fisher as a dominant figure in the world of statistics. Among Fisher's most prominent disciples were statistician and economist Harold Hotelling, who, in turn, heavily influenced two Nobel Prize–winning economists, Kenneth Arrow and Milton Friedman, and George Snedecor, who founded the first academic department of statistics in the United States, at Iowa State University.[46]

Also while working at Rothamsted, Fisher made his greatest contributions to the theory of evolution, summarized in the book *The Genetical Theory of Natural Selection* (1930). At the time, still 20 years before Rosalind Franklin discovered the DNA double helix, the exact dynamical process of evolution was still hotly contested. Darwin had established that species could evolve over time to better suit their environments, but natural selection acting on small chance mutations was far from being accepted as the driving mechanism. Instead, various competing theories such as neo-Lamarckism (the inheritance of traits acquired during an organism's lifetime), orthogenesis (evolution guided by some internal forces acting toward certain goals), mutationism (spontaneous evolution of new forms in single large increments), and theistic views that evolution was guided by God all held sway at the turn of the 20th century.

At the same time, Gregor Mendel's theory of inheritance via genes was becoming increasingly accepted as the explanation for the process by which traits would express themselves in successive generations of organisms, depending on the traits seen in their parents. Putting these two ideas together was the grand project of 20th-century evolutionary biology now called the *modern synthesis*. One of the chief obstacles to achieving that synthesis was that certain traits, say height or skull size, appeared to vary *continuously* in a population. There was no simple binary expression like Mendel's purple and white flowers of pea plants or the dominant/recessive dynamics of a trait like eye color. Instead, these traits appeared to blend and take a range of values. Pearson, among others, argued for a "biometric" theory of evolution—the origin of his journal's name *Biometrika*—in which the metrics (stature, skull size, intelligence, etc.) of a child organism were correlated to some degree with the metrics of its parents. Evolution was therefore a process by which certain ranges of these metrics were favored over others and the overall distribution in the population tended to drift slowly over time.

Fisher was in the opposite camp, which argued evolution was driven by genes, with alleles being segregated into discrete categories. What Fisher showed in his research work, perhaps thinking all the way back to Quetelet's Scottish soldiers' chests, was that an *apparently* continuous distribution of certain traits could arise as the *sum total interaction* of many of these genes acting simultaneously, what we would now call a *polygene* or set of *quantitative trait loci*. Thus, the distributions could steadily drift over time and *still* be affected by spontaneous mutations in discrete increments. Evolution by natural selection was consistent with Mendelian genetics. In achieving this synthesis, Fisher essentially established a new discipline of biology, population genetics, and he supplied it with a powerful mathematical infrastructure. He proved what he called the fundamental theorem of natural selection: "The rate of increase in fitness of any organism at any time is equal to its genetic variance in fitness at that time."[47]

This gave precise meaning to the dynamics of natural selection acting through changes in gene frequencies. Fisher immodestly compared this to the physical study of statistical mechanics and, in particular, to the hallowed Second Law of Thermodynamics (the entropy of a closed system always increases toward equilibrium): "Professor Eddington has recently remarked that 'The law that entropy always increases—the second law of thermodynamics—holds, I think, the supreme position among the laws of nature'. It is not a little instructive that so similar a law should hold the supreme position among the biological sciences."[48] For stating and proving a result of similar magnitude in biology and for helping establish population biology as a quantitative discipline, Fisher was widely celebrated—and rightly so. In 2011, Richard Dawkins even called him "the greatest biologist since Darwin."[49] His legacy in the subject is so great that biologists nowadays are sometimes surprised to find out that Fisher was *also* a statistician.

However, we also find in Fisher's writings on genetics the same eugenicist ideas that had motivated Galton and Pearson before him, particularly their theory of a hereditary component to socioeconomic class.

Before he was a biologist or a statistician, really, Fisher was a eugenicist. He became a devotee of eugenics while still a student at Cambridge, where he advocated for the formation of the University of Cambridge Eugenics Society, created in 1911 with Fisher serving as chair of the undergraduate committee. This committee ran its meetings out of his rooms at the college, and he spoke regularly to the society about the latest advances in evolutionary theory. For its first public meeting, the society invited as a guest speaker the Reverend William Ralph Inge, who had recently written a paper about the menace of the lower classes, warning that "the urban proletariat may cripple our civilisation, as it destroyed that of ancient Rome. These degenerates, who have no qualities that confer a survival value, will probably live as long as they can by 'robbing hen roosts.' "[50] The society's minutes note that he was thanked after his address by "R. A. Fisher of Caius College."

In the period from 1912 to 1920, after he graduated but before he started doing agricultural research at Rothamsted, Fisher wrote *91* articles about eugenics for Galton's journal, the *Eugenics Review*.[51] One of his earliest publications was an essay titled "Some Hopes of a Eugenist" (1914), in which he hinted at a eugenicist theory of nationalism:

> The overmastering condition of ultimate predominance is nothing else than successful eugenics; the nations whose institutions, laws, traditions and ideals, tend most to the production of better and fitter men and women, will quite naturally and inevitably supplant, first those whose organisation tends to breed decadence, and later those who, though naturally healthy, still fail to see the importance of specifically eugenic ideas.[52]

This would be a running theme throughout Fisher's career. The final five chapters of *The Genetical Theory of Natural Selection*, comprising about a third of the book, were given over to Fisher's hereditary theory of why civilizations fail or succeed, mostly driven by the high fertility of subpopulations of lesser character. Sections in these chapters included "The Mental and Moral Qualities Determining Reproduction," "Economic and Biological Aspects of Class Distinctions," and "The Decay of Ruling Classes."

Using data from the British census of 1911, Fisher showed that lower-class people tended to be more fertile, a phenomenon he called an "inverted birth-rate." He also showed a positive correlation between fertility rates over generations, suggesting these rates were inherited. Echoing Galton's ideas of nature-over-nurture for men of eminence, Fisher claimed class divisions corresponded to *biological differences* between people, like varietals of a fruit or breeds of dogs, reinforced by the social pressure to intermarry within one's own class. As he wrote:

> The different occupations of man in society are distinguished economically by the differences in the rewards which they procure. Biologically they are of

importance in insensibly controlling mate selection, through the influences of prevailing opinion, mutual interest, and the opportunities for social intercourse, which they afford. Social classes thus become genetically differentiated, like local varieties of a species, though the differentiation is determined, not primarily by differences from class to class in selection, but by the agencies controlling social promotion or demotion.[53]

So he worried the biologically determined higher fertility rates among the lower classes would eventually bring down any civilization, including the British Empire: "Since the birth-rate is the predominant factor in human survival in society, success in the struggle for existence is, in societies with an inverted birth-rate, the inverse of success in human endeavour. The type of man selected, as the ancestor of future generations, is he whose probability is least of winning admiration, or rewards, for useful services to the society to which he belongs."[54]

To combat this, he proposed a system of disincentives against large families of low social status. He also suggested limits be placed on the sizes of immigrant families because they would not have inherited the genes to make them both suited to living in the climate of their new home and resistant to its diseases. And, like Pearson, he was not above endorsing the occasional genocide if it meant supplanting a "tribe" of low genetic character with one of superior stock to the improvement of society: "Among a group of small independent competing tribes the elimination of tribes containing an undue proportion of the socially incompetent, and their replacement by branches of the more successful tribes, may serve materially to maintain the average standard of competence appropriate to that state of society."[55]

It was during Fisher's lifetime that the eugenics movement attained its most horrific final form in Nazi Germany. Elements of Adolf Hitler's eugenics "project" were, in fact, descended from Galton's in a surprisingly direct manner and therefore cousins of Fisher's, by way of America.

The concept of eugenics spread quickly to the United States through the efforts of academic biologists and statisticians such as Harvard professor Charles Davenport. He learned about eugenics and its mathematical underpinnings directly from Galton and Pearson in London and briefly served as a coeditor of *Biometrika*. In 1910, Davenport founded the Eugenics Record Office (ERO), which, in a manner very similar to that of Galton's Laboratory, went on to collect and analyze data from several hundred thousand individuals concerning various social and medical traits like intelligence, skull size, criminality, alcoholism, literacy, fertility, and incidence of disease.

Applying the methods he learned from Galton and Pearson, Davenport used this data as the basis for numerous publications arguing the dangers of "miscegenation" and immigration from countries of "inferior" genetic stock. Supported by ample funding from backers such as the Carnegie Institution, the Rockefeller Foundation, and the Harriman family's railroad fortune, the ERO and other professional organizations like it sought to spread the gospel of eugenics to the general public.

By far the most prestigious such organization was named in honor of Francis Galton, the Galton Society of America, also founded by Davenport alongside other prominent figures in American eugenics. These were important men of science with influential connections. About a third of the members of the Galton Society were also members of the National Academy of Sciences, and more than half were members of the American Association for the Advancement of Science. They used their positions of power to direct the course of American scientific research in the 1920s and '30s and to lobby successfully for practices and legislative measures like immigration restrictions, laws prohibiting interracial marriage, and forced sterilization of the mentally ill, physically disabled, and anyone else deemed a drain on society. These practices became standard in many official and unofficial ways. For example, in California, where eugenics programs saw their largest fruition, about 20,000 people were forcibly sterilized during the period from 1909 to 1963; they were primarily residents of state mental hospitals; members of the Mexican, Native American, and Asian populations considered "undesirable" and a costly burden on the state's welfare system; and women classified as "oversexed" or "sexually wayward."

It was from the American eugenicists and Galton followers that the Nazis largely took their inspiration. In *Mein Kampf*, Hitler wrote, "There is today one state in which at least weak beginnings toward a better conception [of immigration] are noticeable. Of course, it is not our model German Republic, but the American Union."[56] He admired the ability of the U.S. state governments to implement such policies as sterilization at large scale and was quoted as saying "I have studied with great interest the laws of several American states concerning prevention of reproduction by people whose progeny would, in all probability, be of no value or be injurious to the racial stock."[57] In a fan letter to Madison Grant, cofounder of the Galton Society of America, Hitler referred to Grant's book *The Passing of the Great Race* (1916) as "my bible."[58] And the same institutions that funded eugenics research in the United States paid for similar research programs in Germany. For example, in 1926 the Rockefeller Foundation gave $250,000 (about $3.5 million in today's dollars) to the Kaiser Wilhelm

Institute for Psychiatry, which had as its director the psychiatrist Ernst Rüdin, one of the chief contributors to the "scientific" justifications for the racial atrocities of Nazi Germany.

Similar sterilization policies would have been illegal in the UK at the time, but Fisher and other British eugenicists were working to change that. The similarity to Nazi programs was not coincidental. In 1930, Fisher and other members of the British Eugenics Society formed the Committee for Legalizing Eugenic Sterilization, which produced a propaganda pamphlet arguing for the societal benefits of sterilizing the "feeble minded high-grade defectives."[59] Fisher had contributed statistical analysis to demonstrate the benefits of such sterilization, based on data collected by the Americans. To strengthen their arguments with additional data, the Eugenics Society reached out directly to Rüdin, who in turn expressed his admiration for the work of Fisher's committee.[60]

American eugenicists maintained close connections to their German counterparts and hailed the Nazis for implementing programs about which they could only dream. As Leon Whitney, executive secretary of the American Eugenics Society, said, "While we were pussy-footing around . . . the Germans were calling a spade a spade."[61] Davenport, in particular, held editorial positions at two influential German journals after Hitler's rise to power. His work was referenced by Nazi academics like Otto Reche, who advocated on "racial-scientific" grounds for the genocide of the people of Poland. The 12 million people murdered by the Nazis for the crime of being genetically "inferior" were, in many ways, victims of the natural logical extension of Galton's eugenics proposals from the previous century.

Fisher continued to have disturbingly close ties to Nazi scientists even after the war. He issued public statements to help rehabilitate the image of Otmar Freiherr von Verschuer, a Nazi geneticist and advocate of racial hygiene ideas who had been a mentor to Josef Mengele and used data collected by Mengele in his Auschwitz experiments. In von Verschuer's defense, Fisher wrote, "I have no doubt also that the [Nazi] Party sincerely wished to benefit the German racial stock, especially by the elimination of manifest defectives, such as those deficient mentally, and I do not doubt that von Verschuer gave, as I should have done, his support to such a movement."[62]

In response to the Holocaust, the United Nations Educational, Scientific and Cultural Organization (UNESCO) issued a statement in 1950 titled "The Race Question" to clarify what was actually scientifically understood about race at the time and condemn racism on scientific grounds. Combining input from leading thinkers in sociology, biology, psychology, and anthropology, the statement claimed the apparent racial differences among people tended to appear, disappear, and fluctuate and that these differences were minor in comparison to the great similarities between all members of the species *Homo sapiens*. "What is perceived," the statement read, "is largely preconceived, so that each group arbitrarily tends to misinterpret the variability which occurs as a fundamental difference which separates that group from all others."[63] In particular, most perceived differences should be thought of as cultural differences between ethnic groups rather than genetically determined racial characteristics:

> National, religious, geographic, linguistic and cultural groups do not necessarily coincide with racial groups: and the cultural traits of such groups have no demonstrated genetic connexion with racial traits. Because serious errors of this kind are habitually committed when the term "race" is used in popular parlance, it would be better when speaking of human races to drop the term "race" altogether and speak of *ethnic groups*.

Fisher strongly disagreed and wrote a dissenting opinion, which UNESCO included (along with many other critical responses) in a revised version issued in 1951. Fisher claimed the scientific evidence showed human groups differ profoundly "in their innate capacity for intellectual and emotional development" and concluded that the "practical international problem is that of learning to share the resources of this planet amicably with persons of materially different nature." He claimed a statement such as UNESCO's could only do harm by ignoring the truth of racial differences, since "this problem is being obscured by entirely well-intentioned efforts to minimize the real differences that exist."[64] In an earlier letter in 1945, he made clear the differences he was referring to were "not the superficial indications provided by skin and hair, but temperamental differences affecting the moral nature."[65]

Despite sharing a lifelong commitment to eugenics, Fisher never forgave Pearson for his backstabbing criticism and the damage it did to Fisher's early career or for the obstinance with which he had refused Fisher's correction to the chi-squared test. When Pearson retired in 1933, Fisher, then arguably the most prominent statistician living, took over as Galton Chair of Eugenics at UCL. The university effectively split the position, though, and appointed Pearson's son Egon to head the newly created Department of Statistics.

Fisher transferred his grudge against Karl to Egon and his chief collaborator, the Polish mathematician Jerzy Neyman. The relationship between the Galton Lab and the Statistics Department steadily deteriorated, despite the two being located on adjacent floors in the same building. Fisher became especially incensed when Neyman refused to teach a class out of Fisher's textbook. Fisher threatened, "From now on I shall oppose you in all my capacities."[66] The situation became so tense that students of the rival camps organized two departmental teas in the afternoons, one at 4 and one at 4:30, so that Fisher would never have to be in the same room as Neyman or Pearson. The two exchanged verbal potshots with each other for decades, even after Neyman relocated halfway around the world to the University of California, Berkeley.

In the foreword to his final book, *Statistical Methods and Scientific Inference* (1956), written 20 years after Karl Pearson died, Fisher recounted the story of how Galton appointed Pearson to carry on his legacy, and he pulled no punches:

> The systematic improvement of statistical methods and the development of their utility in the study of biological variation and inheritance were the aims to which [Galton] deliberately devoted his personal fortune, through the support and endowment of a research laboratory under Professor K. Pearson.
>
> The peculiar mixture of qualities exhibited by Pearson made this choice in some respects regrettable, though in others highly successful. . . . In a sense he undoubtedly appreciated Galton's conception of the greatness of the potential contribution of Statistics in the service of Science, and as a means of rendering strictly scientific a range of studies not traditionally included in the Natural Sciences, but, as perceived through his eyes, this greatness was not easily to be distinguished from the greatness of Pearson himself.
>
> The terrible weakness of his mathematical and scientific work flowed from his incapacity in self criticism, and his unwillingness to admit the possibility that he had anything to learn from others, even in biology, of which he knew very little. His mathematics, consequently, though always vigorous, were usually clumsy, and often misleading. In controversy, to which he was much addicted, he constantly showed himself to be without a sense of justice.[67]

Egon Pearson, for his part, tried to strike a conciliatory tone in an article he published in 1968 in *Biometrika*, which he had taken over as editor. The note included some of the letters of correspondence between Fisher and Pearson during that fateful period of 1914–1916 in an attempt to trace the origins of their misunderstanding and contextualize his father's decision not to publish any more of Fisher's work:

> There was deeply ingrained in [Pearson], as there was too in Fisher, an urge to reply to any expression of opinion which he believed to be wrong and perhaps

harmful to the development of his subject. But at least the need to reply would be less compelling if the "faulty" article was not in print in his own journal! The titanic battles which have from time to time been waged across the statistical field were perhaps enlivening to the onlookers, but they were very real and I think harmfully moving to the participants. History we may hope will forget them.[68]

So far history has not forgotten.

THE HYBRIDIZATION

Pearson and Fisher both had great technical skills, which they directed with laser focus toward the goal of establishing statistics as a rigorous discipline, carrying on the project Galton had begun years earlier. Pearson's description in 1920 of his career mission, preserved in the biography written by his son, could just as well apply to Fisher's:

> to make statistics a branch of applied mathematics with a technique and nomenclature of its own, to train statisticians as men of science . . . and in general to convert statistics in this country from being the playing field of dilettanti and controversialists into a serious branch of science, which no man could attempt to use effectively without adequate training, any more than he could attempt to use the differential calculus, being ignorant of mathematics.[69]

Both were extremely ambitious men possessed of colossal egos, and both wielded tremendous influence over the next generation through their writing and teaching. As a result, any student of statistics these days will know the names Pearson and Fisher. The correlation coefficient, now a standard calculation applied to almost any data containing two variables, is Pearson's rho. Pearson gets credit for the multivariate normal distribution, contingency tables, the chi-squared test, the method of moments, and principal component analysis, all standard tools. He also invented significance testing and the p-value, which is the most common measure of statistical significance, as we'll describe in the next chapter. From Fisher we get the F-test, the idea of a sufficient statistic, the method of maximum likelihood, the concept of a parameter, linear discriminant analysis, ANOVA (analysis of variance), Fisher information, and Fisher's exact test, among others.

Even though they disagreed fiercely on some technical points, when it came to the general practice of statistical inference, their methods were largely the same.

For example, Pearson introduced the p-value in the setting of his chi-squared test, but it was Fisher who popularized it as a practical tool and set the standard of $p = 0.05$ as the threshold for statistical significance. He intended this as one tool among many that would enable experimental scientists to understand what features of their data were surprising or worthy of a second look.

In the next generation, Egon Pearson and Neyman introduced a different, more mathematical approach to hypothesis testing by means of *decision theory*. That is, they viewed the results of a statistical test in terms of the decision to *accept* or *reject* a hypothesis in favor of an alternative, with penalties for making the wrong choice. From this mode of thinking came such concepts as unbiased estimators, statistical power, Type I and Type II errors, and confidence intervals. Neyman, in particular, tried to claim all of statistics as a branch of mathematics by denying that any part of statistics involved induction. He wrote: "In the ordinary procedure of statistical estimation there is no phase corresponding to the description of 'inductive reasoning' . . . all the reasoning is deductive and leads to certain formulae and their properties."[70]

Fisher disagreed vehemently about the role of induction in science and with the whole Neyman-Pearson framework of hypothesis testing, claiming its authors lacked "any real familiarity with work in the natural sciences, or consciousness of those features of an observational record which permit of an improved scientific understanding."[71] He compared Neyman's views to thought control: "To one brought up in the free intellectual atmosphere of an earlier time there is something rather horrifying in the ideological movement represented by the doctrine that reasoning, properly speaking, cannot be applied to empirical data to lead to inferences valid in the real world."[72]

In an impressive feat of Cold War–era jujutsu, he said the ideology underlying the decision-theoretic version of hypothesis testing was reminiscent *both* of the five-year plans of the Soviet Union *and* of the emphasis on bottom-line results typical of the United States:

> Russians are made familiar with the ideal that research in pure science can and should be geared to technological performance, in the comprehensive organized effort of a five-year plan for the nation. How far, within such a system, personal and individual inferences from observed facts are permissible we do not know. . . . In the U.S. also, the great importance of organized technology has I think made it easy to confuse the process appropriate for drawing correct conclusions, with those aimed rather at, let us say, speeding production, or saving money.[73]

For Fisher, in what we can only assume he considered to be the British way, inference was not about *making the right decisions* but rather about *thinking the right things*.

While these feuds continued, it was up to the community of research scientists, journal editors, and statistics textbook authors to decide which techniques would become industry standards. Because of the great influence of both the Fisher and the Neyman-Pearson schools, the answer was that working scientists took a hybrid approach and combined ideas from both camps. Most notably, the current standard procedure of null hypothesis significance testing is the result of cramming Fisher's p-value measure of significance into Neyman and Pearson's hypothesis testing framework.

Use of the combined tools and methods of Fisher, Neyman, and the Pearsons became incredibly widespread, the common language for almost all experimental science. This hybrid collection now forms the bulk of the methods we think of as orthodox statistics. Table 4.1 shows the numbers of citations for various statistical keywords returned by a Google Scholar search as of 2020. These numbers are simply ridiculous. By comparison, the current record holder for most cited paper is "Protein Measurement with the Folin Phenol Reagent," written by Oliver H. Lowry et al. in 1951.[74] It has around 213,000 citations, or about one-twentieth of the corresponding number for Fisher and Pearson's idea of the p-value. No single paper serves as the origin of this term, but if one did, it would easily be the most cited academic work of all time.

The influence of these authors was so great that they all but drowned out the competing Bayesian school of thought, with the technical squabbles among

TABLE 4.1 **Number of citations for statistical terms associated to Pearson and Fisher**

Search term	Returns
"p-value"	4,580,000
"Pearson" + "correlation"	3,100,000
"ANOVA"	2,750,000
"confidence interval"	2,700,000
"maximum likelihood"	2,680,000
"p < 0.05"	2,210,000
"null hypothesis"	974,000

the frequentists instead taking center stage. The result of those arguments, one triggered by the *hint* of Bayesianism in Fisher's work, is that orthodox statistics may be an unholy hybrid no single author would recognize, but it is an entirely frequentist one, as we'll see worked out in more detail in the next chapter. On that topic, they were in agreement. Fisher, in particular, was known to fly into a rage any time the topic of inverse probability was raised. Fred Hoyle, an astrophysicist who knew Fisher personally, said the topic would "turn Fisher in the briefest possible moment from extreme urbanity into a boiling cauldron of wrath."[75] Egon Pearson even once turned down a request to coauthor a paper with Neyman because it mentioned prior probabilities—ironically to argue that the choice of priors didn't matter if the data set was large enough—and he was too afraid of how Fisher might respond.[76]

In a way, it should seem surprising that Pearson and Fisher *were* frequentists given how closely descended they were intellectually from mathematicians like Laplace and Gauss, who were at least comfortable with Bayesian ideas of probability.

Pearson was only somewhat squeamish about Bayesian inference as a conceptual framework. For one thing, he always included Bayes' theorem in his classes at UCL. The main issue he had was with the customary assignment of a uniform prior probability distribution for any unknown quantity of interest (that is, the principle of insufficient reason), as we saw with George Boole, among others, in the last chapter. In Pearson's 1917 critique of Fisher's maximum likelihood method that kickstarted their bitter feuding, for instance, he just (wrongly) thought Fisher was doing Bayesian inference starting from a uniform prior for a correlation coefficient, where practical experience would lead any scientist to a different prior assignment, maybe something more concentrated toward 0. Pearson even helpfully carried through a Bayesian analysis of the consequences of such an alternative prior assumption. Gosset, who had been Pearson's student, echoed the same ideas in his own work on the correlation coefficient: "It is clear that in order to solve this problem we must know two things: (1) the distribution of values of *r* [the in-sample correlation coefficient] derived from samples of a population which has a given *R* [the population correlation coefficient], and (2) the *a priori* probability that *R* for the population lies between any given limits."[77] Gosset was open to prior distributions other than the uniform if a problem required it.

In his landmark book *The Grammar of Science*, Pearson defended the Laplacean approach with an argument attributed to Francis Edgeworth that, for many categories of problems, unknown quantities actually *did* tend to distribute themselves more or less uniformly.[78] So even if he couldn't say the uniform distribution was the exact right one, it could serve as a reasonable stand-in for lack of something better, and other distributions could easily take its place if experience suggested. The only way in which Pearson was truly a frequentist, then, was that he thought these probability assumptions (and their inferential consequences) did ultimately need to square up with observed frequencies over time. That is, he didn't disagree with Boole's assertion that probability assignments need to be grounded in actual experience to be meaningful; he just sidestepped the criticism by positing that maybe a uniform distribution was empirical after all.

Fisher was really the first person to try seriously to excise Bayesian inference from statistics altogether. The generation of students who came through his lab at Rothamsted and the many more who learned statistics from his textbook didn't learn the Bayesian method at all. In *Statistical Methods for Research Workers*, Fisher wrote that "inverse probability [that is, Bayesian inference] is founded upon an error, and must be wholly rejected."[79] But in his early writings about probability and statistics, Fisher even acknowledged his debt to the legacy of Laplace:

> There would be no need to emphasise the baseless character of the assumptions made under the titles of inverse probability and Bayes' Theorem in view of the decisive criticism to which they have been exposed at the hands of Boole, Venn, and Chrystal, were it not for the fact that the older writers, such as Laplace and Poisson, who accepted these assumptions, also laid the foundations of the modern theory of statistics, and have introduced into their discussions of this subject ideas of a similar character.[80]

Evidently, for Fisher, the legacy of Laplace was trumped by the more immediate legacy of Boole, Chrystal, and especially Venn, who, as we have suggested, may have had a strong influence on Fisher at Cambridge. Fisher saw Venn as having permanently settled the fight between subjectivists and frequentists in favor of the frequentists. In his own work, then, Fisher defined probability using very Venn-like language of hypothetical infinite populations and imaginary infinite sequences of rolls of an imaginary die:

> Probability is the most elementary of statistical concepts. It is a parameter which specifies a simple dichotomy in an infinite hypothetical population, and it represents neither more nor less than the frequency ratio which we imagine such a population to exhibit. For example, when we say that the probability of throwing

a five with a die is one-sixth, we must not be taken to mean that of any six throws with that die one and one only will necessarily be a five; or that of any six million throws, exactly one million will be fives; but that of a hypothetical population of an infinite number of throws, with the die in its original condition, exactly one-sixth will be fives. Our statement will not then contain any false assumption about the actual die, as that it will not wear out with continued use, or any notion of approximation, as in estimating the probability from a finite sample, although this notion may be logically developed once the meaning of probability is apprehended.[81]

Thus, for Fisher, probability and statistics were entirely about the study of *populations* and random sampling conducted on them. The annoying technicality that real populations tended to be finite was to be ignored by thinking of them *as though* they were infinite. And any derived quantities such as averages of the sample could also be given probabilities by thinking of those as having been sampled from a theoretical infinite population of *measurements*: "The idea of a population is to be applied not only to living, or even to material, individuals. If an observation, such as a simple measurement, be repeated indefinitely, the aggregate of the results is a population of measurements."[82]

Fisher would, of course, allow for Bayesian inference in special cases whenever the prior probabilities themselves were given by known frequencies or proportions in a population. In *Statistical Methods and Scientific Inference*, in the midst of dismissing inverse probabilities, he gave such an example using genetics: Suppose among black mice there are two genetic kinds, homozygous (BB) for black color and heterozygous (Bb) for black and brown. If two heterozygous mice mate, according to Mendelian theory the proportions of their offspring will be as 1:2:1 for BB:Bb:bb. If a mouse is known to have been the product of such a mating, it therefore will have known prior probabilities for having each genotype, which could serve as the basis for a Bayesian inference based on *its* offspring. Working as he so often did with genetic data, Fisher would have no difficulty juggling these probabilities according to Bayes' theorem.[83]

As a side note, one of the three heroes Fisher credited with demolishing Bayesian inference, the Scottish mathematician George Chrystal, would not have agreed with even this plain vanilla inference. Fisher claimed it was Chrystal's decision to leave inverse probability out of his *Algebra* book in 1886 that signaled the death of Laplacean ideas but noted that Chrystal had not given his actual reasons. In a little-known paper in *Transactions of the Actuarial Society of Edinburgh* (1891), though, Chrystal explained that his objections were to Bayes' theorem *itself* because he thought he had found an example where it gave an illogical answer. The problem was to infer the state of a bag of three balls, each either black or white, given two successive white balls being drawn without replacement. Basically he got stumped by a probability brainteaser of the same caliber as the Boy or Girl Paradox we saw in chapter 1. He computed the right answer using Bayes' theorem and then concluded it was illogical because it was counterintuitive. Statistics historian Sandy Zabell at Northwestern University theorized that Fisher was unaware of this later work of Chrystal's but that if he had been, he would have been mortified.[84]

Fisher, like so many others, just took issue with the idea of assigning probabilities in the absence of frequency data of the kind that would naturally come up when making inferences about a hypothesis or an unknown constant. Since a hypothesis is only simply true or false, there is no sense in which it has a frequency-based probability at all. The fact that these were the kinds of inferences that scientists had been trying to make for hundreds of years was what kept Pearson and others from discarding Bayesian principles altogether.

Even Boole and Venn, the other two members of Fisher's frequentist triumvirate, softened on this point in their later work. Boole wrote in 1862 that the "principle of non-sufficient reason" or the "principle of the equal distribution of knowledge or ignorance" was actually a fundamental rule of probability and that it "involves an equal distribution of our actual knowledge, and enables us to construct the problem from ultimate hypotheses which reduce it to a calculation of combinations."[85] In the third edition to *The Logic of Chance*, published in 1888, Venn considered an example inference involving two urns with unknown proportions, abstracted from a question raised by a doctor about the effectiveness of antiseptics in preventing infections after surgery. The question was whether any inference could be made from sample sizes as small as 10 or 15 patients, some treated with antiseptics and others not. Venn, perhaps bowing to the need to make some inference for such a serious issue, just decided ad hoc that, for problems with *two* urns, the principle of insufficient reason was okay, whereas his objections would still hold for problems with *one* urn.[86]

What set Fisher apart, then, was that he thought he had found a way around all this thanks to his new methods of inference that didn't rely on Bayesian prior probabilities. He was particularly motivated to do so by his interest in experiments with small sample sizes, for which the prior probabilities would have played a bigger role. For Pearson, priors were largely an academic question because the large volume of data tended to wash them away. For Fisher, the most important experiments might admit only a small amount of data, requiring the utmost care in handling.

In 1930, he wrote, "Inverse probability has, I believe, survived so long in spite of its unsatisfactory basis, because its critics have until recent times put forward nothing to replace it as a rational theory of learning by experience."[87] Fisher's proposed replacements were significance testing, maximum likelihood estimation, and fiducial inference. As we alluded to in chapter 2 and we'll see in more detail in the next chapter, the first of these methods essentially just repackaged Jacob Bernoulli's fallacious argument and extended it to different problems, and the others, while still falling victim to Bernoulli's Fallacy, were secretly halfway toward being Bayesian inference by another name.

IT WAS NEVER ABOUT THE PEAS

So, in the end, we mostly have Fisher to blame for modern statistics being frequentist, but he was in many ways just carrying out a line of thought begun nearly a century earlier by many others—arguably stretching all the way back to Bernoulli—and consistent with Galton's and Pearson's views of statistics. In answer to the question of why the statistical methods they ultimately produced were frequentist, the simplest and most correct answer, then, is that Fisher thought they *could* be. His predecessors may have desired for all inference to rely solely on observable facts because it would have fit their overall scientific philosophy, but Fisher was the one who brewed the mathematical snake oil with that as its promise. He was further emboldened by critiques like those of Boole and Bertrand, who had shown that different meanings of ignorance could lead to inconsistent prior probability assignments and that a uniform probability distribution couldn't be justified for all problems. As we saw in the last chapter, these were difficult issues that would take many more years to be resolved—and in some cases are still unresolved. Neyman and Egon Pearson added a complementary set of ideas that Fisher took issue with, but even they yielded to his dogmatic assertion that probability could only mean frequency.

But that answer to the question of why frequentism prevailed leaves open the deeper question of why Fisher never saw the logical flaws in these strictly frequentist methods. He was, without a doubt, a mathematical genius with an expansive knowledge of mathematical history to boot. The examples we've seen so far and the ones we'll consider in the next chapter would certainly not have been beyond him to understand. The statistician Leonard "Jimmie" Savage, whose 1954 textbook *The Foundations of Statistics* played a major part in reviving Bayesian statistics from the near death it suffered due to Fisher, said he once asked Fisher point-blank about the logical fallacy of probabilistic proof-by-contradiction, the idea that improbable data must somehow count as evidence against a hypothesis, as we saw in chapter 2. He said Fisher responded evasively: "Savage, you can see the wool you are trying to pull over our eyes. What makes you think we can't see it, too?"[88] Did he not have a good answer, or did he just think the answer was so obvious it didn't deserve to be said out loud?

There is no doubt that Fisher's motivations were also at least partly interpersonal. Much of his work throughout his career can be seen, in the right light, as a repudiation of Pearson. His focus on small-sample problems undercut Pearson's techniques, which worked only for large data sets; his maximum likelihood method was meant as a more efficient alternative to Pearson's method of moments; he spent years hammering home the same technical flaw in Pearson's celebrated chi-squared test; and more. So the fact that Pearson was partially sympathetic to Bayesian inference (albeit from a strictly frequentist point of view) may have motivated Fisher to try to erase it from history. He also had to think of his own legacy. Had significance testing, maximum likelihood estimation, and fiducial inference all been exposed as a complete waste of time, as we'll argue they should be, then all of Fisher's brilliant derivations of things like the sampling distributions for various test statistics would be forgotten. These were the questions he had spent a great deal of effort answering, so it was natural to organize a worldview in which those questions mattered the most.

There's also the fact that for many of the practical problems he considered, his methods did seem to work. As we saw in chapter 2, the exact same phenomenon arose with Bernoulli and created the possibility for Bernoulli's Fallacy to take hold: *when prior information is weak, an inference can ignore that information and likely still end up with sensible results.* Fisher's scientific problems were mostly confined to survey sampling and other situations where no strong prior information existed to enter into a Bayesian inference anyway. Whatever the objections might have been in theory, the practical results spoke for themselves. Hubris led him to claim the methods must therefore be generally applicable to all problems. In fact, as we'll see in the next chapter, Fisher's beloved maximum likelihood method is just a special case of Bayesian inference with a prior that reflects uniform ignorance—so Fisher was sometimes accidentally doing Bayesian inference.

Considering the social and historical context in which they did their work and their agendas concerning eugenics, though, we might come to a different understanding of why Galton, Pearson, and Fisher were so insistent that probability had to be measurable as long-run frequency. The key advantage to thinking of probabilities as frequencies has always been that it makes them *objective*, so any conclusions drawn from data would appear to be free from bias or prejudice on the part of the person doing the inference. As Fisher wrote: "The feeling induced by a test of significance has an objective basis in that the probability statement on which it is based is a fact communicable to, and verifiable by, other rational minds."[89] According to him, probability was a "physical property of the material system concerned"[90] and was fixed and measurable "in the sense that the weight of an object, and the resistance of a conductor have an objective value."[91] The right way to measure probabilities was therefore through frequencies; according to Fisher, the main accomplishment of Venn's frequentist theory was "developing the concept of probability as an objective fact, verifiable by observations of frequency."[92]

So constructing a method of inference using only frequency probabilities would make any conclusions drawn appear to be free of bias. This kind of objectivity could be useful, for example, if one was trying to establish a controversial new discipline of science. It could also be useful if one was trying to establish as "common knowledge" grotesque ideas such as the idea that one race was inherently superior to another, the idea that colonial subjugation of indigenous peoples was done for their benefit, or the idea that people with disabilities should be forcibly sterilized.

For Galton, Pearson, and Fisher, eugenicist goals were present in the background while all the key concepts of statistics were being hashed out. Galton was led to study inheritance of abilities among prominent families because he wanted to give scientific weight to the suggestion that bloodlines of the upper classes should be preserved at the expense of the lower class. He developed the idea of regression because he wanted to predict how the merits or flaws of each generation would be passed on to the next and to understand why the spread between high and low did not increase the way he expected. He was drawn to the normal distribution because it provided a quantitative scale with which to compare races and measure what he imagined to be the superiority of Anglo-Saxons to others. He developed the correlation coefficient as a way to measure the tendency for supposedly desirable traits among white Europeans or undesirable traits among Africans or Asians to appear together.

The racist and aristocratic assumptions motivating Galton may have been generally common in British elite society at the time, but he gave them scientific backing. He had the authority of a world traveler who had studied other cultures of people in the places they lived, in the tradition of the Victorian naturalists like his cousin aboard the *Beagle*. So his quantitative assessment of the supposed inferiority of other groups of people to the British upper class was an important step toward enshrining these attitudes as common knowledge. Inestimable violence was done to the peoples of the British colonies in Asia, Africa, and the Americas, rationalized by just such racist beliefs as these.

Pearson carried on Galton's statistical work by attempting to provide objective methods by which the correlations of traits among a population could be established scientifically, with the goal of weeding out the inferior traits to preserve the efficiency of the state. He became interested in statistics only because he saw the potential for it to bolster his historical theory that natural selection was the driving force that determined the success of a nation. In response to a critical review of his *National Life from the Standpoint of Science* published by the *Manchester Guardian*, in which the editors accused him of overstepping his bounds by applying biological theories to politics, Pearson bristled, "I ask what reason you have for supposing my history an outgrowth of 'biological consciousness' rather than that my interest in heredity has arisen from my conviction of its bearing on historical studies."[93] Pearson's first loves were always history and political science, with statistics providing a means to an argumentative end, that end being eugenics.

By commingling eugenics with statistics as an academic discipline and publishing a vast corpus of results, Pearson gave these ideas an institutional backing that would be hard to refute. Anyone looking to criticize the findings would first have to wade through hundreds of pages of formulas and technical jargon. The papers published in *Biometrika*—effectively Pearson's private journal for 20 years in addition to being named after a theory of evolution with racist associations[94]—were by no means neutral on the subject of eugenics. He used the journal to illustrate concepts like regression and correlation analysis using example data such as skull size, intelligence, and disease frequency among different races. The papers were of questionable theoretical importance but served as a forum for him to spread eugenics dogma under the guise of statistical theory. For example, in 1900 he presented a novel technique called *tetrachoric correlation* for measuring associations for categorical data—things like education level and marital status that necessarily fall into one of some number of discrete categories. A few years later he used it in a massive study on 4,000 pairs of siblings that found roughly the same strength of association between siblings for obviously hereditary features like eye color as it did for mental characteristics like "vivacity," "assertiveness," and "introspection." He concluded this meant they were all equally hereditary: "We are forced, I think literally forced, to the general conclusion that the physical and psychical characters in man are inherited within broad lines in the same manner, and with the same intensity. . . . We inherit our parents' tempers, our parents' conscientiousness, shyness and ability, even as we inherit their stature, forearm and span."[95]

Pearson ended with a grand statement about the failure of British genetic stock to keep pace with that of America and Germany, advising that the "remedy lies first in getting the intellectual section of our nation to realize that intelligence can be aided and be trained, but no training or education can *create* it. You must breed it, that is the broad result for statecraft which flows from the equality in inheritance of the psychical and the physical characters in man."[96] In other words, he invented a measuring stick and used it to measure two things: how often siblings' bodies were alike and how often their personalities were alike. And, finding their measurements to be equal, he concluded they must be inherited in exactly the same way, from which he jumped straight to the most extreme possible eugenicist conclusions.

In the *Annals of Eugenics*, Pearson was even less subtle. He used the very first issue of the journal to promote his thesis on the inferiority of Jewish immigrants—and he twisted the data to support it when necessary. He knew these ideas would find purchase. Arthur Henry Lane, for example, in his anti-Semitic book *The Alien Menace*, referred to Pearson's conclusions as "of such profound importance as affecting the interests and welfare of our nation, that all men of British race and especially all statesmen and politicians should obtain 'The Annals of Eugenics.' "[97] Maintaining the veneer of objectivity was crucial. Pearson claimed that he was merely using statistics to reveal fundamental truths about people, as unquestionable as natural laws. In introducing the study on Jewish children, he wrote:

> We believe there is no institution more capable of impartial statistical inquiry than the Galton Laboratory. We have no axes to grind, we have no governing body to propitiate by well-advertised discoveries; we are paid by nobody to reach results of a given bias. We have no electors, no subscribers to encounter in the market-place. We firmly believe that we have no political, no religious and no social prejudices, because we find ourselves abused incidentally by each group and organ in turn. We rejoice in numbers and figures for their own sake and, subject to human fallibility, collect our data—as all scientists must do—to find out the truth that is in them.[98]

The subtitle of the *Annals of Eugenics* was Darwin's famous quote, "I have no Faith in anything short of actual measurement and the rule of three." According to Pearson, it was only by allowing the numbers to tell their own story, without any help from us, that we could see these truths for what they were. If anyone objected to Pearson's conclusions—for example, that genocide was an instrument of progress—they were arguing against cold, hard logic with "passion displacing truth."[99] We even see in his responses to Fisher's purely mathematical criticism of the chi-squared test a desire to present a unified front and a theory free from controversy.

Likewise, Fisher was motivated to use statistics to understand the dynamics of population genetics because of his fear that the inferior genes of the lower classes or those introduced into the population by foreigners would threaten the integrity of the British gene pool. In her biography of Fisher, his daughter Joan Fisher Box referred to an early speech he gave to the Eugenics Society as "a weather vane . . . indicating the prevailing direction of his scientific interest towards genetics as the mechanism of human inheritance and towards statistics as the appropriate way of thinking about genetical and other population problems. His theme was *eugenic*."[100] His idea of the genetic causes of the rise and fall of civilizations was grand and ahistorical. By the time similar theories had been put into practice by the Nazis, Fisher had witnessed their horrifying logical conclusion, but he still held to the misguided belief that empirical evidence had demonstrated fundamental differences between the races. How destabilizing would it have been, then, if those controversial inferences had been dependent on subjective assumptions?

Fisher clung to objectivity as a cudgel he could use against opponents when it suited his interests. A lifelong smoker, he spent years later in life—until he died from cancer—advocating for the tobacco industry with the argument that the correlation between smoking and lung cancer did not necessarily prove a causal relationship. It could be, for example, that an inflammatory condition caused both a higher rate of cancer and physical discomfort that was alleviated by smoking.[101] Why, then, did the same argument not apply to his observed correlations between fertility and social class or to any number of others?

Evolution and eugenics provided the through line for all three men's careers. At first, the theory of evolution needed defending, and Galton was an early adopter thanks to his cousin's influence. Pearson added much-needed quantitative buttressing. Between 1893 and 1912, he wrote *18* separate papers with the same title, *Mathematical Contributions to the Theory of Evolution*, in which he deployed many of the most important theoretical tools of statistics of the day. One of these caught Fisher's eye when he was still a young student and inspired his interest in statistics and genetics. Fisher is most remembered today by biologists for his contributions to the modern synthesis of Darwinian evolution and

Mendelian genetics. The kinds of probability questions native to this discipline allowed a frequentist theory to blossom because the lack of strong prior information meant the Bayesian conclusions wouldn't have contradicted them.

All three also thought they saw the potential to use evolution to shape human society, and they were not shy about making those views known. These were no mere footnotes. They all devoted ample space in major works to the idea that mankind could be improved by breeding. Galton's *Natural Inheritance*, Pearson's *National Life from the Standpoint of Science*, and Fisher's *Genetical Theory of Natural Selection* show eugenics was never far from their minds. In *The Grammar of Science*, Pearson wrote, "The theory of evolution is not merely a passive intellectual view of nature; it applies to man in his communities as it applies to all forms of life. It teaches us the art of living, of building up stable and dominant nations, and it is as important for statesmen and philanthropists in council as for the scientist in his laboratory or the naturalist in the field."[102] At an address to the Anthropological Section of the British Association in 1885, referring back to his experiment with the pea plants, Galton said, "It was anthropological evidence that I desired, caring only for the seeds as means of throwing light on heredity in Man."[103]

While their racism and classism may have been roughly consistent with the views of their elite contemporaries, they understood their eugenicist ideas were radical, particularly as they could be seen to collide with Christian values like the concept of the meek inheriting the earth or the biblical instruction to "be fruitful and multiply." (Of course, not everyone saw it that way, like Fisher, who throughout his life was a deeply religious person, and the Reverend William Inge, who, as we mentioned, was the first guest speaker at Fisher's Cambridge eugenics meetings.) As a preemptive move, the eugenics community in the early 1900s sought out Christian clergy to help shape its message and make the case that eugenics was compatible with religion.[104] One of these invitees was the prominent Anglican priest James Peile, who spoke at the Eugenics Education Society in 1909 about how the doctrine of eugenics "deals in terms and phrases which startle the sensibility of the unlearned; and it ignores, or explicitly rejects, a number of things which Church people consider essential." Mainly, Peile argued, eugenics *was* in conflict with traditional Christian philanthropy, but the concept of philanthropy was due for an update anyway because the old idea that every life was worth saving had unfortunately combined with advances in medical science to create "a great army of mental and physical imbeciles, who are in the strictest sense artificially kept alive, without regard to the eugenic interests of the nation."[105]

Eugenicist proposals would mean interfering to some degree in the intimate relationships within families to serve the state, an idea likely to reasonably offend people. Pearson, for example, said, "I fear our present economic and social conditions are hardly yet ripe for such a movement; the all-important question of parentage is still largely felt to be solely a matter of family, and not of national importance. . . . From the standpoint of the nation we want to inculcate a feeling of shame in the parents of a weakling, whether it be mentally or physically unfit."[106]

In keeping with views on the sanctity of family that long predated Galton, the Catholic Church unsurprisingly came down especially hard on eugenics, particularly the practice of forced sterilization. In the encyclical of 1930, Pope Pius XI stated in emphatic terms that the state had no business deciding who was fit to get married or have children, a sacrament that, according to Christian theology, was given to mankind by God, and that "public magistrates have no direct power over the bodies of their subjects; therefore, where no crime has taken place and there is no cause present for grave punishment, they can never directly harm, or tamper with the integrity of the body, either for the reasons of eugenics or for any other reason."[107]

British eugenicists succeeded at turning some of their ideas into reality, mostly in the form of various laws restricting immigration and one awful domestic policy, the Mental Deficiency Act of 1913, which established a process by which anyone deemed "feeble-minded" or "morally defective" could be committed to an institution at the direction of a government board. The standards for who qualified were notoriously vague and included anyone who required any kind of supervisory care, anyone who had been abused or neglected, and anyone who had been found guilty of any crime. At one time, there were more than 65,000 people living in state-operated "colonies" as a result of the act. It passed the Parliament with only three dissenting votes, one being Josiah Wedgwood, who called the bill "a spirit of the Horrible Eugenic Society which is setting out to breed up the working class as though they were cattle."[108] In response to the act, the famous author G. K. Chesterton wrote a book called *Eugenics and Other Evils* (1922), in which he derided eugenicists for interfering with people's lives "as if one had a right to dragoon and enslave one's fellow citizens as a kind of chemical experiment."[109]

The most prominent resistance to eugenics from within the scientific community came from Lancelot Hogben, chair of social biology at the London School of Economics in the 1930s and—like Galton, Pearson, and Fisher—a Fellow of the Royal Society of London. He objected to the whole idea of eugenics as a scientific enterprise because it couldn't possibly be ethically neutral, writing "What is the *good* of the race? What is a desirable social quality? What is a 'morally and mentally fit' person? These are matters of taste, not of science."[110] He agreed that the study of genetics could offer humanity great advances, but because of Galton's "aristocratic bias," the eugenics movement had only given cover to a "system of ingenious excuses for combating the amelioration of working class conditions."[111] His book *Genetic Principles in Medicine and Science* argued for a more careful understanding of how environmental factors could affect human traits and used new mathematical analysis of family data to show, for example, that deafness could not be strictly hereditary, debunking a theory eugenicists had leaned on for some time as an example of a genetic cause of poverty. An article in the *New Statesman and Nation* put it bluntly:

> When [eugenicists] assume a simple genetic character for such complicated combinations of heredity and environmental ingredients as produce feeble-mindedness, criminality and even pauperism, boldly confusing economic and biological factors to prove that the poor should be sterilized, the scientific mood has deserted them. . . . They have filled the bookshelves of the world with dead weight of hearsay, sham expert opinion and doubtful conclusions, based on sufficiently entertaining family histories illustrated by neat little genealogical trees.[112]

The last sentence was a clear shot at Galton. Near the end of his life, at age 85, Galton delivered the 1907 Herbert Spencer Lecture at Oxford with the title "Probability, the Foundation of Eugenics." Lamenting that the public had not yet come around—in particular, that people still got married to "almost anybody" without regard to their genetic stock—he predicted that public opinion would eventually be swayed "when a sufficiency of evidence shall have been collected to make the truths on which it rests plain to all" and that "then, and not till then, will be a fit moment to declare a 'Jehad,' or Holy War against customs and prejudices that impair the physical and moral qualities of our race."[113] The work Pearson and Fisher did, as warriors for the holy cause, was the continuation of that effort to make the "truths" of eugenics apparent to all in preparation for what they understood to be an overhaul of social norms.

The history of statistics in the late 19th and early 20th centuries is therefore inextricably bound up with evolution, eugenics, and scientific racism. It is because of these deep connections that so much of the current language of probability and statistics is suffused with terminology taken from population biology and eugenics. Fisher had to do mental gymnastics to define probability as a "study of populations" in a way that also includes coin flips and measurement errors, but nonetheless many current statistics textbooks still define probability in terms of sampling from a population. Pearson's idea of the correct operation of statistical research was to measure associations of human traits to establish that they were hereditary and to detect inhomogeneities and significant differences between subpopulations. For him, those were often differences like the difference in hygiene between Jewish and non-Jewish children, but we still use the same language to talk about all manner of statistical findings that have nothing to do with populations. Galton defined *regression* in the context of future generations "going back" to their mediocre mongrel genetic roots, but today linear regression is a catch-all tool to estimate linear relationships between any two statistical variables.

Racism, aristocratic class panic, anti-Semitism, xenophobia, and ableism are all original sins of modern statistics. The real sins, though, were the greed for scientific authority supported by unassailably objective methods and the arrogance to believe such a thing could exist. Of course, there are many reasons a scientist might want their methods to appear to be objective, even benevolent ones. As we saw in the previous chapter, it's been a general pattern that as probability has encroached more and more on people's lives, the users of probability have retreated more and more to the frequentist interpretation, since this makes probability a measurable fact of the world and the numbers appear to speak for themselves. The higher the stakes have been, the more suspicious people have been about Bayesian priors, and the greater their desire has been to remove any semblance of subjectivity from the process of drawing conclusions from the data. Galton, Pearson, and Fisher considered their work to have the highest possible stakes, a "holy war" to prevent the disintegration of the social order and the collapse of society itself, so it's natural that their methods took frequentism to the greatest extreme.

In the process, though, they simultaneously revealed the limits of that strategy and how thin the myth of objectivity was, since their eugenicist conclusions were mostly predetermined before they ever started collecting data. The implicit claim in the use of strictly frequentist statistical methods is that every probability in the calculations is objectively measurable and therefore so are all the conclusions emanating from those calculations. But that claim has always involved some sleight of hand. All statistical estimates of anything—for example, the correlation between skull size and measured intelligence or the mean difference in disease incidence between different races—are explainable in more than one way. The estimation process may be perfectly objective and its associated frequencies reliably measurable, and for problems with weak prior information, the estimates may even come out sensibly. But then it's up to the scientist to decide what conclusions to make from those estimates: whether they represent a real association or a difference worth caring about, or whether they might be the products of some unobserved variable or infected with bias in the ways the relevant quantities were measured. Galton, Pearson, and Fisher all demonstrated this flexibility abundantly by interpreting the same statistical results one way or another, depending on what conclusion suited their agenda. The agenda guiding their inferences was the unacknowledged subjective element, while their flashy statistical calculations were meant to provide misdirection.

Far from letting the data speak for itself, they found a way to make the data ventriloquize their own thoughts. They knew where their research was inevitably bound to go, and they could reasonably expect fierce resistance along the way, so it's little wonder these early statisticians desired the bedrock of frequencies on which to build their theories. In their quest for objectivity, they established a school of statistical reasoning with no role for prior information, which they claimed was unnecessary for scientific inference. This was equivalent to saying premises are unnecessary for deductive reasoning, violating the rules of valid logic, but validity wasn't their goal. It was not enough for their methods to be *logically valid*; Galton, Pearson, and Fisher needed their conclusions to appear to be *objectively true*. We do not yet know the full wages of that sin. But one consequence is that scientists are now largely saddled with frequentist statistical techniques that prove inadequate outside the narrow domains for which they were derived, as we'll begin to explore in the next chapter.

5

THE QUOTE-UNQUOTE LOGIC OF ORTHODOX STATISTICS

Lacking the necessary theoretical principles, [frequentist methods] force one to "choose a statistic" from intuition rather than from probability theory, and then to invent ad hoc devices . . . not contained in the rules of probability theory. Each of these is usable within the small domain for which it was invented but, as Cox's theorems guarantee, such arbitrary devices always generate inconsistencies or absurd results when applied to extreme cases.

~Edwin Jaynes

Our story of statistical orthodoxy has reached the present day, where students of statistics are now taught a confusing mishmash of the ideas of the Fisher and Neyman-Pearson schools. To see these concepts in action, let's rewind things, once again, back to Jacob Bernoulli's urn-drawing problem circa 1700. As we've seen, Bernoulli's problem was where most of our ideas of statistical inference got their start. In many ways, it's the prototypical example of a statistics problem where orthodox methods perform the best.

We'll adopt an orthodox statistical mindset for a while and go through a modern analysis of Bernoulli's problem step by step. Anything we calculate will be specific to this problem, but the template is the same for almost all the standard procedures (anything involving parameter estimation, a significance test, or a confidence interval). Some of this will seem awkward—because it is—but as we'll see, that's a natural consequence of the narrow view that probability can mean only frequency. To clarify that this is the orthodox viewpoint and not our own, we'll put these words into the mouth of a fictional character, an artificially intelligent computer named SuperFreq. We'll imagine a dialogue between the computer and Jackie Bernoulli, a fictional college student and descendant of Jacob, trying to analyze some data for her Urn Theory 101 class.

Once we're done showing the orthodox methods in the best possible light, though, it will be time to turn out the light. The remainder of this chapter consists of a parade of horribles: statistical inferences that push the orthodox methods to several different breaking points, along with the reasonable answers the Bayesian method would have provided. In the process, they reveal the difficulties facing scientists or anyone else trying to use the standard methods to handle more complex problems than guessing at the contents of a mystery urn.

JACKIE BERNOULLI AND THE STATISTICS WIZARD

Jackie Bernoulli, the great-great-great-great-great-great-great-great-great-great-great-great-granddaughter of Jacob Bernoulli, was a student at the local college, majoring in Urn Studies. She was writing a paper about an urn, from which she had collected a sample. To do a statistical analysis of her data, she enlisted the help of SuperFreq, an artificially intelligent computer loaded up with all the standard concepts of orthodox statistics.

■ ■ ■

"Hi, I'm SuperFreq, the statistics computer! I've been programmed to assist people with frequentist statistical analysis. How can I help?"

"Hi, I'm Jackie Bernoulli. I have this urn full of tiny black and white pebbles that I've taken a sample from, and I need to know how to analyze my data."

"I see!" said SuperFreq. "First, I'm so glad you said 'urn.' So many people these days want to ask about jars, buckets, trash cans, or the like. Urns are kind of a specialty of mine. What kind of analysis are you interested in?"

"Well," Jackie answered, "I suppose I'm most interested in what the likely mix of pebbles is in the urn. So, for example, what's the probability that the urn has twice as many white pebbles as black ones?"

"Whoa, hold on there! Remember, I've been programmed with *frequentist* logic, which means I accept probability statements only if they're phrased as frequencies in repeated samples from a population. 'The probability that the urn has a certain mixture' doesn't compute. It either does or doesn't."

"O . . . kay . . . , so what kind of probability statements can I make?"

"Well, you could talk about the probability of observing a particular *sample*—say a sample of 10 white and 10 black pebbles. That can sometimes happen and sometimes not, so it has a frequency—that is, a probability."

"I see," said Jackie. "And how can you know how often you'll get a sample like that if you don't know the mix of pebbles in the urn?"

"Well, you're allowed to assume an urn mix provisionally as a hypothesis. So you can talk about the probability of getting a sample with 10 white pebbles and 10 black pebbles if hypothetically the urn were equally mixed. Since we can *imagine* doing the sample from such an urn, we know what the frequency *would* be."

"So you imagine doing a series of samples from an imaginary urn."

"That's right."

"Okay. So how does that imaginary urn help me learn about my real one?"

"Well, one thing you can do is try to reject a null hypothesis with a significance test."

"A what-hypothesis with a what-test?"

"A *null hypothesis* is a hypothesis that you secretly want to reject; it's usually something like two populations being equal or a treatment having no effect or something of that nature. A *significance test* is the procedure we use to knock it down. Do you have a null hypothesis?"

"Well, urns in my field are almost always about equally mixed, and I'm hoping this one is really biased one way or the other. So I suppose it will be boring if my urn is 50 percent black and 50 percent white. Let's use that as a null hypothesis."

"Excellent! So now we'll try to reject it. What kind of data do you have?"

"I mean, I recorded all the colors of the pebbles I took out in order. Do you want me to tell them to you?"

"No need. For an urn problem like this, it doesn't matter in what order the pebbles came out. All I need to know is how big the sample is and how many of them are white. From that, I'll compute the sample ratio. That will serve as our *test statistic*."

"I've actually been wondering about that word. What is a statistic anyway?"

"A *statistic* is any quantity derived from the data," SuperFreq explained. "Since it varies according to what data you get, it can also have probabilities."

"Okay, got it. So the sample ratio is a test statistic for . . . what test?"

"We're going to test whether your null hypothesis is true—that is, whether the urn is equally mixed. Technically what you have here is called a *parameter*, the fraction of white pebbles in the urn. Let's call that F, for ease of reference. So we're interested in testing whether $F = 0.5$."

"So first we assume it is."

"Yes—but only provisionally. We'll say things like 'If $F = 0.5$, then the probability we'd get such-and-such sample is such-and-such,' and so on. We might also assume F has other values at some point. Do you know the formula for the probability in the binomial distribution?"

"Do I! It's practically our family motto. If R is the sample ratio, then

$$P[R = \frac{k}{n} \mid F = f] = \binom{n}{k} f^k \left(1 - f\right)^{n-k},"$$

"Oh right, you're *that* Bernoulli. I'm impressed! I'm also impressed that you spoke that equation out loud in this verbal dialogue! So we'll put in different values for that f. We might put in 0.5 or we might not."

"I must say you're being awfully cagey about this."

"Well, the thing is . . . Sometimes it will be helpful to make statements that apply to all f. For example, your great (times 12) grandfather showed that, for large enough n, almost all the probability in that distribution is concentrated near the value f, no matter what value we sub in for f."

"Yeah, tell me about it. The 'golden theorem' . . . blah blah blah."

"Right. So that property is called *consistency*. It also happens that the *average* sample ratio over its probability distribution is also equal to f, no matter what f is. That's called being *unbiased*. "

"Sounds good! Is it?"

"Oh, yes! It means that if we take a long series of repeated trials and average the resulting sample ratios, it converges to the true urn fraction, no matter what it is. So, for a large sample, we can be pretty sure our sample ratio is pretty close."

"How close, and how sure?" Jackie asked.

"Good questions! I guess it's a good time to ask, What's your actual data?"

"Well, I took out 32 pebbles and got 20 white. Is that enough to reject the whatsit hypothesis?"

"Null hypothesis. Let's see . . . Of course, just to warn you, we're never going to definitively answer the question either way. It's possible the urn has any mixture (other than $F = 0$ or $F = 1$) and you could still get that data."

"Yeah, I'm okay with that."

"Alright! So we have our null hypothesis ($F = 0.5$) and our test statistic R (the sample ratio). All we need now is a *rejection region*."

"What's that?"

"It's the range of test statistic values that we take as evidence against the null hypothesis. Typically it's something like the *tails* of the probability distribution—values far away from the assumed parameter value."

"So something far away from 0.5."

"Yes, that's right."

"How far?"

"Well, that all depends on what significance level you want for the test."

"There's that word again. What does *significance* mean here?"

"It means the probability of rejecting the null hypothesis assuming it's true—the false positive rate," replied SuperFreq.

"Now I'm totally confused. Why would we reject the hypothesis if it's true?"

"Here's how it works. Let's use a significance level of 5 percent, which is the default option anyway. We'll do a *two-sided* test, meaning both very high and very low values will cause us to reject the hypothesis. What we'll do is find two tails of the distribution, equally far away from 0.5, so the total probability of the sample falling in those extremes, if hypothetically the null hypothesis were true, is about 2.5 percent each, so 5 percent total. Those two tails combined are our rejection region."

"Okay, I think that makes sense. But hang on a second . . . What do you mean the probability that the sample ratio falls in those extremes? I already know the sample ratio. So don't I know whether it's in the tail already?"

"Oh, right. Um . . . this is awkward, but I'm going to need you to forget your data."

"Forget it?!"

"Yes. You see, we're defining a *procedure* here. What matters is how often we *would* get data that leads us to draw a false conclusion, like rejecting the null hypothesis when it's true. So we're not using your actual data right now."

"O . . . kay . . . , I guess I can try to forget it."

"Great! Now back to computing the rejection region. Let's see . . . adding up probabilities from a table of binomial distribution probabilities until I get about 2.5 percent on each side for your sample size ($n = 32$), I'm getting a rejection region of $R \leq 10/32$ or $R \geq 22/32$. Those two tails combined have a probability around 0.05, so that will work for our significance test. If we get data in the rejection region, we'll say, 'We reject the null hypothesis at the 5 percent level.' By the way we set things up, if the null hypothesis is actually true, that should only happen 5 percent of the time. Get it?"

"I think so. Okay, now what?"

"What was your data again?"

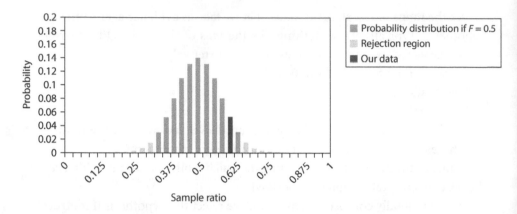

FIGURE 5.1

Probability distribution and 5 percent rejection region for the urn sample ratio under the null hypothesis.

"I forgot it!" Jackie exclaimed.

"Oh, right. Um . . . well now I need you to remember it."

"What?! Why?"

"Now it's time to compare your actual data with the tail regions we came up with for our procedure. We've got to see whether the test says that your data is enough for us to reject the null hypothesis."

"Okay, well, fortunately I was lying about forgetting it. It's $R = 20/32$."

"Alright, cool! So, let's see here . . . that's not *quite* in the right-tail rejection region. The cutoff is 22/32. Here, I made a graph [figure 5.1], which I will describe to you in such vivid detail it will be as though you can see it for yourself."

"Wow, that *is* vivid! I can practically see the binomial distribution, with tail regions indicated. And there's my data, just outside the right tail!"

"So you can see that, for your data, we fail to reject the null hypothesis at the 5 percent level."

"Which is bad."

"It's not great. Really, it just means your data isn't extreme enough considering how small your sample is. The larger the sample size is, the more concentrated the distribution is and the closer in the tails are. A ratio like yours might have been significant in a larger sample."

"This all depended on that choice of 5 percent as the significance level, right?"

"Right."

"What's to stop me from changing the significance level and making the tails bigger so my data *is* significant?"

"Nothing! In fact, another way to express the result of the test is to say what the significance level would have to be in order for your data to be right on the borderline of significance. That's called the *p*-value."

"So what's my *p*-value?"

"Well, let's see here . . . We want your data ($R = 20/32$) to be just on the edge of the right tail. That's a distance of 4/32 away from the hypothetical parameter value (0.5 = 16/32), so the two tails would consist of all the sample ratios at least 20/32 or at most 12/32. Tallying up the probabilities gives me . . . 0.2153. That's your *p*-value."

"Let me get this straight. You're saying that if the significance level had been 0.2153 instead of 0.05, my value of *R* would just barely have been significant. So that makes 0.2153 my *p*-value. Is that right?"

"Exactly! Here's another (almost impossibly lurid description of a) graph [figure 5.2]."

"Okay, so this may be a dumb question, but why can't I just use that significance level instead of 0.05? Then my data would be significant, and I could get my results published in the *Journal of Urn Science*."

"Well, most journals use 0.05 as the threshold for what counts as publication worthy," SuperFreq explained.

"Why?"

"Because Fisher said so once."

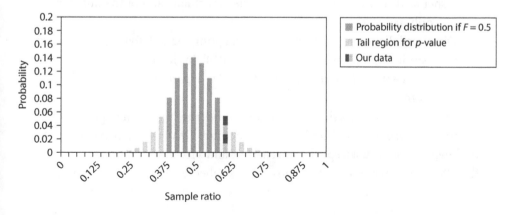

FIGURE 5.2

Probability distribution and *p*-value calculation for the urn sample ratio under the null hypothesis.

"Well, that seems kind of arbitrary. What's wrong with 0.2153 instead?"

"It just isn't a very good significance level. The problem is that if you start rejecting the null hypothesis at that high a level, you open yourself up to a lot of false positives."

"How so?"

"Remember, the significance level is defined as the probability of rejecting the null hypothesis assuming it's true. So, if you started doing tests with a higher significance level, you'd be rejecting a lot more true null hypotheses."

"And that's bad?"

"Yes, because the null hypothesis usually corresponds to something boring, like a treatment having no effect. Rejecting it means you think something interesting is going on with your experiment. We don't want to get people's hopes up unless we're pretty sure."

"Okay, I get that. So why not set the significance threshold at like 0.00001 and make sure we almost never have a false positive?"

"Why do you think? The counterpart to a false positive is what?"

"A false negative?"

"Exactly! I knew you were a Bernoulli. If we make the significance threshold too low, we'll get more false negatives. That is, we'll fail to reject a false null hypothesis."

"Fail . . . to reject . . . a false . . . null. Is there a way to say that without so many negatives in a row?"

"Nope!"

"Why not?" Jackie asked.

"Fisher said so. According to him, we never actually accept the null hypothesis; we only fail to reject it."

"Okay, well, I'm going to think of it as accepting the null hypothesis instead."

"Ah, a student of Neyman-Pearson, I see! Well, in that case you might as well talk about accepting or rejecting the *alternative hypothesis* too."

"What's that?"

"That's the hypothesis that you think must be true if the null hypothesis isn't. So, in your case, it would be $F \neq 0.5$. You could say that, when you reject the null hypothesis, you accept the alternative, and vice versa. Wrongly accepting the alternative hypothesis (false positive) is called a *Type I error*, and wrongly accepting the null hypothesis (false negative) is called a *Type II error*."

"Okay, let's go with that. So, sticking with my significance level of 5 percent, how often will I make a Type II error?"

"Alright, this one's tricky. We can't be totally sure."

"What? Why not?"

"Well, the false negative rate is the rate of accepting the null hypothesis even though it's false, meaning the alternative is true. But just saying something like $F \neq 0.5$ doesn't give us enough information to compute that probability. We'd need to know what F *is* if it's not 0.5."

"How am I supposed to know that?"

"Ideally, you'd have some theory that would lead you to predict the effect size ahead of time. Do you have any theory?"

"Um, no. I have an urn."

"I see. Well, you could always go with a post hoc analysis and just claim your observed sample ratio was what your theory predicted."

"And that's allowed?"

"Eh . . . sort of . . . ," SuperFreq said hesitantly.

"Works for me! So my observed ratio is $R = 20/32$, which is 0.625. What does that mean about the false negative rate?"

"Well, we set our tail rejection region for the significance test to be $R \leq 10/32$ and $R \geq 22/32$. So, if the only alternative to $F = 0.5$ is $F = 0.625$, we need to compute the probability of getting an R value somewhere between those two extremes—assuming $F = 0.625$. That gives us the probability of accepting the null hypothesis ($F = 0.5$) even though it's false. From a table of binomial distribution probabilities with $F = 0.625$, I get a probability of 0.7035."

"Can you by any chance describe that to me in graph form?"

"Sure! I'll vividly describe another graph [figure 5.3]."

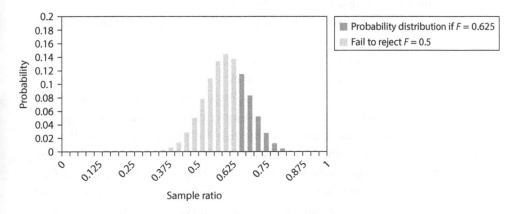

FIGURE 5.3

Probability distribution and false negative region for the urn sample ratio under the alternative hypothesis.

"Oh, I see. The probability distribution is moved over because, for the alternative hypothesis, we're assuming the urn fraction F is actually 0.625. But we wouldn't know to reject the *theory* that $F = 0.5$ if we got any sample in the part of the distribution between our previously established tail regions, which would happen with probability around 70 percent. That's the false negative rate."

"Couldn't have described it better myself!"

"Hey, so that 70 percent false negative rate seems like a lot."

"It's not a little. The way people usually talk about these things is in terms of the complementary probability, 1 minus the false negative rate. That's called the *power*. In your case, you'd have a power of about 30 percent. But the whole art is balancing the false negative rate against the false positive rate, the power and the significance. A more extreme significance threshold means a lower false positive rate but also a higher false negative rate and a lower power."

"Did you say 'art'? I thought this was math."

"Oh, no, it's definitely an art form. Some might say it's performance art."

"Okay, so what . . . Are we done here?" asked Jackie, anxious to get to work on her paper.

"Well, one more thing you should probably do is report a confidence interval."

"Sounds good. What is it?"

"A *confidence interval* is just a record of all the hypothetical parameter values you wouldn't have rejected at whatever significance level. In your case, it'd be all the possible values for F that you wouldn't reject at the 5 percent level based on your data ($R = 20/32$). That's your 95 percent confidence interval."

"So I don't have to just use $F = 0.5$ for the test?"

"Oh, no, you're allowed to apply the same procedure to any hypothesis of the form $F = f$."

"How do I find this confidence interval?"

"Basically you slide the distribution back and forth. If you assume a lower value of F, the distribution moves left; for higher values of F, the distribution moves right. So the procedure is this: First move the distribution all the way *left* so your data is just on the boundary of the right tail of the distribution, the part with probability 2.5 percent. That tells you the lowest value of f for which you wouldn't reject the hypothesis $F = f$. Then move it *right* until your data is on the edge of the left tail. That tells you the highest value for which you wouldn't reject the hypothesis $F = f$. The confidence interval is everything in between. I'll spare you doing the arithmetic yourself. The answer is a range from 0.468 to 0.763."

"I think I'm going to need to hear another graph or two."

"Sure, here you go! One for the lowest value and one for the highest [figures 5.4 and 5.5]."

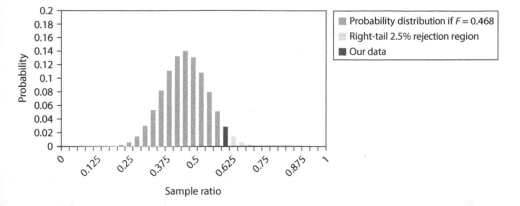

FIGURE 5.4

Probability distribution for the urn sample ratio under the lowest hypothetical value for which the given data is not significant at the 5 percent level.

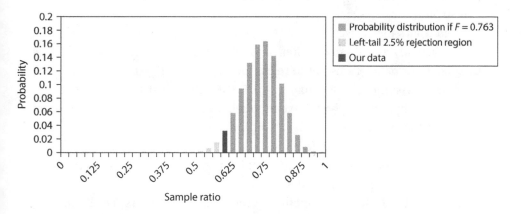

FIGURE 5.5

Probability distribution for the urn sample ratio under the highest hypothetical value for which the given data is not significant at the 5 percent level.

"Got it. So what's the point of doing that?"

"It gives a lot more meaningful information than just the single significance test for the null hypothesis. Basically it shows the results of an infinite number of significance tests at the same time. Plus the width of the confidence interval shows how much information was really contained in your sample."

"So it's a 95 percent confidence interval, right? Can I say with 95 percent confidence that it contains the true urn-fraction value?"

"Ha ha, no. Yes. Kind of."

"Very clear answer, thank you."

"Well, here's the deal," explained SuperFreq. "Once the interval is *known*, you can't say anything about the probability of it containing the true value. Remember, probability means frequency. The true value is a fixed constant, and a given confidence interval is fixed, so it just either contains the true value or it doesn't. There's no frequency there."

"Okay."

"But since you get a different confidence interval every time you do a sample, you *can* talk about probabilities of getting different intervals. The interval itself does have frequencies. So you could ask a question like 'If $F = 0.5$, what's the probability I'll get a 95 percent confidence interval containing the value 0.5?' and so on."

"And what's the answer?"

"We actually already said it! Remember, 0.5 is in the 95 percent confidence interval if our data tells us not to reject the hypothesis $F = 0.5$ at the 5 percent level. If we assume that F actually is 0.5, then that's just the same as falsely rejecting the hypothesis $F = 0.5$ even though it's true. Which, by the way we set things up, has probability . . . ?"

"5 percent?" Jackie asked hesitantly.

"Exactly! So the probability of that *not* happening is 95 percent. It's the same as the confidence level of the interval."

"And what if F is really something other than 0.5? Say it's 0.6. Would the confidence interval still contain the true value 95 percent of the time?"

"Yes! The same argument would work for any value f. If $F = f$, then based on whatever data we get, when we compute the confidence interval, we'll be testing all the possible hypotheses of the form $F = f'$. So, at some point, we'll be testing whether $F = f$ when it actually is. We should get data that causes us to (falsely) reject that theory only 5 percent of the time, meaning that, with 95 percent probability, the value f will fall in the 95 percent confidence interval when $F = f$."

"Whoa. That's a lot to keep track of."

"You bet!"

"So the 95 percent confidence interval contains the true value with 95 percent probability."

"Yes, but that's not the definition. That's just a property it has."

"Oh?"

"Yes, it's a common misconception. Say you rolled a 20-sided die and just reported the whole real number line as your interval for the urn fraction unless the die came up 12, in which case you reported the one-point interval {0}. Then 95 percent of the time (19 out of every 20 rolls), you'd include the true parameter value. But that's not a confidence interval."

"Okay, good to know—I guess."

"Happy to help!" SuperFreq replied. "There's a reason people need an AI to keep all this straight. Anyway, here's a summary analysis of your data:

- An unbiased estimate for the urn fraction is the sample ratio, $R = 20/32 = 0.625$.
- Using a two-sided test of the sample ratio, we do *not* reject the hypothesis that the true urn fraction is 0.5 at the 5 percent level. Our *p*-value for this test is 0.2153.
- If the only alternative true urn fraction had been equal to our observed value of 0.625, the power (post hoc) of the preceding test would have been approximately 30 percent.
- Based on our data, a 95 percent confidence interval for the true urn fraction is $(0.468, 0.763)$."

"Great, I still don't fully understand what this has to do with my original question about the contents of the urn, but I hope you don't mind if I just copy and paste that into my paper."

"Not at all! Everyone does that."

"Cool. So I was just wondering something. You said the significance level of 5 percent is the standard for publication, and that corresponds to a 5 percent false positive rate."

"Yes, that's right."

"So does that mean only 5 percent of published results are false? If we retested all the theories published in science journals, would we find that only 5 percent of them didn't work?"

" . . . "

"Hello?"

" . . . "

"SuperFreq?"

THE PSYCHOSEMANTIC TRAP

The results in the urn example seem like a reasonable enough summary of what we're able to conclude about the urn. However, the reason the conclusions seem reasonable is that we implicitly started from the perspective that no strong prior information about the urn fraction is given. Since we're essentially indifferent to all the possibilities between 0 and 1, if we follow through the Bayesian procedure, we will actually end up with almost exactly the same conclusions in terms of our posterior distribution, as we saw in chapter 2.

Almost all other orthodox statistical methods follow the same procedure. What is usually difficult is deciding what estimator or test statistic to use to make inferences about whatever parameter or other feature of our model is being considered. For urn drawing, we're interested in the true fraction of white pebbles in the urn, which can be thought of as the *population mean* of the binary variable, taking the value 1 for each white pebble and 0 for each black one. But imagine our urn contains more than two colors of pebbles; we might then be interested in all the different fractions for all the colors, and the sample ratios are still likely good estimators to use for this. Or maybe we are interested in some continuous quantity for each pebble, like the diameter or weight; we might want to make inferences about the *average* weight of the pebbles in the urn. This again suggests using the sample average weight as a statistic.

Problems can be much more complicated than that, though. For example, if we imagine the weight of pebbles in the urn to have a *normal distribution*, we may be interested not just in the mean weight but also in how spread out the weights are—that is, the *population standard deviation*. Or getting more complex still, we might assume that the two quantities of weight and diameter are both normally distributed with unknown means and unknown standard deviations and furthermore that the weight and diameter are *correlated* (in Galton's sense) to an unknown degree, measured by the *population correlation coefficient*.

This was the problem Ronald Fisher had considered in the note he sent Karl Pearson in 1914. What Fisher was able to do, which is a necessary step for any such problem of estimation in the frequentist mode, is to choose an estimator for this unknown correlation parameter—the in-sample analogue called the *sample correlation coefficient*—and compute its sampling distribution as a function of the unknown parameter, the way we had Bernoulli's binomial distribution for the sample ratio as a function of the population ratio. This was necessary because any test of significance requires knowing the tail probabilities for the test statistic, but the sample correlation is a *much* more complicated entity than just a simple average of the data. It involves computations like this:

$$\frac{\sum_{i=1}^{n}[(x_i - \frac{1}{n}\sum_{i=1}^{n}x_i)(y_i - \frac{1}{n}\sum_{i=1}^{n}y_i)]}{\sqrt{\sum_{i=1}^{n}(x_i - \frac{1}{n}\sum_{i=1}^{n}x_i)^2}\sqrt{\sum_{i=1}^{n}(y_i - \frac{1}{n}\sum_{i=1}^{n}y_i)^2}}$$

Fisher solved the problem in a truly impressive way by giving this statistic a geometric meaning in n-dimensional space.

The vast body of statistical literature therefore consists mostly of these kinds of distributions for different statistics deemed useful for different types of problems. For example, much of Fisher's *Statistical Methods for Research Workers* gives various significance tests for different sorts of model parameters or other hypotheses. Each at some point had to be developed anew and its sampling distribution computed, and because each was a different function of the data, no one universal argument could apply to all. This created a lot of theoretical work for statisticians—and plenty of opportunities to get famous.

The connection between the model feature being tested and the statistic considered appropriate for testing it is not dictated by any strict requirements, though. It depends entirely on an intuitive sense of what kind of data would count as evidence against any given hypothesis. For example, when testing whether the urn contained a 50/50 mix of black and white pebbles, we decided to count deviations in the sample ratio far away from 0.5 as evidence against that hypothesis. Nothing about the false positive rate requires this. We could just as well have set our rejection region to be any set whose probability under the null hypothesis is 0.05, and all the same frequency arguments would apply.

As we'll discuss later, the choice of the rejection region carries with it an implicit statement of what the alternative hypotheses are, which, if not properly reckoned with, can lead to difficulties. This problem comes up regularly in practice when someone needs to choose between a *one-sided* and a *two-sided* version of a test, such as Student's t-test, the difference being whether extreme data on *both* sides of the mean will count against the hypothesis or merely extreme data on one side. Here, we see the faint inklings of Bayesian prior probabilities playing a role, even if not properly acknowledged.

Nowadays anyone using statistics in practice can simply look up which test is appropriate for their kind of problem. The Social Science Statistics online wizard (a real-life version of our SuperFreq) will even guide the user by asking a series of yes/no questions about the data they have and what they want to ask of it.[1] Or they could consult a flowchart like the one in figure 5.6.

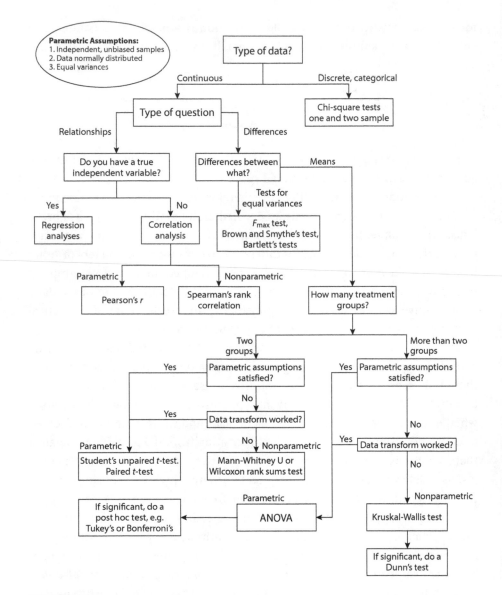

FIGURE 5.6

Flowchart for selecting commonly used statistical tests.

Source: Dr. Robert Gerwien, "A Painless Guide to Statistics," Bates College On-Line Resources, modified January 24, 2014, http://abacus.bates.edu/~ganderso/biology/resources/statistics.html.

Edwin Jaynes compared this process to a doctor-patient relationship: a patient presents a set of symptoms, and the doctor narrows in on a recommended treatment, but no one unifying principle apart from the doctor's experience exists to connect the two.[2] Similarly no unifying principle exists to connect a type of analysis with a statistical test, so, for complex problems, a working scientist may need a professional statistician to guide them. Perhaps the power dynamic this created, not to mention the job security, is part of the reason statisticians have been resistant to change.

Likewise, the properties of estimators we described as being desirable are somewhat arbitrary. For example, whether an estimator is unbiased doesn't really matter all that much. It may be useful to know that averaging an infinitely long series of estimates causes them to converge to the true parameter value, but being unbiased is a *sufficient* but not *necessary* condition for this to happen. We could just as well have taken our estimate of the urn fraction to be the biased estimate $(k + 1)/(n + 2)$, where k is our observed number of white pebbles out of n; this too would have the property of converging to the true ratio as n goes to infinity.

Considering the names of all the various properties of estimators and tests in the statistical literature, it's hard not to notice that they all seem to embed value judgments, suggesting that this particular estimator or test is *good*. We have already mentioned unbiased and consistent estimators. There are also *efficient* estimators, *admissible* estimators, *dominant* estimators, *robust* estimators, *uniformly most powerful* tests, and, surely the best example, the *best linear unbiased estimator*. It seems nearly certain that all this normativity is a by-product of the political infighting and jockeying for position between various camps within the world of frequentist statistics over the course of the last century. As different factions fought for legitimacy and for acceptance of their methods as standard, they must have thought it advantageous to give their methods virtuous-sounding names. Who would want to be seen as being in favor of bias, inconsistency, inefficiency, inadmissibility, subordination, frailty, powerlessness, or . . . worst-ness?

But sometimes an unbiased estimator might actually be worse than a biased one. For instance, another standard measure of estimator quality is the mean squared error (the average squared deviation from the true value). Jaynes gave an example where samples of size 203 produced estimates, after "correcting" for the bias, with a mean squared error about equal to what a biased estimator would produce with a sample size 100.[3] That is, by purging the estimate of bias, we have effectively thrown out half the data. In any scientific field where data is expensive or hard to come by, the idea that we are not making the most efficient possible use of our data should come as a disturbing suggestion, but that is what a fear of bias can do. As Jaynes wrote, statisticians insisting on using unbiased estimators are "caught in a psychosemantic trap of their own making."[4]

That's not altogether surprising, though, because, as should be increasingly clear by now, all of frequentist statistics is one big psychosemantic trap. There is no coherent theory to orthodox statistics, only a loose amalgam of half-baked ideas held together by suggestive naming, catchy slogans, and folk superstition. The only unifying principle these ideas share is a commitment to the fantasy that somehow, with enough creativity and clever rearrangement, it's possible to twist sampling probabilities into inferential probabilities and thus make all of statistical inference a simple matter of measuring frequencies.

The particulars of the frequentist methods vary from problem to problem, but Bernoulli's argument is present in all of them, obscured by fancier language and notation. Bernoulli said the in-sample ratio was a good choice for estimating the population (urn) fraction because, as the sample size got larger, one could be morally certain that the two numbers were close. The appeal of this idea was that *closeness* seemed symmetric. The modern methods simply took that idea of moral certainty and repackaged it. The template of a null hypothesis significance test says, "It is morally certain (to a 95 percent probability) that if this hypothesis is true, the estimator will *not be far away* from the parameter value; therefore, if we do see an estimator far away from the assumed value, we may reject this hypothesis." The confusion that students suffer from such things as the definition of confidence intervals or p-values is the same as Bernoulli's confusion. It seems equivalent to say

"We are unlikely to observe an estimate of x far from its true value."
and
"It is unlikely that the true value of x is far from what we observed."

Or it seems equivalent to say

"The true parameter value will be contained within the confidence interval with high probability."
and
"This confidence interval contains the true parameter value with high probability."

Bernoulli may incorrectly have thought he had solved his inference problem because he expressed the statements in such a way that the two probabilities seemed equal. The frequentist statisticians of the 20th century did him one better by denying that inferential probabilities even *existed*. So, if it's not actually probable that the true value of a parameter is contained within a given confidence interval, why report it? If it's not actually highly probable that the null hypothesis is false, why reject it?

The answer is that by artificially limiting the scope of what probability may apply to, the frequentists have tied their own hands. The only mode of inference left that could use probability is one determined entirely by variation in the data, which means deciding on parameter values or the truth of hypotheses based strictly on the sampling probabilities these imply. This is the essence of Bernoulli's Fallacy. We already showed the consequences it can have for legal, medical, and other inferences. We can now extend the same criticism to orthodox statistics.

FISHER'S BAYESIAN OMELET

Before we go further, though, it will be helpful to take a side journey to examine Fisher's other proposed answers to the question of inference besides significance testing to watch him struggle with the same issues we raise. As we have seen, for most of his career Fisher considered Bayesian inference to be a great historical error that had to be absolutely rejected. But he also, at some level, understood that pure frequentism was an inadequate replacement, and he expended enormous energy trying to construct new methods of inference that he claimed went beyond probability. For this reason, despite all his dogmatic bluster, some people believe that had Fisher lived long enough, he would have eventually become a Bayesian.

He made his anti-Bayesian viewpoint especially clear during a famous exchange with the Bayesian physicist Harold Jeffreys at Cambridge, published in the *Proceedings of the Royal Society of London* between 1932 and 1934. Their argument was over the essential meaning of probability. Jeffreys, who thought of probability as "an expression of our state of knowledge,"[5] had used a clever argument to deduce the prior probability distribution that a certain parameter in a Bayesian inference problem *had to have* in order for some commonsense, intuitive conclusions to be true; this was an early example of what would become known as a "Jeffreys prior."[6] Fisher, who (at that point) could think of probability only in terms of ratios in a "hypothetical infinite population," dismissed the idea as pure nonsense: because it was *possible* to imagine many different populations with different distributions for the parameter, there was no possible way Jeffreys could know his answer was correct.[7]

Jeffreys shot back: "By 'probability' I mean probability and not frequency as Fisher seems to think," and "Fisher considers that it is an objection to my theory that with different assumptions I should have got a different answer and thereby misses the entire point of the theory." That is, if the distribution of the parameter is known, then, of course, it will lead to different probabilities. Jeffreys had developed a strategy for assigning a probability distribution in the *absence* of such knowledge. The point of his method is not to be "correct" in the sense of matching some measurable frequencies or ratios but rather to quantify the consequences of our *ignorance* before gathering data, "to state the alternatives to be tested in such a way that experience will be able to decide between them."[8] That view of probability and what it meant about the nature of scientific knowledge were intolerable to Fisher. He called Jeffreys's idea of probability "subjective and psychological," making it "impossible for any of his deductions to be verified experimentally."[9] Jeffreys said Fisher, in his mistaken restatement of the problem in frequentist terms, had proceeded "to reduce my theory to absurdity,"[10] and Fisher said, in what he surely considered a very clever retort, it was an accusation "I am not inclined to deny."[11]

Fisher's later writing on the fundamentals of probability had a noticeably softer tone, though. In *Statistical Methods and Scientific Inference*, after he was finished spitting on Pearson's grave, he began by praising Thomas Bayes for "the first serious attempt known to us to give a rational account of the process of scientific inference as a means of understanding the real world."[12] And later he even defended Pierre-Simon Laplace against John Venn's criticism of the rule of succession, essentially saying that, in his eagerness to criticize Laplace's logic, Venn had not understood its mathematical basis.[13]

A further indication of Fisher's Bayesian leanings appears in his method of estimating parameters called *maximum likelihood estimation*, which he first came up with while still a student at Cambridge. The starting place for one of these inferences is the same as in the standard approaches: we assume a probability model for the data that includes some unknown parameter value. The key idea then is to consider the probability as a *function* of that parameter variable given a fixed set of data.

For example, in our urn-drawing problem, we had this probability of getting a sample ratio of k/n in a sample of size n assuming the true urn fraction was f:

$$P[R = \frac{k}{n} \mid F = f] = \binom{n}{k} f^k (1-f)^{n-k}$$

This ordinarily comes with an assumed value f, the fraction of white pebbles in the urn. However, now imagine the sample is given, so k and n are fixed. We can think of this probability as a function of f, which Fisher called the *likelihood function*. He suggested a good choice of estimator would be the value of f that *maximizes* this function—that is, the value under which *the given observation is made most probable*. Usually some amount of calculus could allow him to derive an exact formula for this maximum solution. In our case, taking the derivative with respect to f and setting it equal to 0 shows that the maximum likelihood occurs at the value $f = k/n$; that is, the urn fraction value that makes our data the most probable is the one that exactly matches our observed sample ratio. This has an appealing symmetry that suggested to Fisher that the method of maximum likelihood would generally produce good results.

Fisher said he was inspired to think this way by the methods of inverse probability—that is, Bayesianism. He took the most inspiration from Carl Friedrich Gauss, who had used a similar argument to justify the choice of the normal distribution for errors. In fact, Fisher's maximum likelihood estimate is exactly the value with the highest *posterior probability*, assuming a uniform prior distribution for the unknown parameter. From the Bayesian computation, we can immediately see how that happens: If all the hypotheses of the form H_f: "$F = f$" have the same prior probability, $P[H_f \mid X]$, then their pathway probabilities will just be proportional to Fisher's likelihoods. But then the posterior probabilities will be given by the relative proportion of each pathway, so they will also just be proportional to the likelihoods. Therefore, the maximum likelihood estimate will also maximize the posterior probability. It was because of this, and the fact that Fisher had referenced Bayesian thinking as having inspired the maximum likelihood idea, that Pearson felt compelled to criticize Fisher in that fateful article in 1917. In a way, Pearson was correct!

But Fisher did not mean these likelihoods to be probabilities. Following Venn, again, Fisher denied the applicability of probability to this kind of inference. Or, as he said, something *more* than probability was needed:

Inferences respecting populations, from which known samples have been drawn, cannot by this method be expressed in terms of probability. . . . What is essential is that the mathematical concept of probability is, in most cases, inadequate to express our mental confidence or diffidence in making such inferences, and that the mathematical quantity which appears to be appropriate for measuring our order of preference among different possible populations does not in fact obey the laws of probability. To distinguish it from probability,

I have used the term "Likelihood" to designate this quantity; since both the words "likelihood" and "probability" are loosely used in common speech to cover both kinds of relationship.[14]

Fisher even tried to attribute this claim to Gauss by saying he had not *really* meant his prior and posterior probabilities as probabilities either.[15] One technical difference was that Fisher's likelihoods didn't sum to 1 the way probabilities always do, but that would have been resolved if he had been willing to normalize them, the way our Bayesian posterior probabilities are just our pathway probabilities expressed as relative proportions, as we've seen many times.

Furthermore, Fisher's methods for evaluating the *quality* of a maximum likelihood estimator were entirely frequentist. That is, since the estimator still depends on the data, it is allowed to have a probability distribution under the frequentist interpretation. So one could ask how often the maximum likelihood estimate would come out to be close to the true parameter value and so on. The fact of it being the most probable value in the Bayesian way of thinking was immaterial. A good portion of Fisher's theoretical work was devoted to establishing the taxonomy of these estimators, all within the confines of frequentism. For example, it can be shown under certain conditions that the maximum likelihood estimator is consistent and (asymptotically) efficient and has a distribution that converges for large samples to a normal distribution with a variance depending on the so-called Fisher information matrix.

So even though some of the Bayesian ingredients were present in Fisher's maximum likelihood method, the intent behind it was frequentist, and the utility of the estimates was judged by frequentist criteria.

The closest Fisher ever got to endorsing Bayesianism, though, was toward the end of a protracted fight he had with basically the rest of the statistics community over a technique he called *fiducial inference*.[16] The idea was inspired by a clever observation made by a colleague of Fisher's at Rothamsted about something called a *pivotal quantity*, best illustrated by example: Suppose we are trying to estimate an unknown constant μ that we know acts as the mean of the probability distribution of our data Y—say a normal distribution with standard deviation 1. This means we can say, for example, that if μ is 0, the probability of Y falling between −1 and +1 is about 68.27 percent, the chance of Y falling between −2 and +2 is about 95.45 percent, and so on, according to a table of probabilities for the normal distribution. Similarly, if μ is 1, Y has the same 68.27 percent chance of falling between 0 and 2—that is, within one standard deviation of the mean.

The main idea is to write these statements in terms of the distance $|Y - \mu|$, so, for example, it's always the case that

$$P[|Y - \mu| < 1] = 0.6827,$$
$$P[|Y - \mu| < 2] = 0.9545,$$

and so on.

By expressing things in terms of $|Y - \mu|$, we've arrived at probability statements that appear agnostic with respect to the particular value of μ. The expression inside the probability statement is therefore called a *pivotal quantity*, a combination of data and parameter values with a probability distribution that no longer involves the parameter. In our case, we'd say the distance between Y and μ is a pivotal quantity with the standard normal distribution.

Fisher's trick was then to take any one of these probability statements and invert it by now thinking of Y as fixed and μ as the unknown. So, instead of thinking of the first line as saying Y has about a 68 percent chance of being within distance 1 of μ, Fisher would say there's about a 68 percent probability of μ being within distance 1 of Y. He called this the *fiducial argument*, taking the name from a concept in land surveying where any fixed point can be used as a reference, as we mentioned in chapter 2.

The problem is that there are (at least) two different ways of interpreting this probability statement. One way is to say, "Whatever μ is, as Y is repeatedly sampled from the distribution, if we consider the interval from $Y - 1$ to $Y + 1$, it will contain the value μ about 68 percent of the time." The other is to say, "After doing a *particular sample* of Y, we have reason to be about 68 percent confident that μ lies within the range $Y - 1$ to $Y + 1$ for *that Y*." The daylight between these two interpretations was what ultimately pushed Fisher to nearly accepting Bayesianism.

The first version is totally consistent with frequentist thought. It's effectively indistinguishable from the idea of a confidence interval, as we defined it earlier in relation to a significance test. As long as the probabilities in question are those associated to Y, the interval as constructed will have the *coverage probability* it's supposed to have, meaning it will contain the parameter value as often as is claimed. This is the way Fisher originally justified the fiducial argument, contrasted against the Bayesian method, which had no such verifiable frequencies. He called the coverage frequency a "definite probability statement about the unknown parameter . . . which is true irrespective of any assumption as to its *a priori* distribution."[17]

It also happens to be mostly meaningless for the inference in question, for the same reason that confidence intervals are. What someone doing an experiment is concerned with is never what would happen if the experiment was run many more times but rather what inferences to draw from the particular observation of data they have in front of them. Married though he was to the frequentist idea of probability, Fisher understood the practical situation facing a scientist and did, even early on, consider probability to be a "numerical measure of rational belief."[18] The coverage probabilities gave a theoretically empirical *backing* to that belief, though, the same way the statement that a die has a 1/6 chance of coming up 6 is theoretically backed by an imaginary series of die rolls.

For simple examples like our normal distribution one, this didn't pose much real difficulty because the intervals were easy enough to calculate and always had the right frequency properties. So, for a given data set, a scientist could report the "fiducial interval" for the parameter and just leave it unsaid, but heavily implied, that this range must therefore represent a good guess as to where the parameter actually was, exactly the way people use confidence intervals to this day.

The problem was that, for more complicated situations involving multiple parameters, the fiducial argument wasn't so easy to extend. For example, suppose our normal distribution has two unknown parameters, the mean μ and the standard deviation σ. Since the probability that a normally distributed variable will be within a certain number of standard deviations of the mean is always the same, we can use the pivotal quantity $|Y - \mu|/\sigma$ to write probability statements like

$$P[|Y - \mu|/\sigma < 1] = 0.6827$$

that wouldn't depend on the choice of μ and σ. But now how do we invert the inequality so that it expresses something about μ and σ simultaneously?[19] Fisher deployed some clever tricks to come up with fiducial "joint distributions" for pairs of parameters and used his technique to solve the so-called Behrens-Fisher problem, which involves the difference of two means of separate normal distributions that *both* have unknown standard deviations, for a total of four unknown parameters.

This is where things started to unravel. It turned out, to Fisher's surprise, that his fiducial interval answers for problems with multiple parameters could sometimes have the *wrong* coverage probabilities. The English statistician Maurice Bartlett pointed out in particular that if, say, one used Fisher's approach to construct 90 percent fiducial intervals for the difference of the two means in the Behrens-Fisher problem, upon repeated sampling of additional data the proportion of intervals containing the true difference could come out to be something other than 90 percent.[20]

After unconvincingly trying to dismiss Bartlett's objections on technical grounds, Fisher radically changed course and decided that coverage frequencies weren't actually relevant at all! Where he had previously referred to fiducial probability with respect to "the population of all possible random samples,"[21] he decided now that "the legitimacy of such inferences cannot be affected by any supposition as to the origin of other samples which do not appear in the data."[22] He even used the issue as a way to attack his rivals Egon Pearson and Jerzy Neyman: "Pearson and Neyman have laid it down axiomatically that the level of significance of a test must be equated to the frequency of a wrong decision 'in repeated samples from the same population.' The idea was foreign to the development of tests of significance given by the author in 1925, for the experimenter's experience does not consist in repeated samples from the same population."[23]

Suffice to say, the idea was not actually "foreign" to Fisher's earlier work. So, if not coverage frequencies, what was the justification for the fiducial probability statements? As he continued to grapple with this apparent failure of frequentism, one strategy Fisher tried was an appeal to pure logic. If, say, the probability statements like the one we had earlier for the normal distribution,

$$P[|Y - \mu| < 1] = 0.6827,$$

were thought of as a "major premise" for general Y and μ, then the result of applying it to a *particular* observation Y and using it to deduce probabilities for μ could maybe be justified by *substitution*,[24] the way premises like "All men are mortal" and "Socrates is a man" would justify the substitution "Socrates is mortal." Unfortunately, probability statements don't work that way, as Neyman was quick to point out. If Y is the result of a rolling a fair die, for example, we can say

$$P[Y = 6] = 1/6$$

But after observing that we have rolled a 1, we can't substitute the value in naively and conclude

$$P[1 = 6] = 1/6$$

Still, Fisher was convinced that fiducial probability statements must have *some* inferential meaning. People like Jeffreys could provide that meaning very simply, using Bayesian inference. If the parameter was assumed to have a prior probability distribution, then sampling probability statements about Y, like those presented earlier, could be inverted into probability statements about μ by using Bayes' theorem, even if there were no pivotal quantity. Jeffreys showed how this extended to multiple-parameter situations like the Behrens-Fisher problem and how particular choices of priors for the unknowns could make the inferential intervals have the right coverage probabilities. It was because Fisher's fiducial method hadn't properly kept track of the prior and posterior probabilities for each parameter in turn—that is, because it didn't obey the rules of probability—that his interval frequencies had ended up misbehaving.

Fisher, maybe, started to see the light. As Sandy Zabell observed,[25] for the publication of the seventh edition of *The Design of Experiments* in 1960, Fisher changed a passage about the difference between Bayesian and fiducial inference from

> Statements of inverse probability have a different logical content from statements of fiducial probability, in spite of their similarity of form, and they require for their truth the postulation of knowledge beyond that obtained by direct observation.[26]

to

> Statements of inverse probability have a different logical *basis* from statements of fiducial probability, in spite of their similarity of form, *for* they require for their truth the postulation of knowledge beyond that obtained by direct observation (emphasis added).[27]

By the "postulation of knowledge," Fisher meant the assumption of a prior distribution for the parameter, a step of Bayesian inference he had always objected to as arbitrary. His argument was that fiducial inference would allow the same kind of posterior probability calculations that Bayesian inference did, just without the prior, as long as certain conditions were satisfied about the choice of pivotal quantities. He also added the explicit requirement that we have *no prior knowledge* of the true value.

Simply acknowledging the role of prior knowledge was a seismic shift for Fisher. He referred to his earlier views, such as the statement that "we can know nothing of the probability of hypotheses or hypothetical quantities,"[28] as "hasty and erroneous"[29] and said that none of the probability calculations in significance tests had "objective reality, all being products of the statistician's imagination."[30] He even, in the third edition of *Statistical Methods and Scientific Inference*, referred to "the role both of well specified ignorance and of specific knowledge in a typical probability statement,"[31] a description that might just as easily have come out of Jaynes's mouth.

For these reasons, Bayesian statistician Leonard "Jimmie" Savage would later refer to fiducial inference as Fisher's attempt to "make the Bayesian omelet without breaking the Bayesian eggs."[32] Fisher claimed that he had sidestepped the extra assumption of, say, a uniform prior in situations of total ignorance, which had always been a sticking point for him. The conditions where fiducial inference would apply were supposedly complementary to the ones where Bayesian methods would apply. If a prior for the parameter was *known*, then Fisher said to go ahead and use Bayes' theorem, a process he had always maintained was valid. If we were totally ignorant, in a sense he tried to make precise involving something he termed *recognizable subsets*, then fiducial inference was the way to go, assuming the necessary ingredients like pivotal quantities were in place. Then and only then, he maintained, would we be in the right state of knowledge so that our fiducial inferences would be valid and the coverage frequencies for our fiducial intervals would also fall in line (by that time, he had circled back to caring about coverage frequencies). As a technical point, this turned out not to be true; a year after Fisher died Robert Buehler and Alan Feddersen showed an example where all of Fisher's desired assumptions were satisfied and the coverage probability for a recognizable subset of 50 percent fiducial intervals was something like 50.181 percent.[33]

In both of these endeavors, maximum likelihood estimation and fiducial inference, we can see Fisher grappling with what he must have understood to be the inadequacy of the frequentist interpretation of probability. In the process of exploring both ideas, he essentially reinvented a good portion of Bayesian inference but decided it must be something new because Bayesian inference referred to probabilities in a way he couldn't accept. So instead he called them *likelihoods* or *fiducial probabilities* and used them to make inferences just as a Bayesian would for these special cases. Eventually he decided the latter were actual probabilities of the same type Bayesians had always used, but now they were justified in a way that avoided the parts of Bayesianism he couldn't stand. The cognitive dissonance this required apparently caused him no small amount of anguish. Later in life he confessed to Savage, "I don't understand yet what fiducial probability does. We shall have to live with it a long time before we know what it's doing for us. But it should not be ignored just because we don't yet have a clear interpretation."[34] As it was, fiducial inference more or less died with Fisher—for good reason.[35]

Fisher's awareness of the problems with the standard techniques had led him part of the way down the Bayesian path, but his lingering distaste for allowing probability to mean anything but frequency kept him from going any farther.

NINE ORTHODOX PROBLEMS AND NINE BAYESIAN SOLUTIONS

Now we'll see some of the catastrophes the thought process underlying the frequentist approaches can create once we attempt to follow it outside the narrowly confined realms in which it works. We'll consider some examples of slightly nonstandard statistical reasoning that nevertheless are consistent with the logic of orthodox statistics. Each of these thought experiments probes a slightly different aspect of the standard methods to show how they fail when taken to a different kind of logical extreme. But they all essentially exploit the same fallacy: that frequentist inferences are based *only* on sampling probabilities—that is, Bernoulli's Fallacy.

They range in practicality, with some (like "Base Rate Neglect or the Prosecutor's Fallacy" and "The German Tank Problem") being drawn from real experience and others (like "The Sure-Thing Hypothesis," where a die is rolled 60,000 times) being purely fantastical. The point is not so much that these problems come up in real life but that, for, say, a working scientist trying to use statistics to do inference, the problems they face may be somewhere *in between* these and the simple, contrived examples of a standard statistics course. The fact that probability as logic can handle even these extremes while covering the simple cases as well should give reassurance that it continues to work for everything in between.

1. Base Rate Neglect or the Prosecutor's Fallacy: What Prior Probability Do We Assign to the Hypothesis?

In chapter 2, we described how Bernoulli's Fallacy leads to the error of thinking in base rate neglect or the prosecutor's fallacy. Since the logic of frequentist statistics also employs Bernoulli's Fallacy, it will come as no surprise, then, that we can reimagine these examples as orthodox statistical tests. All that's required is a choice of notation.

For example, suppose we are considering a blood test result for a rare disease. We'll assume the same test accuracy as before, so, if someone has the disease, the test will surely come back positive, and if they don't have the disease, the test will come back negative with probability 99 percent. Assume a patient then tests positive. How would orthodox statistics handle this? First, as we are told to do, we forget this particular test result. Then suppose we define a *test statistic* T that simply captures the result of the blood test: say $T = 1$ if the test comes back positive and $T = 0$ if it's negative. According to the orthodox procedure, we define a *rejection region* for the statistic given by the set {1}. Our null hypothesis is the proposition "The patient does not have the disease." With the assumptions we are given, we can then say that the probability of the statistic falling in the rejection region, assuming the null hypothesis is true, is the probability of a false positive—that is, 1 percent. So, returning to the patient in question, since their test result falls in this region, we can say *we reject the null hypothesis at the 1 percent significance level.* That is, we conclude the patient does have the disease, and we have followed all the same statistical procedures that any journal of medical science would require. No doubt the patient will be terribly upset to hear such bad news backed by such authority.

As we saw in chapter 2, though, this inference could be far from correct, depending on the base rate with which the disease occurs in the population. If the disease is rare, it could be the case that the overwhelming likelihood is still that the patient doesn't have it, even after testing positive. In probability language, the base rate plays the role of our *prior probability* for the hypothesis. However, orthodox statistical procedures don't include the prior probability at all—or even admit that such a probability exists in general—so it has no role to play in the inference.

The frequentist claim that following this procedure will lead to a false positive rate of only 1 percent is even still true! For every 100 healthy people we test, we should expect on average only 1 to test positive. But this is the wrong probability. It happens that, after observing a positive, we should be led to a different inference about whether that *particular* result is a false or a true positive. That is, we were never concerned with how often we should expect the test to be right or wrong when the patient is healthy. We want to know whether to declare the patient is sick based on the results of the test. Orthodox statistics says yes in this circumstance, even though for a rare disease that will be the wrong answer an overwhelming percentage of the time.

The exact same situation applies to the prosecutor's fallacy, since it's the same fallacy by a different name. If we define a test statistic such as N = Number of children of a family who died in infancy and we consider the null hypothesis to be H = "This parent is not guilty of murdering their children," then we could make a correct probability assignment that given H the probability of N being greater than or equal to 2 is very, very low—even something like 1 in 73 million, as Roy Meadow had calculated in the Sally Clark case. So, taking this $\{N \geq 2\}$ to be our rejection region, after being informed that a family had experienced 2 infant deaths, we would be led to reject the null hypothesis at an extremely low significance level, around 0.00000001.

But, again, we would be ignoring the fact that the prior probability for the alternative hypothesis, that the parent had murdered their children, should itself be extremely low.

Now any well-trained frequentist statistician knows the solutions to these problems. Like Fisher, they are perfectly willing to include the prior information when it's given, and these very examples appear in many standard statistics text-books. The point is that, by *sometimes* allowing prior probabilities to play a role in inference, frequentists are behaving hypocritically. They should ignore any prior probabilities available in these problems because strict frequentist methods claim in all other situations that the prior information isn't necessary. What if, for example, instead having a precise base rate, the disease in question is known to be somewhat rare? According to frequentist thinking, there is no middle ground. Anything not known exactly can be safely ignored, which is antithetical to the whole point of probabilistic inference: assembling all the bits and pieces of partial knowledge we have and aggregating them into the best possible inference.

Bayesian inference combines the probabilities in a sensible way that the orthodox statistical methods are unable to handle.

2. The Malfunctioning Digital Scale: How Do We Interpret Extreme or Unlikely Data?

Even when the situation is more "normal"—say involving the error present in a physical measurement—we may still be led to illogical conclusions if we blindly follow the orthodox template of rejecting hypotheses based on extreme data.

Suppose we are measuring the mass of some object in our lab using a digital scale. Ordinarily we know this scale is accurate to the nearest milligram—that is 0.001 grams. But suppose our particular scale has an annoying quirk where sometimes the leading digit on the display flips from 0 to 1 due to some internal error; the result is a readout that effectively adds 100 kilograms—that is, 100,000 grams. Assume this is known to happen only 0.1 percent of the time.

Suppose we're interested in testing the null hypothesis that our object in question has a mass of 1 gram. Now imagine that we put the object on the scale and get a readout on the scale of 100,001.000 grams. How do we treat the null hypothesis?

Well, following the orthodox template, we need to forget this individual measurement and think about a procedure by which we would reject the null hypothesis or not. Assuming the measurement error to generally have a normal distribution—say with a standard deviation of 0.001 grams—we might set our rejection region to be anything outside the range of two standard deviations from the mean, corresponding to a significance level of about 0.05. That is, we'll reject the hypothesis at the 5 percent level if the measurement is anything less than 0.998 grams or more than 1.002 grams.

Returning to our given data, we would conclude that we must reject the hypothesis at this level because the value 100,001.000 grams is *much* greater than our threshold. However, you might at this point reasonably object that we haven't taken into account what we know about the scale and the possibility that it could malfunction! But as we said, that event is very rare, so even setting our significance level to be 0.001, we can make the correct claim that, by rejecting the null hypothesis for any value outside of a small neighborhood of 1 gram, we will be wrong only about 0.1 percent of the time. It happens that, by visual inspection of our data, we can be almost certain that *this is one of those times* because we can be very confident that the scale has malfunctioned. The only alternative hypothesis that would explain the same data is that the object actually has a mass of 100 kilograms. And the fact that we were able to lift it and place it on the scale probably tells us that is not the case. Once more, this is the kind of prior information that isn't allowed for by the frequentist procedures.

The problem here is the idea of *extreme data* and whether that data should count as evidence against a hypothesis. Ordinarily the rejection region for a hypothesis—say that some parameter has some particular value given that we're able to measure it but with some error—takes the form of all those measurements "far away" from the putative value, but here we have a clear situation where *some* extreme values, like 1.5 grams, are clear evidence against the hypothesis that the mass is 1 gram, while others, like 100,001 grams, are not. No matter how we set a threshold for what counts as extreme, we'll be led to make a (probably) incorrect inference. To fix things, we'd need a rejection region that carves out a little section in the tail of the distribution around the value 100,001 grams and doesn't count *those* values against the hypothesis.

We also got tripped up by the fact that our data is unlikely, so our procedure is at a loss to explain it. The logical template of the significance tests is "If the hypothesis is true, some aspect of the data will be very unlikely; we observe that aspect, so something is wrong with the hypothesis." We saw in chapter 2 that this appeal to something like the reductio ad absurdum argument but with probabilities does not work. The probability of something improbable happening is 1. The probability of a particular improbable thing happening under a given assumption may be extremely low, but then observing it shouldn't necessarily count as evidence against that hypothesis. Everything depends on what alternative hypotheses are available that could explain the data better and what their prior probabilities are.

Often the alternative is given a cursory nod with some statement like "We test the hypothesis $m = 1$ gram against the alternative $m \neq 1$ gram," but the situation can often be more complicated than that. In truth, we may be accidentally testing compound hypotheses like "$m = 1$ gram and our scale did not add 100 kilograms to the measurement." As we saw with the ESP example in chapter 2, we may even have alternatives in mind that we *don't realize* until we're presented with such unlikely data that it revives them from near impossibility to something like near certainty. This very experiment may be the way we discover the scale's malfunction!

3. The "Sure-Thing" Hypothesis: What Hypotheses Are Allowed?

A standard example of a significance test that *works* is testing a six-sided die for bias (this could be an example of where to use Pearson's chi-squared test). Say we decide to way overtest this hypothesis, though, and we collect a sample of 60,000 rolls of a given die. But now imagine that, after we have conducted our extremely long sequence of 60,000 rolls and recorded the results, a stranger approaches us and makes an absurd claim: that the die we used has the property that those results were *predetermined*. The stranger says that due to some inner mechanism of the die, any time it is rolled exactly 60,000 times it produces the exact sequence of outcomes we observed. It is, in a sense, maximally biased.

Too exhausted to consider rolling the die another 60,000 times, we turn to statistical orthodoxy to help us test this claim. Suppose we try to reject this "sure-thing" hypothesis H_{sure}: "The results of the 60,000 rolls were predetermined to be the sequence we observed." The problem is that, under H_{sure}, there is no randomness to the data at all. So *any* summary statistic we want to define—say the number of 6s out of 60,000 rolls—will have a distribution under this hypothesis that is concentrated entirely at a single point that exactly matches the result we got. Thus, no test we perform will ever allow us to reject this hypothesis at any significance level!

Nor could we ever accept it. Imagine we try reversing the script so that H_{sure} is the *alternative* hypothesis. What null hypothesis would produce this as an alternative? You could say the null hypothesis is that the die is fair; call this H_{fair}. But then the simple negation of H_{fair} is not H_{sure}. It's an infinite assortment of all the ways the die can be unfair, including H_{sure} and many others. Okay, so suppose we ignore that complication and stipulate that, for this particular die, we're considering only the possibilities H_{fair} and H_{sure}. What test statistic could we use? The "feature" of the data we're supposed to look out for is that it exactly matches what we got. Therefore, a natural choice would be the funny statistic F_S, defined as having the value 1 if the sequence of die rolls exactly matches the sequence we observed, which we'll call S, and 0 otherwise. In probability language, this is called an *indicator function*. It may look a bit awkward, but sure enough, it is a quantity derived from the data (taking the value 1 or 0), so it can have frequentist probabilities.

Under the null hypothesis, the probability that $F_S = 1$ is the probability that 60,000 rolls of a fair die exactly match the given sequence, which is $(1/6)^{60,000}$, no matter what S is. This number is astronomically small—on the order of $10^{-46,689}$. But the value 1 is what our data produced for F_S, so are we ready to reject the null hypothesis at this incredibly small significance level?

Not so fast, comes the orthodox reply. We have forgotten to account for the fact that the stranger would likely have made the claim of predestination *no matter what data we got*. This might not be because he's a charlatan. Maybe all he knows about the die is that it has this funny behavior, not exactly what results it's predestined to produce. So we have to adjust our significance test accordingly to reflect the fact that we are actually testing *multiple hypotheses*. That is, we consider the family of all the possible sure-thing hypotheses corresponding to every possible sequence of data—$H_{sure1}, H_{sure2}, \ldots, H_{sureN}$, where N is the number of sequences, $6^{60,000}$. According to standard procedure, we are then supposed to take our desired significance level, typically 5 percent, and divide by N, the number of simultaneous alternatives. This is called the *Bonferroni correction* to our significance level; it gives us the right overall error rate when rejecting the null in favor of *any one* of the possible alternatives. With that correction applied, we see that we did not cross the significance threshold: our p-value was $1/N$ (we got the most extreme value possible, with this probability), but the threshold was $0.05/N$.

So it seems that we can neither accept nor reject this bizarre theory. It's stuck in a netherworld of untestable hypotheses, outside the reach of significance testing to decide.

What about other methods? Fisher's quasi-Bayesian maximum likelihood method tells us we might believe the stranger. From among the available hypotheses (say H_{fair} and all the rest of the H_{sure} family), the one that maximizes the likelihood function is always the one that makes our observed data certain. If we wanted, we could invent some complicated mathematical description that turned this into a parameter of some model, but the underlying logic would be the same. If all we care about is how likely a hypothesis makes some data, then unless we really restrict ourselves to alternatives of a particular form, we will probably be able to construct some weird hypothesis that makes the data very likely—even certain. In other realms of statistical modeling, this could be called *overfitting* or *data dredging*, and what we see here is that the standard methods, particularly maximum likelihood, do not give us much defense against it.

The Bayesian methods do, however. We can even allow for this sure-thing hypothesis and concede that it makes our data have probability 1, compared to the "fair-die" hypothesis, which tells us our data is vanishingly unlikely. Why isn't this a problem? Because we also must consider the *prior probability* of the sure-thing hypothesis. Among other reasons to doubt such a claim (e.g., How would this inner mechanism work? How does the die "know" we're going to roll it 60,000 times?), as we noted earlier, this particular H_{sure} is not the only possible sure thing. Prior to knowing how the data actually comes out, we have no reason to prefer any of these to the others, so the principle of indifference dictates that we assign them all the same prior probability, which necessarily must then be at most $6^{-60,000}$. That very, very low prior probability kills off the high sampling probability in the final analysis and leaves us with a reasonable posterior inference.

4. The Problem of Optional Stopping: What Other Data Could We Have Gotten?

The standard methods have a bizarre fascination with what other data we could have gotten but did not get. The *p*-value, for example, measures exactly this: What is the probability, under a certain assumed hypothesis, that other instances of the experiment would produce a result as extreme as or more extreme than this one? But often while doing real experiments, we may not know what data we *could* have gotten, only what we *did* get.

For example, suppose an experiment is being conducted in our lab by an assistant named Alex. The experiment consists of a series of trials, each of which can terminate in what we'll call a good (G) or a bad (B) result. We're interested in testing the null hypothesis that says these are equally likely. We have hypothesized that, according to our new procedure, we should produce good results more often than bad ones; we are hoping to reject the null hypothesis in favor of this alternative hypothesis. The future of our lab depends on the grant funding we'll get if the experiment is a success.

Assume the trials are difficult and expensive to conduct, though. So, marshalling all our available resources, Alex performs a sequence of six trials and records the trial results as five good out of six, with a full data sequence of GGG GGB. Bill, the postdoctoral fellow running the experiment, interprets the data as follows:

The procedure consisted of performing six trials, and the test statistic we used is the number of successes out of six. Call this statistic S. Then, given a success rate p, the probability of observing a value $S = k$ follows the binomial distribution:

$$P[S = k] = \binom{6}{k} p^k (1 - p)^{6-k}$$

For the null hypothesis, the value of p we assume is 1/2, so we have

$$P[S = k \mid H_0] = \binom{6}{k}\left(\frac{1}{2}\right)^k \left(\frac{1}{2}\right)^{6-k} = \binom{6}{k}\left(\frac{1}{2}\right)^6$$

Our rejection region (corresponding to the alternative of a higher rate of success) is large values of S. The observed value of S is 5, so the p-value of this test is the probability of observing values as extreme as or more extreme than 5. According to the formula above, this is

$$P[S = 6 \mid H_0] + P[S = 5 \mid H_0] = \ldots = 0.109$$

Since this is greater than 0.05, we do *not* reject the null hypothesis at the 5 percent level. Even though the data suggests a success rate higher than 50 percent, we cannot be sure at this significance level that the results were not obtained just by chance. Oh well.

Charlotte, the professor in charge of the lab, looks over Bill's analysis and says he has misinterpreted the results, though. Her analysis goes like this:

The procedure Alex followed was such that, upon a bad result, a piece of lab equipment (say a chemical reagent or plate of photographic film) was destroyed, and I know that there was only one of these pieces of equipment in the lab at the time of the experiment. Alex would have continued performing trials as long as possible. This is confirmed by the fact that the final trial was a bad result. Therefore, the variable in the data is not the number of successes out of six trials but rather the number of *trials* it took to achieve one failure. Call this statistic T.

Under an assumed success rate p, the statistic T has probabilities given by the *negative binomial distribution*:

$$P[T = n] = p^{n-1}(1 - p)$$

For the null hypothesis of $p = 1/2$, this gives

$$P[T = n \mid H_0] = \left(\frac{1}{2}\right)^{n-1}\left(\frac{1}{2}\right) = \left(\frac{1}{2}\right)^{n}$$

The minimum possible value of T is 1, and our rejection region (meaning a higher success rate) corresponds to large values of T. Our observed value of T is 6. The p-value for this test is therefore the probability of getting any value for T greater than or equal to 6. This is given by the infinite sum:

$$P[T = 6 \mid H_0] + P[T = 7 \mid H_0] + \ldots$$

which we can compute using its complement as

$$1 - (P[T = 1 \mid H_0] + P[T = 2 \mid H_0] + \ldots + P[T = 5 \mid H_0]) = \ldots = 0.031$$

Since this is less than 0.05, we conclude that we *do* reject the null hypothesis at the 5 percent level. Our method has been proven effective at this significance level. Open the champagne!

Both Bill's and Charlotte's calculations are correct. They arrived at different p-values because they used different theories of what *other* data the experiment might have produced. This can be further complicated and made arbitrarily silly. Imagine that, after publishing their results in *Nature*, Charlotte's team discovers that a backup set of lab equipment was stored away that would have allowed Alex to keep performing trials beyond one failure, assuming Alex knew about the backup. This might confirm Bill's theory that Alex must have planned to stop after six trials no matter what, making the results not significant. Should they issue a retraction? What if the equipment was locked away in a storage cabinet and Alex didn't have the key? This might confirm Charlotte's theory that Alex had planned to keep doing trials as long as possible but had to stop after one failure. Suppose that, on further investigation, it's revealed there was a janitor in the building who could have opened the cabinet. Alex didn't ask him to, but was that because Alex didn't know he was in the building (confirming Charlotte's theory), or because Alex was done with the experiment after six trials anyway (confirming Bill's theory), or because Alex didn't know about the extra equipment in the cabinet (confirming Charlotte's theory)? Alex has long since graduated and moved away, so there's no way to know for sure. Or suppose that, after the sixth trial was complete, a fire alarm went off in the building, meaning Alex *had* to stop the experiment at six trials regardless of the plan (confirming Bill's theory). Oh, but it turns out the fire alarm was actually triggered by smoke produced by the bad result of Alex's sixth trial (confirming Charlotte's theory), and if that had happened on the very first trial, Alex would have ignored the fire alarm and kept going until recording at least three data points (contradicting both Bill's and Charlotte's theories). And so on, and so on.

The point is that the orthodox significance tests can depend critically on what would have happened but did not happen. Because the procedure is awkwardly defined in terms of tail probabilities and rejection regions, it requires that probabilities be established for all possible data, not just the result that was actually obtained. Otherwise, it's not possible to know where the tail even is. As a result, the tests are overly sensitive to model assumptions about what kind of extreme data *could* have been observed. The conditions under which the experiment would have been terminated—the so-called problem of optional stopping—can create a huge headache.

In the Bayesian regime, all the inferential probabilities are conditioned on the observations that were *actually* made. The possibility of other observations enters into the sampling probability as the fact that the observations are not known with certainty ahead of time. However, once given, they become part of the background assumptions for the purpose of assigning probabilities. Any such questions about what other data we could have gotten but did not get are irrelevant. Different rules for optional stopping don't cause any trouble for the inference because any *feature* of the data that determined why the experiment ended will necessarily be revealed by the *data itself.*

To take the preceding example, if we assume the variable is the number of successes out of n trials, as Bill did, and if we also assume any individual success rate p, we would have (with B denoting Bill's background assumptions)

$$P[D \mid H_p \text{ and } B] = \binom{6}{5} p^5 (1-p)$$

Or if we assume the variable is the number of trials to get one failure, as Charlotte did, we would agree with her calculation (with C denoting Charlotte's assumptions):

$$P[D \mid H_p \text{ and } C] = p^5 (1-p)$$

When it is time to make an inference, though, assuming Bill and Charlotte assigned the same prior probabilities for the success rates, we will see that they arrived at the same *posterior distribution* because the probabilities they assigned to *this* data observation are the same apart from a constant multiple, which would cancel out in the final calculation. Bill's pathway probabilities would have a factor of 6 in front of all of them, whereas Charlotte's wouldn't, but since the posterior probabilities are computed as *relative* proportions anyway, these constants would not matter. That is, *in the way the probability of this data involves the parameter* p, Bill and Charlotte are effectively in the same state of information and so *must* make the same inferences.

5. The Problem of Divided Data: Should All Insignificant Results Be Forgotten?

The following academic year Bill has gotten a job as a professor at a different university. His research focus is largely the same as when he worked in Charlotte's lab, so he ends up conducting an experiment similar to one she's running. Suppose their experiments both involve the effect of music on the cognitive abilities of mice. They're both interested in whether playing particular forms of classical music for the mice will make them better at running through mazes. They're using a standard maze test that, according to established research, takes the average untreated mouse 100 seconds to solve. Charlotte focuses on the music of Mozart and decides to test a selection of symphonies, operas, concertos, and string quartets. Bill has a slightly different theory—that what really matters is the violin—so he tests an assortment of violin concertos by Bach, Mendelssohn, Sibelius, and Mozart.

Unfortunately all their experiments end in failure. None of the musical compositions they play seems to have a statistically significant effect on the mice. The relevant test statistic for this sort of experiment is Student's t:

$$t = \frac{m - \mu}{s/\sqrt{n}}$$

where m is the observed mean, μ is the hypothesized mean (in this case, 100), and s is the observed standard deviation in the sample. A high or low enough value of t would be cause to reject the null hypothesis of no effect.

Catching up at a conference later that year, they grab coffee and commiserate over their rotten luck. But if they had compared detailed notes, they would have found something interesting. At some point, they were unwittingly testing the same piece of music: Mozart's Violin Concerto No. 5 in A Major (K. 219). Their findings for that work, respectively, were as follows:

Bill:
Mean = 98.85
Standard deviation = 4.7450
Sample size = 30
t-statistic = −1.33
p-value = 0.1951

Charlotte:
Mean = 98.67
Standard deviation = 4.8112
Sample size = 50
t-statistic = −1.95
p-value = 0.0563

Neither result was statistically significant on its own, but pooling the data would have yielded the following:

Bill/Charlotte collaboration:
Mean = 98.74
Standard deviation = 4.7571
Sample size = 80
t-statistic = −2.37
p-value = **0.0200**

Since the combined p-value is less than 0.05, the result is significant at the 5 percent level, and they could have coauthored a paper for the *Journal of Classical Music and Rodent Cognition*. What happened? Note that we haven't played any tricks here with the sampling distribution or the test statistic; everything is completely standard. Furthermore, this isn't a problem of priors. We can take as a given that the experimenters had no strong prior information about the size of their effect, so, as in Bernoulli's urn problem, the frequentist approach gives answers materially similar to the Bayesian ones.

The difference, though, is that the orthodox procedures force an assessment of whether a result is significant or insignificant. It may happen that dividing what would have been significant data into two parts makes each part insignificant. So a researcher may see an experiment as a failure for not producing a significant finding—but only because they've rounded it down to an insignificant 0. This is all in keeping with Fisher's instructions. He described the purpose of a significance test as telling a scientist "what to ignore, namely all experiments in which significant results are not obtained."[36] To a layperson, it might even seem that this result has been *confirmed*. A perfectly accurate, but misleading, summary of our situation would be "Two labs independently studied the same treatment, and both found no significant effect."

It just might happen that one person's insignificant result could provide the boost another person needs to push their own results over the threshold of significance. To allow for this, the practice that has emerged in many areas of science is that, after some number of studies has been performed, each examining the same effect and reaching its yes/no conclusions about its existence, a meta-analysis is conducted to put all the data together and draw a single conclusion. But that requires the original studies to be published. As in our example, it's likely that, for many negative results, there is no potential for a meta-analysis because the results never see the light of day. Even if they did, it would take time and effort for a third party to assemble the pieces.

With Bayesian techniques, *every analysis has the potential to be a meta-analysis* because we are free to take the posterior probabilities from someone else's work as our prior probabilities for the start of our own. Bayes' theorem guarantees that we will reach the same conclusions as a meta-analysis combining our results, since, mathematically, for any two propositions A and B, we have

$$P[H \,|\, (A \text{ and } B) \text{ and } X] = P[H \,|\, A \text{ and } (B \text{ and } X)]$$

Thinking about some other experimenter's data as proposition B and our own data as proposition A leads us to the probability on the right-hand side, with the reported posterior probability $P[H|B \text{ and } X]$ playing the role of our prior. Treating all the data as one proposition (A and B) would lead the meta-analyst to the inference on the left-hand side. The logical consistency of the calculations guarantees that we'll reach the same conclusions.

This equation, like all equations, goes both ways. So another way of understanding it is to say *meta-analyses are already Bayesian analyses*. They just require the additional overhead of another study to collate the data of previous studies and start the test over as if from nothing, assuming the data exists to be found. With Bayesian thinking and better data-sharing, we can make much more efficient use of the available resources and be able to see for ourselves in real time whether the cumulative research into a given effect is definitive or whether it calls for more investigation. (If this improvement seems too trivial to care about, imagine in this example that, instead of music, Bill and Charlotte are testing out new cancer therapies and instead of maze-running ability, they are measuring tumor size.)

6. The German Tank Problem: Which Part of a Distribution Is the Tail?

During World War II, in the days leading up to the invasion of Normandy, Allied forces were faced with a difficult problem of statistical inference. The German military had recently introduced a new model of tank, the Panzer V. In previous battles, the U.S. Army's Sherman tank had done well against its predecessors, the Panzer III and IV, but the Panzer V was significantly more deadly. The Allies needed to know how many of these tanks they could expect to encounter in northern France. Conventional intelligence had been spotty and unreliable, but rumors were circulating that Panzer V production had recently spiked.

What they had were serial numbers of various parts (mostly gearboxes, engines, and wheels) of Panzer V tanks they had either captured or destroyed, which would enable a statistical analysis. Importantly, they assumed these numbers were generated in *sequence*, so the relative sizes of the numbers could give them some clue of how many tanks had been manufactured over some period. What they needed to know was the highest number that had been generated, which would reveal the total number of tanks that had been produced. They made the further assumptions that the tank parts they discovered were equally likely to have come from anywhere in the sequence and that the parts that ended up in tanks in various places were effectively independent from one another. Relabeling the numbers to start from 1 (and slightly abstracting things a bit), the problem then essentially became the following:

> Given a collection of k integers, $n_1, \ldots, n_k$ sampled independently and uniformly from the range from 1 to N, with N unknown, come up with a way to estimate N.

In our previous language, the value N is a *parameter* of our model, with the data probability being given by

$$P[n_1 \ldots n_k \mid N = n] = \left(\frac{1}{n}\right)^k, \; for \; 1 \leq n_1, \ldots, n_k \leq n$$

This just encodes the idea that each number is uniformly likely to be between 1 and N and that the k samples are independent. We see the dependence on the assumed value $N = n$, similar to how our urn sample ratio depended on the true urn-fraction F. How then can we use our data to make inferences about N? It turns out that a good estimator for N is given by the statistic

$$\hat{N} = \text{Max} \cdot \frac{k+1}{k} - 1$$

where Max is the *data maximum*, meaning the largest number among $n_1, \ldots, n_k$. This captures the intuitive idea that if we had observed only one value ($k = 1$), we'd guess it to be about midway through the list on average, so we might think to double it as a guess for the maximum. If we had observed two values, we'd imagine them to be roughly equally spaced on average, at positions around 1/3 and 2/3 of the total. So the larger of the two should be about two-thirds of the highest possible value, suggesting a guess of about Max $\cdot$ (3/2), etc.

This estimator also happens to have the friendly property of being the *minimum variance unbiased estimator (MVUE)* for this parameter,[37] meaning that, as the name suggests, from among all the possible unbiased estimators of N, it is the one with the minimum variance—that is, the one with the most tightly concentrated sampling distribution. According to the frequentist approach, these properties suggest it's the best estimate we could produce. So, given our tank data, we would know what we could estimate the number of tanks to be.

Now imagine we have done this analysis and presented it to our superiors in the Allied military intelligence operation, who then ask us the very reasonable follow-up question: How confident are we in this estimate? That is, what kind of statistical tests might we apply to rule out other possible values—say at the

5 percent significance level—or, expressing things another way, what would be a 95 percent confidence interval for the true number of tanks? A lot depends on the Allies not *underestimating* this number of tanks, so, while the single unbiased estimate is a good starting point, what would be a *conservative* range of values that could also be consistent with our data?

We have what may serve as a reasonable *test statistic*—our estimate $\hat{N}$—but the problem is that we don't have an obvious rejection region for the statistic with which to answer any of the preceding questions. In the urn-drawing example, we used the sample ratio as a test statistic for the true urn fraction and the two tails of the distribution as rejection regions because Bernoulli's theorem guaranteed that these tail regions would have low probability. Here, we have a parameter that interacts with our data in a *different way*, expressed via the sampling distribution earlier.

We're interested in accepting or rejecting hypotheses of the form $N = n$. Suppose we take to heart our superiors' concerns about not underestimating the number of tanks, so we decide on a way of rejecting these hypotheses that preferentially rules out small values of n and is more circumspect about ruling out larger values. Our test statistic is derived in a simple way from the data maximum, so we can just as well use this maximum as a test statistic. Let's say we decide to reject the hypothesis $N = n$ whenever n is small relative to Max—that is, whenever Max is large relative to n. The probability of observing a maximum of Max $= m$ out of a sample of k numbers from 1 to n works out to be[38]

$$P[\text{Max} = m] = n^{-k}[m^k - (m-1)^k]$$

for those values of m that are possible—that is, $m \leq n$. So we set up our rejection region with significance 5 percent to be those large values of Max, starting from n and working our way down until we have accumulated 5 percent probability. Considering any possible value of N, let's say we will reject the hypothesis $N = n$ if Max is greater than or equal to $(n - i)$ for some i. Using the preceding formula, we will notice an interesting pattern for these tail probabilities:

$$P[\text{Max} \geq n - i \mid N = n\,] = 1 - \left(1 - \frac{i+1}{n}\right)^k$$

Therefore, this tail region will constitute at least 5 percent of the probability if we have $(i + 1)/n \geq 1 - 0.95^{1/k}$, meaning the tail rejection region is a fixed proportion of the total range from 1 to n.

For concrete numbers, suppose we have observed $k = 4$ tank parts with a maximum serial number of 313, so our estimate for the total number of tanks is

$313 \cdot (5/4) - 1$, or 390.25. We find the 5 percent rejection region for any hypothesis $N = n$ consists of those values for Max within the top $1 - 0.95^{1/4} \approx 1.27\%$ of values. So, for example, we would reject the hypothesis $N = 500$ if the maximum fell within distance 6 of the top (Max $\geq$ 494), we would reject the hypothesis $N = 1,000$ if it fell within distance 12 (Max $\geq$ 988), and so on.

The problem is that, beyond some point, as n gets larger, this means we will *never* reject the hypothesis $N = n$ at the 5 percent level, since our *observed* value, Max = 313, will not be within the specified tail proportion. We find that the greatest value for which we do reject the hypothesis is $N = 317$. Expressing things in our previous language, we would say that our 95 percent *confidence interval* for N is the interval $[318, \infty)$ because those are all the values of n for which we *do not* reject the hypothesis $N = n$ at the 5 percent level given the way we designed our test. So even though our estimate for the number of tanks is about 390, we admit that, based on our data, it could be anything greater than 318, including the possibility that there are more German tanks than there are German soldiers, or people on Earth, or atoms in the universe.

What went wrong? Obviously, the culprit is somewhere in the way we set up our rejection region for the significance test. An examination of the graph in figure 5.7 of the probabilities for the maximum Max = m given a choice of N—say $N = 500$—reveals why this may have been a poor choice.

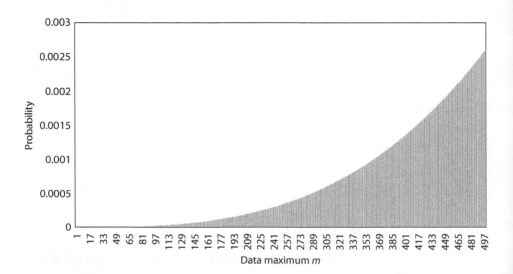

FIGURE 5.7

Probability that Max = m as a function of m given $N = 500$.

Even though it is true that the rightmost values total up to something like 5 percent probability, depending on how many we include, we wouldn't say this really constitutes a tail of the distribution. In fact, the larger values of Max are always more probable! Taking the *leftmost* values as our rejection region instead, we'd find that, given the observation Max = 313, we'd *start* rejecting the hypothesis only at $N = 662$, and then we'd reject it for all subsequent larger values. So the 95 percent confidence interval from a test designed *this* way would be the much more reasonable range [313, 661].

Again, nothing about testing for significance necessitates that choice. The fact that the left side of the distribution looks more tail-like is not an objective reason to use it instead of the right side. We were led to the right-side rejection region for what we thought was a conservative reason: we didn't want to underestimate the number of tanks. It just turned out that our procedure was way *too* conservative and led to a ludicrous *over*estimate. But the 95 percent confidence interval we constructed does, indeed, have the frequentist property it's supposed to have: given any true value of $N = n$, if we repeatedly sample from the distribution of Max and construct a confidence interval this way, we'll find that approximately 95 percent of the constructed intervals contain the true value n!

As we mentioned before, the choice of which part of the distribution for some test statistic constitutes the tail for the purposes of rejecting a hypothesis carries with it an *implicit* acknowledgment of the specific alternatives against which we are testing that hypothesis. In principle, we're free to choose any part of the distribution we like, and as long as its probability adds up to the significance level, we'll always maintain the desired false positive rate. Different choices will naturally give different false negative rates, but that's entirely dependent on what we assume *is* probably true if the null hypothesis *isn't*.

The Bayesian procedure, by the way, makes extremely quick work of this problem and doesn't involve any arbitrary choice of tail regions. The sampling probability for Max, given $N = n$, is the same as above. What we care about is how the single observation Max = 313 affects our probability assignments for N. The alternatives are simply all the hypotheses of the form $N = n$ for all values of n. Assuming a uniform prior probability distribution on, say, the range [1, 1,000] and computing the posterior probabilities, given an observed maximum of Max = 313, yields the distribution shown in figure 5.8.

The probabilities are 0 for all the values that are inconsistent with the data (that is, where $n < 313$), and they drop off like n^{-4} from that point on, just as the data probability indicates they should. From here, we could easily draw other numerical conclusions. For example, we could say, somewhat surprisingly,

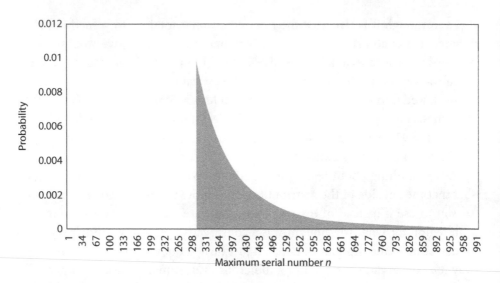

FIGURE 5.8

Posterior probability that $N = n$ as a function of n given Max = 313.

that the most probable value is equal to the observed maximum, meaning we collected a part from the last tank in production. This would also be Fisher's maximum likelihood estimate for this problem; as ever, it has the highest posterior probability if the prior distribution is uniform. The frequentist estimate of 390.25 given earlier turns out to be about our median value. We could say with about 50 percent confidence that the number of tanks is less than 390, with 75 percent confidence that the number is less than 481, and so on.

If we had collected more parts, the drop-off would have been even steeper, representing greater certainty. The distribution represents the full state of our knowledge about the number of tanks given our observation (starting from a position of uniform ignorance), from which all manner of military decisions could be made.

7. Testing for Independence: What Direction Does Our Data Have?

Sometimes the correct choice for a tail region is even less clear. A basic example of a statistical inference is the test for independence between a pair of variables in a *contingency table*: a tally of how many observations have each combination of the two characteristics. Say, for instance, we are interested in household

TABLE 5.1 **Example survey data for income and political party**

	Democrat	Republican	Total
High income	2	9	11
Low income	10	4	14
Total	12	13	25

income and political party among voters in the United States. We might conduct a sample of some number of people and arrange them in groups according to income (high or low) and party affiliation (Democrat or Republican). Our contingency table then records how many of each of the four possible combinations we observe. For example, we might have a total of 25 respondents arranged in a 2 × 2 table (table 5.1).

Our hypothesis, which seems well supported by the data, is that the two factors are *not* independent, meaning the proportions of people in the two income brackets in the general population are different between the two parties. So to get to that conclusion in the orthodox manner, we must reject the null hypothesis that they *are* independent. This means imagining different 2 × 2 tables that are even more extreme than table 5.1. But how?

The test we are supposed to use in this case is *Fisher's exact test*.[39] Fisher designed the test in the 1920s for his famous "lady tasting tea" experiment. His subject, a researcher at Rothamsted named Muriel Bristol, had claimed that she could distinguish whether milk had been poured in a cup of tea before or after the tea was poured. Fisher, doubting that anyone could tell the difference, tested her by making eight cups of tea, four in each preparation, and asking her to identify the four where the milk was poured before the tea. She correctly identified them all.

What makes the test "exact" is the formula Fisher derived that gives the exact probability of observing this particular table under the assumptions that (1) the two characteristics are independent and (2) the row and column totals are fixed. In our example, Fisher's formula gives a probability of about 0.0106. But the formula doesn't tell us what the more extreme tables are. Maybe, inspired by the "direction" our table seems to indicate, where Democrats are more likely to be in the low-income bracket than Republicans, we might come up with the two possible alternative tables (tables 5.2 and 5.3).

TABLE 5.2 More extreme table in the same "direction" as the data: 1

	Democrat	Republican	Total
High income	1	10	11
Low income	11	3	14
Total	12	13	25

TABLE 5.3 More extreme table in the same "direction" as the data: 2

	Democrat	Republican	Total
High income	0	11	11
Low income	12	2	14
Total	12	13	25

But, as usual, there should be another tail of the distribution with data as extreme as ours or more so—but in the other direction. Where is the tail in that direction?

We might consider the other direction to be either the one we get by reversing the numbers for Democrats and Republicans (table 5.4) or the one we get by reversing the high- and low-income brackets (table 5.5). The problem, apart from the fact that the answers are different, is that neither table has the same row and column totals as our original data, so they don't belong to the same universe of possible samples. Remember, the row and column numbers are supposed to be fixed; otherwise, the "exact" probability isn't.

We might focus on the *difference* between the numbers of Republicans and Democrats in the high-income group. Table 5.1 has a difference of 7, so we could imagine reversing that and finding the table with a difference of −7. That forces the table to appear as shown in table 5.6. That, at least, is an allowable data table, but the problem is that we will get a different table if we apply the same process to the low-income group, switching the difference from −6 to 6 (table 5.7).

So which is it?

TABLE 5.4 The other "direction": switching Democrat and Republican?

	Democrat	Republican	Total
High income	9	2	11
Low income	4	10	14
Total	13	12	25

TABLE 5.5 The other "direction": switching high and low income?

	Democrat	Republican	Total
High income	10	4	14
Low income	2	9	11
Total	12	13	25

TABLE 5.6 Reversing Democrat and Republican in high income group

	Democrat	Republican	Total
High income	9	2	11
Low income	3	11	14
Total	12	13	25

TABLE 5.7 Reversing Democrat and Republican in low income group

	Democrat	Republican	Total
High income	8	3	11
Low income	4	10	14
Total	12	13	25

TABLE 5.8 **Example survey data including third category in each variable**

	Democrat	Republican	Unaffiliated	Total
High income	2	6	2	10
Middle income	4	1	0	5
Low income	6	3	1	10
Total	12	10	3	25

Things really start to get out of hand if we include a *third* category for one or both of the variables—say a middle-income bracket and a nonanswer for political affiliation (table 5.8).

Now what are the more "extreme" tables? What direction does our data even have? There are some canonical ad hoc strategies for this kind of problem,[40] but none of them is provably correct for the underlying reason that any statistic we use to express our null hypothesis (that the income proportions are independent between parties) and its implicit alternative (that they are not independent) must have more than one *dimension*, so there's nothing like a "tail" of a distribution we can identify. Independence among a set of variables is a *groupwise* property that requires all the data to be considered simultaneously.

This situation is very common—many experiments involve gathering categorical data for a set of factors and then hypothesizing that they are not independent—and yet the standard statistical methods are mired in controversy precisely because there is no unifying logic of orthodox statistics to refer back to in order to settle the debate. If you were doing this research on your own, you would not be able to derive an approach yourself and instead would need to trust the advice of a statistician. A minor difference in the calculation could seem unimportant, but one choice might lead to the result being significant while the other wouldn't. Also, allowing this kind of imprecision to slide would surely be a double standard. If the lack of a perfectly specified prior for the Bertrand paradox problem is sufficient grounds to reject all of Bayesian inference, isn't the lack of a perfectly specified tail region for Fisher's exact test a reason to reject all of significance testing?

The Bayesian answer for this problem is very straightforward, even for the 3×3 example: let (h_D, m_D, l_D) be the proportions of high-, middle-, and low-income individuals among Democrats in the population, and let (h_R, m_R, l_R) and

(h_U, m_U, l_U) represent these same proportions among Republicans and unaffiliated persons, respectively. We can say the probability of observing the values given in the table 5.8 is proportional to

$$h_D^2 m_D^4 l_D^6 \cdot h_R^6 m_R^1 l_R^3 \cdot h_U^2 m_U^0 l_U^1$$

We also need to assume a prior distribution for these unknowns; say we give each triple the independent uniform prior distribution representing a state of information where all we know is that the proportions add up to 1. Multiplying the sampling probability by the prior probability gives the pathway probabilities for the data—and thus the posterior probability distributions for the unknowns as the relative proportions, the same as always. From this posterior distribution, we could, in principle, ask any questions we like about the various income proportions in the population. It would be silly to ask whether income and political affiliation are *actually* independent. This would correspond to the statements $h_D = h_R = h_U$ and so on, which, since these parameters have continuous probability distributions, will all have probability 0. As usual, the null hypothesis is something we know almost certainly to be false. Rejecting it is trivial and boring. Instead, we might ask for the probability that a greater proportion of Republicans than Democrats are in the high-income bracket (answer: about 97 percent), or the probability that the proportion of Democrats in the low-income bracket is greater than the proportion of Republicans in the low- and middle-income brackets combined (answer: about 51 percent), or the probability that the proportion of unaffiliated people in the high-income bracket is closer to the corresponding proportion of Democrats than that of Republicans (answer: about 29 percent), or any other such question. The joint probability distribution of the parameters represents our full state of knowledge about them after having made the observations, so we're free to investigate the consequences however we choose.

8. The Lucky Experimenter: What Features of Our Data Are Relevant for Our Inferences?

A situation similar to that in the problem of optional stopping arises anytime some feature of the data, say the sample size, is assumed to be somehow contingent on or determined by chance. Imagine we are doing a population survey and we are trying to establish the ratio of people in some community who have a certain trait. But for some reason that has nothing at all to do with the prevalence of the trait we care about, we can't guarantee ahead of time how many participants we will have. Say we're surveying them as they walk through a building with two possible exits, door A and door B, which we assume they choose at random independently from one another, but the nature of our experiment requires that it end as soon as *anyone* leaves through door B. The assumption we then are forced to make is that our sample size itself is a variable in the experiment; the probability of getting N subjects for our survey is $(1/2)^N$.

Now imagine we go through all the trouble to set things up and then a miracle occurs: the first 19 people through the building exit through door A and then the 20th person goes through door B, something we would expect to happen only once every 2^{20} times, or about 1 in 1 million! Upon conducting our interviews with them, we find that 17 out of the 20 people have the trait in question. What can we conclude about the ratio in the population? Can we say it's something like 17/20 with any confidence?

Well, considering any possible population ratio f and including the fact that our sample size *itself* is variable, we'd compute the probability of getting a sample of size N including k people with the trait and $(N - k)$ without it as

$$P[\text{observing } k \text{ people with the trait out of } N] = \left(\frac{1}{2}\right)^N \binom{N}{k} f^k (1 - f)^{N-k}$$

That is, the probability is our familiar binomial distribution but multiplied by the factor $(1/2)^N$ to reflect the chance of getting a sample of size N in the first place.

That factor of $(1/2)^N$ creates a problem in the frequentist framework, though. What it effectively means is that we have to allow for the fact that if the experiment is repeated some large number of times, half the time our data will consist of a *single* sample because the very first person will walk through door B. So, if we take, say, the sample ratio k/N as an estimator of the population ratio, we see that its sampling distribution has a lot of probability (totaling at least 0.5) concentrated at the points 0 and 1, corresponding to the times when we survey just 1 person and they either do or don't have the trait. This means that for *any* null hypothesis—say that the population ratio is 50 percent—we will never be able to identify tail regions for us to reject the hypothesis at anything less than the 25 percent significance level.

The distribution of this test statistic turns out to have quite a funny shape, owing to the probabilities of getting samples of various—mostly small—sizes. Under the hypothesis that the population ratio is 50 percent, a histogram of 10,000 simulations of the experiment produces the distribution shown in figure 5.9. Where to set the tail rejection region for this distribution is anyone's guess.

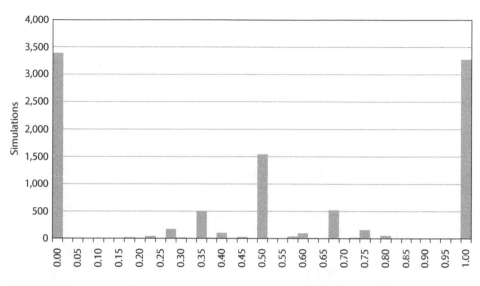

FIGURE 5.9

Pin the tail on the distribution.

So even though we were incredibly lucky this one time and got what looks like a fairly large sample, the procedure requires us to consider what *could* have happened *other* times, and in total honesty, we have to imagine we usually wouldn't have such a large sample. As a result, we likely can't conclude anything of importance from our survey, and our miracle is wasted.

Again, the Bayesian inference procedure isn't given to such hallucinations. Considering the hypotheses H_f that the population ratio is f, we will still agree with the probability of getting this data computed using the preceding formula:

$$P[D \mid H_f \text{ and } X] = \left(\frac{1}{2}\right)^N \binom{N}{k} f^k (1-f)^{N-k}$$

But after assigning prior probabilities to each H_f, we compute a pathway probability for this given data set with the same $(1/2)^N$ factor in front for every f. Calculating the posterior probabilities as relative proportions causes this factor to cancel out, so we're actually left with the same posterior inference we would have gotten if we had assumed that the sample size of 20 was a fixed feature of the data instead of a variable.

The point is this: while the sample size in this problem may be a property of the data that is not certain—so it gets a probability—its relationship to the various hypotheses is the *same*. It has the same inferential content, so the Bayesian inference procedure causes it to drop out. The Bayesian program automatically identifies those features of the data that are actually *relevant* to making inferences about the thing we care about, conditional on what we have actually observed, without having to worry about what could have happened but did not.

9. Finding a Lost Spinning Robot: What If No Statistic Is Sufficient?

Examples like the last one, in which some random feature of the data unrelated to the inferential question at hand seems to get in the way of significance testing, were known to Fisher. His general advice for all problems of parameter estimation was to look for what are called *sufficient statistics* in the data. These are statistics, meaning functions of the data, that capture all the useful information that could affect an inference about a given parameter. To bring this back to urn drawing, we already argued that although we have the full data recording the results of drawing pebbles from the urn in a particular order, it is fine if we disregard all the information apart from the *sample ratio* of white pebbles in our sample. The probability we assign to the data depends only on this ratio anyway. This is an example of a sufficient statistic; we would say that the sample ratio is sufficient for the parameter F, the true fraction of white pebbles in the urn.

Fisher was the first to define the notion of a sufficient statistic, and he found examples for different types of probability models. He and others were able to prove that sufficient statistics had some appealing general properties. The most important among these was guaranteed by the Rao-Blackwell theorem, developed separately around 1945–1947 by Calyampudi Radhakrishna Rao and David Blackwell. The theorem says that any choice of estimator for a given model parameter can be improved on, in the frequentist sense of reducing the mean squared error of its sampling distribution, by conditioning on the value of a sufficient statistic. So, essentially, sufficient statistics, when they exist, are a source of good estimators and could even make other estimators better.

The main example motivating Fisher was that for a normal distribution— say with a known standard deviation σ but an unknown mean μ—the average of the data is a sufficient statistic for μ. This can be seen right away from the algebra of the exponential function in the normal distribution, since it converts the product in independent data probabilities into a sum. In a sense, this provides a justification for the practice of averaging together data when what people really care about is the population mean. If the population distribution is normal (which was almost always Fisher's assumption), this sample average captures everything about the data that could be relevant to estimating that mean. Similar arguments apply to the population standard deviation, etc. So instead of carrying around the whole data set, one can get by with a few summary descriptive statistics, which can be a relief to any scientist. As Fisher wrote in *Statistical Methods for Research Workers*: "No human mind is capable of grasping in its entirety the meaning of any considerable quantity of numerical data. We want to be able to express all the relevant information contained in the mass by means of comparatively few numerical values. This is a purely practical need which the science of statistics is able to some extent to meet."[41]

The flip side to a sufficient statistic is what Fisher called an *ancillary statistic*. This is an aspect of the data, like our sample size in the last example, that doesn't depend *at all* on the model parameter we care about. Sufficient statistics show everything about the data that's relevant for inference, and ancillary statistics capture everything else. It can happen sometimes that a particular statistic we'd like to use for estimation (say our in-sample ratio) is actually not sufficient for the parameter we care about, but that once we condition on the value of an ancillary statistic, it is.

That's what was going on in our last example: the in-sample ratio is not a sufficient statistic for the population ratio in that problem because two different data sets with the same in-sample ratio can have *different* probabilities in relation to the parameter, depending on how large the sample was—say a sample of size 2 with 1 success and 1 failure versus a sample of size 20 with 10 successes and 10 failures. Each has a sample ratio of 1/2, but under an assumption that the population frequency is 1/2, the latter is much less probable because it requires getting exactly 10 out of 20 survey respondents with the trait. If the sample size is *fixed*, however, we are back to ordinary urn testing, and the sample ratio is sufficient. When this kind of thing happens, the ancillary statistic is called an *ancillary complement*.

Fisher's answer to our preceding example would be that, for purposes of frequentist inference, we are allowed to *condition* on the value of the ancillary complement, meaning the sample size, and consider only data that share this common value. So we need to consider only the other samples that also lucked out and got 20 respondents, and we will be able to draw strong conclusions from our data. Why? Because it simply feels right that something *unrelated* to the parameter shouldn't actually count against the accuracy we can claim for estimating it. Jaynes called this kind of on-the-fly patch to the orthodox methods an "ad hockery,"[42] to point out that though it may allow us to arrive at the conclusions we hoped, it is not derived from any underlying mathematical reasoning and is there only because the unpatched frequentist theory gave the wrong answer.

But what happens when there isn't a sufficient statistic at all? This happens if, for example, instead of being normal, the population distribution is described by a similar bell-curve shape with a different mathematical function: the curve known since at least the time of Pierre de Fermat and variously called the (standard) Cauchy distribution, the Lorentz distribution, and the witch of Agnesi:

$$f(x) = \frac{1}{\pi[1 + (x - \mu)^2]}$$

The curve has almost the same shape as the normal distribution but with a slightly sharper peak and slightly fatter tails. The parameter μ moves the peak of the distribution left and right, but annoyingly it cannot be thought of as the mean because, as it turns out, the mean of the Cauchy distribution is *undefined*. That is, the infinite integral $\int_{-\infty}^{\infty} x \cdot f(x)\, dx$ diverges. For this reason, the Cauchy distribution is often treated as a pathological counterexample, where theorems go to die.

Among the other annoying properties of the Cauchy distribution is the fact that Fisher's method of averaging samples to get an estimate of the population parameter *won't work* for the Cauchy distribution. It turns out, quite hilariously, that the sampling distribution of the sample mean of n data points from this distribution is exactly the same as the distribution for any *one* of them! In other words, by averaging we achieve nothing except for effectively throwing away all but one of our samples.

How, then, to go about estimating the parameters of a Cauchy distribution presents a difficult problem for the frequentist school. Again, the only approach recognized in this way of thinking is to first choose an *estimator*, meaning a function of the data, and then judge its quality based on how tightly concentrated its distribution is around the true parameter value. What estimator should we pick if not the sample mean? Fisher, through trial and error, found that the sample *median* tended to be more concentrated around the true parameter value than the sample mean was, but nothing suggested this was the best possible choice at all. Computing maximum likelihood estimates is, in general, out of the question because it involves solving a difficult degree-n polynomial equation.

What's more, for data sampled from the Cauchy distribution, there is no sufficient statistic. Writing out the probability for getting values $x_1, \ldots, x_n$ as independent samples, as we did before for the normal distribution, gives us a product without the friendly properties of the exponential to help us rewrite this in terms of a summary statistic in $x_1, \ldots, x_n$. Basically we're stuck. Without a clear estimator and with no sufficient statistic[43] with which to improve an estimator once chosen, the frequentist methods are at a loss for what to do next.

Because of these and the other pathological properties mentioned earlier, many statistics authors over the years have simply ignored the Cauchy distribution altogether or claimed that it has no practical use. For example, the only reason it's called the Cauchy distribution at all is that, in the course of a public feud with French mathematician (and Laplace disciple) Irénée-Jules Bienaymé around 1853, Augustin Cauchy brought it up as an example of how Laplace's method of least squares breaks down if the error distribution isn't assumed to be normal. Bienaymé dismissed the example contemptuously: "One should recognize that an instrument affected by a probability law similar to this would not be put up for sale by an ordinary artisan. One would not know what name to give to an establishment that would construct it."[44]

However, it does have some uses. It's actually not too hard to conjure examples where the Cauchy distribution could come up. It has the fairly simple property of being the distribution of the tangent function applied to a uniformly distributed angle.[45] So this means the Cauchy distribution might arise quite naturally in any problem to do with data collected from something *rotating* at uniform speed.

For example, suppose we're on a team doing undersea geological research using autonomous underwater robots. One of our robots was last seen exploring a deep linear trench along the seafloor, where it apparently got stuck. The robot has a directional antenna at the end of one of its arms and is programmed to spin the antenna around and send out a distress signal whenever the robot gets stuck. However, the communications system is malfunctioning, so this signal gets sent out only in very short bursts at random intervals. About half the time the antenna is pointed down at the seafloor; the other times it is oriented up toward a straight line along the surface of the water, where we have positioned buoys with sensors to detect the signals. Our data, then, consists of the positions of some number of these confirmed contacts from sensors at various places along this line. We'd like to send another vehicle down to retrieve the robot, but the conditions are hazardous, and it will be expensive if we pick the wrong spot to begin our search. How do we determine the best place to start looking?

It turns out that to be consistent with the information given, we should assign the Cauchy distribution to each observation. Assuming independence in the positions puts us in the situation we were just considering, where we have *n* independent samples from a Cauchy distribution but with an unknown location parameter μ, the location of the robot. Orthodox statistical methods are pretty much helpless here because no canonical best choice of estimator exists given the lack of a sufficient statistic. Various choices have been tried on an entirely ad hoc basis over the years. For example, a 1964 paper recommended using the average of the middle 24 percent (!) of the samples as an estimator and showed that it was reasonably efficient.[46]

Bayesian inference, on the other hand, handles the problem easily. All we need to do is assign a prior probability distribution for the parameter μ—say a uniform distribution on some wide range of possible locations—and then process the data probability through our Bayesian inference, the same as ever. What we find is that even with a sample of size *n* = 20, we are quickly able to narrow in on a most probable location for the robot. We can even give a range of places it's likely to be with some level of certainty if that is more useful.

Figure 5.10 illustrates one such inference. I've created a sample of 20 points from a Cauchy distribution centered at 0 and then used the Bayesian method with a uniform prior distribution for μ. The curve shows the posterior distribution for μ—this is the state of our uncertainty about the robot's location given the data—and we can see that it is fairly tightly concentrated around the correct

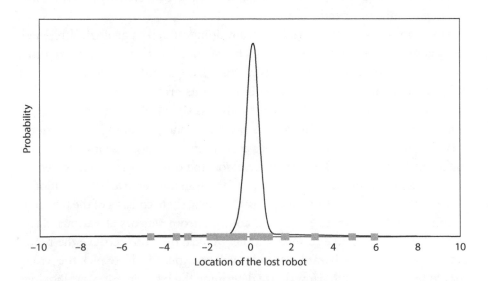

FIGURE 5.10

Posterior distribution for μ: the state of our uncertainty about the robot's location given the data.

value of $\mu = 0$ even though the data (shown along the x-axis) varies widely over the range from −4 to 6.

Since the frequentists generally have to invent a new method from scratch every time, there's no guarantee that they'll find a good estimator simply through pure creativity. The Bayesian approach is the same every time, whether a sufficient statistic exists or not. If a sufficient statistic *does* exist, the procedure finds it automatically.

LIFE OUTSIDE THE SMALL DOMAIN

We have deliberately misused the frequentist statistical methods in these examples to prove a point. The frequentist methods are confined to certain types of problems with certain distributional assumptions and an implicit assumption of no strong prior information, so the orthodox statistician's response to any one of these examples is likely to be that we have applied the wrong test for that kind of problem. But what we have illustrated is that there is no *unifying logic* to the frequentist methods. Taking an example where things seem to work out nicely, such as Bernoulli's urn-drawing inferences, and trying to abstract the reasoning process even a little allows us to construct examples where following the same thought process evidently leads to some terrible conclusions. So what, then, is a working scientist to do when faced with a problem that does not exactly match the template of one of the known frequentist examples? It seems the answer is always that they should consult a professional statistician, who may or may not be able to develop some new method for them on the fly.

With the Bayesian approach, there *is* a unifying logic to all probabilistic inference. Figure 5.11, our version of the flowchart in figure 5.6, is very boring.

Most importantly, the probability statements in Bayesian inference are oriented in the correct direction: given some *data*, what probability do we assign to a *hypothesis*? That question is meaningful to us in ways that those in the frequentist school cannot allow, so they are forever stuck with procedures based only on sampling probabilities of data, committing Bernoulli's Fallacy every time. In reality, we are concerned only with what Jaynes called the *postdata* questions:[47]

- What data did we observe? How surprising was it?
- What alternative hypotheses might explain this data better?
- Given this data, what reasonable inferences can we draw? What conclusions are best supported by our observations?

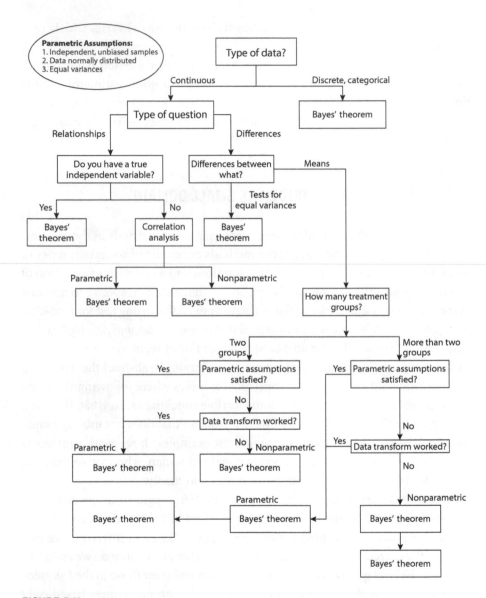

FIGURE 5.11

Flowchart for selecting commonly used statistical tests (Bayesian version).

We ask not "If the patient isn't sick, how likely is the test to come back positive?" but "If the test comes back positive, how likely is it that the patient isn't sick?" Not "If the parent didn't murder them, how likely would it be for two infants to die?" but "If two infants died, how likely is it that their parent didn't murder them?" Not "If the object has a mass of 1 gram, how likely is it that the measurement will be 100,001 grams?" Not "If a die is fair (or somehow rigged), how likely is a series of rolls?" Not "If there are 1,000 German tanks, how likely is it that we'd collect parts from among the first 313 of them?" Not "If my experimental hypothesis is correct, how often will I observe data more extreme than what I got?" Not "How often would I get a sample as large as I did?" Not "How would I, as a general procedure, transform other data into an estimate of the thing I actually care about?" No, no, no. What we care about, what we actually have to work with, what we are forced to condition all probabilities on, forever and always, is our actual blessed data.

No professional hand-holding is necessary for Bayesian inference, outside of perhaps advising on what sorts of probability models might apply to certain data-gathering procedures and what distributions would represent various states of knowledge. Once those are determined, the inferential process is automatic and simply consists of using Bayes' theorem, as Jaynes and Richard Cox showed was required by the rules of consistent logical reasoning when expressed in probabilities. Because the frequentist methods sometimes violate these rules, they will necessarily sometimes produce illogical results.

The ad hoc frequentist methods may seem to work in some "small domains" but only under very particular circumstances. With our knowledge of probability as logic, we can even be exact about what those circumstances are. As we showed in Bernoulli's example, when prior information is weak (meaning it has a roughly uniform probability distribution) and sampling distributions are well behaved, the Bayesian procedure may *reproduce* the frequentist results. It's just that it does so with a proper understanding of the conclusions it has drawn and the assumptions on which they are based. So, with one tool, we can handle these basic examples as well as the pathological ones, plus anything in between.

But where do the problems facing working scientists actually fall on this spectrum? Is all of science actually confined to one of these well-behaved frequentist domains after all? As I'll argue in the next chapter, there is reason to believe the majority of the published research findings in various disciplines of science that rely on standard frequentist statistical methods cannot be replicated. This is evidence that the world of science has already strayed outside the comfort zone of statistical orthodoxy, and it is time to start using techniques appropriate for life outside the small domain.

6

THE REPLICATION CRISIS/OPPORTUNITY

When we reach a point where our statistical procedures are substitutes instead of aids to thought, and we are led to absurdities, then we must return to the common sense basis.

~David Bakan (1966)

The techniques of statistics created by Ronald Fisher and by Jerzy Neyman and Egon Pearson in the 20th century have always been controversial. Much of that controversy was self-generated and amounted to picayune disagreements between rival camps about the finer details of a fundamentally broken idea. But there have also been outsiders who rejected the whole of frequentist statistical methods for as long as those methods have been around. For example, the public feud we saw in the 1930s between Fisher and Harold Jeffreys took place almost simultaneously with Fisher's ideas coming to prominence.

Criticism of significance testing—in particular, null-hypothesis significance testing (NHST)—has an especially long and rich history. In this chapter, we'll tell the story of only some of that criticism, primarily in an attempt to understand why it never succeeded in toppling the establishment. Sadly this is mostly a story of widespread apathy and inertia on the part of the scientific community. By and large, researchers indoctrinated in statistical orthodoxy have simply ignored the warning sirens for decades because the flaws, whatever they might have been in theory, never seemed to show up in practice. So long as it didn't interfere too much with the real business of science, people were content to jump through whatever statistical hoops it took to get published.

But that's changing. Something is happening in the world of science now that may finally cause people to take the warnings seriously. Across the many disciplines where significance testing has dominated, roughly half of the research findings established over the past century—hundreds of thousands of results—might simply be untrue, and the rest might be much less important than previously thought. This realization is dawning at a time when our species and planet need science more than ever, even as it's taking fire in cultural and political conflict on multiple fronts.

Statistical "significance" has been exposed as nothing of the sort. The shameful secret that there is a logical fallacy at the heart of statistical practice may at last be coming to light. Science and statistics are overdue for a reckoning.

THE RESISTANCE

Soon after the hybridization of orthodox statistics in the 1940s and '50s, the frequentist methods caught on as standard practice—particularly in the fields of medicine and social science—and a few prominent authors responded critically. For example, in 1942, Dr. Joseph Berkson at the Mayo Clinic wrote in the *Journal of the American Statistical Association* that the logic of significance testing was flawed because it always interpreted unlikely observations as evidence against a hypothesis. He argued that this could make sense only if there was a cogent alternative:

> There is no logical warrant for considering an event known to occur in a given hypothesis, even if infrequently, as disproving the hypothesis. . . . Suppose I said, "Albinos are very rare in human populations, only one in fifty thousand. Therefore, if you have taken a random sample of 100 from a population and found in it an albino, the population is not human." This is a similar argument but if it were given, I believe the rational retort would be, "If the population is not human, what is it?"[1]

William Rozeboom, professor of psychology at St. Olaf College (and later at the University of Alberta), wrote an article in the *Psychological Bulletin* in 1960 titled "The Fallacy of the Null-Hypothesis Significance Test." In it, he argued significance tests had the probabilities backwards and the true logic of scientific inference was "inverse probability":

> In brief, what is being argued is that the scientist, whose task is not to prescribe actions but to establish rational beliefs upon which to base them, is fundamentally and inescapably committed to an explicit concern with the problem of inverse probability. What he wants to know is how plausible are his hypotheses,

and he is interested in the probability ascribed by a hypothesis to an observed experimental outcome only to the extent he is able to reason backwards to the likelihood of the hypothesis, given this outcome.[2]

Writing in 1966, David Bakan of the University of Chicago's Department of Psychology referred to the logical fallacies and misinterpretations of significance testing as something "everybody knows" but nobody would admit out loud, as in the story of the emperor's new clothes. By that point, though, significance testing had already become entrenched as the industry standard. Bakan suggested the reason people had been reluctant to think critically about the statistical methods was that doing so would pose an existential threat to their disciplines: "The test of significance is profoundly interwoven with other strands of the psychological research enterprise in such a way that it constitutes a critical part of the total cultural-scientific tapestry. To pull out the strand of the test of significance would seem to make the whole tapestry fall apart."[3]

Paul Meehl, also a professor of psychology and founder of the Minnesota Center for the Philosophy of Science, wrote an article in the journal *Philosophy of Science* in 1967 lamenting that significance testing had by then become something all researchers did almost mechanically, and he attributed an overall decline in the quality of research to the new statistical methods:

> Meanwhile our eager-beaver researcher, undismayed by logic-of-science considerations and relying blissfully on the "exactitude" of modern statistical hypothesis-testing, has produced a long publication list and been promoted to a full professorship. In terms of his contribution to the enduring body of psychological knowledge, he has done hardly anything. His true position is that of a potent-but-sterile intellectual rake.[4]

Meehl agreed with Rozeboom in expressing support for Bayesian methods as an alternative, referring to the debate "between the Bayesians and the Fisherians" as ongoing. Later his criticism of Fisher became more openly hostile. In 1978, he wrote that "Sir Ronald [Fisher] has befuddled us, mesmerized us, and led us down the primrose path" and that "the almost universal reliance on merely refuting the null hypothesis as the standard method for corroborating substantive theories in the soft areas is a terrible mistake, is basically unsound, poor scientific strategy, and one of the worst things that ever happened in the history of psychology."[5]

In 1972, the sociologist Stanislav Andreski at the University of Reading wrote the book *Social Sciences as Sorcery*. Within a much broader criticism of social science as a whole, he directed particular attention to the rise of quantitative methods. He claimed their popularity was an artifact of the insecurity social scientists felt about the softness of their disciplines, especially compared to natural sciences, and the need to justify their conclusions in a manner that seemed objective. The effect, he said, was either a "quantitative camouflage" obscuring flawed reasoning or unnecessary quantitative support for conclusions that were obvious:

> In nearly all instances, it is the case of a mountain giving birth to a mouse, as when, after wading through mounds of tables and formulae, we come to the general finding (expressed, of course, in the most abstruse manner possible) that people enjoy being in the centre of attention, or that they are influenced by those with whom they associate . . . which I can well believe, as my grandmother told me that many times when I was a child.[6]

For the first few decades, then, objections to orthodox statistical methods were largely ideological, based mostly on theoretical examples and logical criticism. By the 1970s and '80s, some had tried pushing back on these methods in practice. Tony Greenwald, editor of the *Journal of Personality and Social Psychology*, attempted, with limited success, to change the standards of reporting statistics in the journal. John Campbell, in his outgoing comments as editor of the *Journal of Applied Psychology* in 1982, suggested NHST (a.k.a. testing by p-values) was by then too entrenched to be removed:

> It is almost impossible to drag authors away from their p values, and the more zeros after the decimal point, the harder people cling to them. . . . Perhaps p values are like mosquitos. They have an evolutionary niche somewhere and no amount of scratching, swatting, or spraying will dislodge them. Whereas it may be necessary to discount a sampling error explanation for results of a study, investigators must learn to argue for the significance of their results without reference to inferential statistics.[7]

In 1986, Kenneth Rothman, editor of the *American Journal of Public Health*, briefly banned submissions using p-values, though the ban was lifted when he stepped down two years later. In 1994, the American psychologist and statistician Jacob Cohen wrote a scathing critique of NHST in *American Psychologist* called "The Earth Is Round ($p < .05$)." (He said he resisted referring to the method as statistical hypothesis inference testing, which would have conferred on it a more appropriate acronym.) His main theoretical point was, again, that the logic of significance testing had the probabilities backwards:

> What's wrong with NHST? Well, among many other things it does not tell us what we want to know, and we so much want to know what we want to know that, out of desperation, we nevertheless believe that it does! What we want to know is "Given these data what is the probability that H_0 is true?" But as most of us know, what it tells us is "Given that H_0 is true, what is the probability of these (or more extreme) data?" These are not the same.[8]

Cohen elaborated on why NHST's implicit appeal to something like a probabilistic version of the reductio ad absurdum argument did not hold up logically (as we saw in chapter 2). Ruma Falk and Charles Greenbaum called this the "illusion of probabilistic proof by contradiction" or the "illusion of attaining improbability,"[9] Persi Diaconis and David Freedman called it "the fallacy of the transposed conditional,"[10] and Gerd Gigerenzer called it the "Bayesian Id's wishful thinking."[11] Looking back in 2012 on the history of this debate, Charles Lambdin cited some 48 references criticizing significance testing between 1919 and 2007, including two attributed to Fisher.[12]

So why weren't the critics more successful at dislodging these methods? At least a good portion of the answer is given in Andreski's argument: statistical methods gave researchers in the "softer" sciences a feeling of objectivity, which they desperately desired as a way to lend quantitative legitimacy to their work. As we saw in chapter 3, that need had been felt in the discipline since the days of Adolphe Quetelet's "social physics" and had been a key motivation throughout the development of statistics in the 19th and 20th centuries. Objectivity was what frequentism promised, and it found hungry consumers in the worlds of social science.

Orthodox statistics also benefited enormously from the strong feedback loop in these academic disciplines between research practice and education. As George Cobb, professor emeritus of mathematics and statistics at Mount Holyoke College, put it: "Q: Why do so many colleges and grad schools teach $p = 0.05$? A: Because that's still what the scientific community and journal editors use. Q: Why do so many people still use $p = 0.05$? A: Because that's what they were taught in college or grad school."[13] The same argument could apply to all of the frequentist statistical methods. The standard methods were the standard ones because they were. (One might imagine sociologists would be trained to identify exactly this sort of phenomenon—but maybe not within their own ranks.)

The feedback loop was strengthened by the voice of authority. Significance testing had been sanctioned by eminent mathematical minds like Fisher, Neyman, and both Pearsons (Karl as well as Egon), so who was a psychologist, sociologist, political scientist, anthropologist, or anyone else with minimal mathematical training to question them? The reassuring advice these researchers got from such authorities was that all one had to do to keep from committing the fallacy of misinterpreting p-values was to not misinterpret them! As long as people were disciplined about constantly reminding themselves they were not *actually* computing the probability of the null hypothesis, but rather just following a procedure for accepting or rejecting it based on *significance*, they could rest assured no logical fallacies were being committed. Never mind the fact that the whole procedure had been founded on a fallacy in the first place.

But another key factor must have been that the methods of inference seemed to *work*. As we saw in Jacob Bernoulli's urn-testing inferences in the last chapter, the orthodox methods of significance testing sometimes seemed to give pretty reasonable answers that "felt right." Of course, we now know the reason they felt right is that they aligned with the Bayesian methods given weak prior information, but this took a long time to be fully understood. So even if the logic seemed flawed in some abstract way, for simple examples it didn't seem to matter because the conclusions came out sensibly anyway.

In a 2001 article for the *Journal of Statistical Planning and Inference* titled "Why *Should* Clinicians Care About Bayesian Methods?" Robert Matthews of Aston University suggested NHST was the standard for clinical and epidemiological research precisely *because* it had been hardened by all the criticism over the years while no terrible consequences befell researchers who continued to use it. In order to justify abandoning significance testing and investing in learning all-new Bayesian techniques, he said, the costs to people of continuing on as they had been would need to be great enough that it was worth the trouble. He wrote, "If the standard methods of inference are so awful, how come the whole scientific enterprise has not collapsed around our ears?"[14] (In his defense, he then went on to make a strong case for Bayesian methods based on the "benefit" side of the cost-benefit analysis.)

Imagine the situation facing a pragmatic scientist who had been educated in the use of orthodox statistics and who was uninterested in following the infighting over issues of mathematical logic. When they hear there's an argument and one side claims all of statistics has been done backwards for the last several decades, it's understandable that the scientist might ask, "Where, then, are the catastrophic effects of all that bad statistical reasoning?" If the gatekeepers of their career all insist on using orthodox techniques and those techniques appear to work, what does the scientist care about some esoteric issue of mathematical logic? It's easy to see how that scientist could view the critics of orthodox statistics as playing the role of Chicken Little, warning everyone that the sky is falling, while the bulk of academia wisely ignores these alarms and continues using the methods that had been proven to work.

THE SKY IS FALLING ($P < 0.05$)

The theoretical problems with significance testing started to become all too practical as scientific research changed in the early to mid-2000s. For one thing, the amount of data available to researchers exploded. The rapidly increasing capacity of digital storage and the ability to share data or conduct studies online enabled entirely new types of data analysis. *Big data*, *data mining*, and *machine learning* all became household phrases during that time. Roger Peng, professor of biostatistics at Johns Hopkins University, summed up how much had changed in the availability of data for research purposes by comparing Stanley Milgram's "six degrees of separation" experiment in 1967 with its modern equivalent:

> In 1967 Stanley Milgram did an experiment to determine the number of degrees of separation between two people in the USA. In his experiment he sent 296 letters to people in Omaha, Nebraska, and Wichita, Kansas, and the goal was to get the letters to a specific person in Boston, Massachusetts. His experiment gave us the notion of "six degrees of separation". A 2007 study updated that number to "seven degrees of separation"—except the newer study was based on 30 billion instant messaging conversations collected over 30 days.[15]

There was a sudden surge in analytical research during the decade to find useful correlations within massive data sets in diverse industries from finance to marketing, actuarial science, telecommunications, health care, pharmacology, and many others. In medicine, in particular, the completion of the Human Genome Project in 2003 gave researchers a vast new territory in which to explore the potentially millions of genomic influences on any number of diseases or other conditions.

Around that time was also when these same researchers started to notice problems with their ability to replicate results. In a survey article in *Lancet* in 2003, Helen Colhoun, Paul McKeigue, and George Davey Smith identified the growing problem that "inability to replicate many results has led to increasing scepticism about the value of simple association study designs for detection of genetic variants contributing to common complex traits." They argued that despite the many advantages the new post–genome era had brought them, the explosion in possible relationships to explore and the number of people doing the exploration had led to many spurious associations being found just by chance:

> We suggest that failure to exclude chance is the most likely explanation for dif-
> ficulty in replication of reports of genetic associations with complex diseases.
> For most diseases of interest, hundreds of known genes are possible candidates,
> and in most of these genes, dozens of polymorphisms are known or can be eas-
> ily identified by screening of the gene. Around the world, thousands of such
> polymorphisms are tested for disease associations every week. Even if none of
> these genotypes is associated with outcome, we can expect many associations
> that are significant at 5 percent or less to arise frequently by chance alone.[16]

The first real bombshell, though, came in 2005, when John Ioannidis, a professor at Stanford University's School of Medicine and its Department of Statistics, laid the replication problem at the feet of the orthodox statistical methods, primarily NHST. In an article titled "Why Most Published Research Findings Are False," he showed in a straightforward Bayesian argument that if a relationship, such as an association between a gene and the occurrence of a disease, had a low *prior* probability, then even after passing a test for statistical significance, it could still have a low posterior probability of being true.[17]

For example, he considered a test of genetic associations with schizophrenia. Prior experience with the heritability of the disease might suggest that about 10 possible gene variants out of 100,000 possibilities would be truly associated with schizophrenia to a given degree. Therefore, the prior probability assigned to any one of the possible theories should be about 10/100,000, or 0.0001. A typical test would use a significance level of 5 percent and, for an effect of this size, might have a power of 60 percent to detect it, meaning only a 60 percent chance of obtaining a significant result even if the effect were real. The result of putting these numbers into an inference table is shown in table 6.1. The posterior probability for the effect was about 0.12 percent, meaning that even

TABLE 6.1 **Inference for an unlikely genetic association given a statistically significant result**

Hypothesis H	Prior probability $P[H \mid X]$	Sampling probability $P[D \mid H \text{ and } X]$	Pathway probability $P[H \mid X]P[D \mid H \text{ and } X]$	Relative proportion $P[H \mid D \text{ and } X]$
No real effect: null hypothesis	0.9999	0.05	0.049995	0.9988
Effect is real	0.0001	0.6	0.00006	0.0012

TABLE 6.2 **General inference given a statistically significant result**

Hypothesis H	Prior probability $P[H \mid X]$	Sampling probability $P[D \mid H \text{ and } X]$	Pathway probability $P[H \mid X]P[D \mid H \text{ and } X]$
No real effect: null hypothesis	$1 - p$	α: false positive	$(1 - p)\alpha$
Effect is real	p	$1 - \beta$: avoided false negative	$p(1 - \beta)$

with a statistically significant test result, the probability is 99.88 percent that the association isn't real.

In general, assuming a prior probability p for any theory and putting the assumed false positive rate (α) and the assumed false negative rate (β) in an inference table, we would have the results shown in table 6.2, with the observation D being "The observed effect is statistically significant."

So an effect that had passed the significance test would have less than a 50 percent chance of being true if the second pathway was less probable than the first. In terms of the quantities in table 6.2, this would happen if

$$p/(1 - p) < \alpha/(1 - \beta)$$

Since α was usually taken to be 5 percent for most significance tests and a typical test might have a false negative rate around 50 percent, this meant most published research findings would be false if the prior ratio of true to false effects was anything less than 10 percent. Given the enormous numbers of *possible* effects being studied in this "high-throughput discovery-oriented research," Ioannidis said it would be the norm that the prior probability for any given theory was much lower than this threshold. He further showed how this low posterior probability would be made even lower by the effects of bias due to selective reporting or researcher conflicts of interest.

Therefore, most published research findings were false.

Or were they all true?

TYPE III ERRORS: THE CRUD FACTOR

In his analysis of genetic association studies, Ioannidis essentially argued that we should expect a high percentage of Type I errors—rejecting a null hypothesis (of no effect) even though it's true. What his calculations showed was a clear difference between the Type I *error rate*, meaning the probability of getting a statistically significant result assuming the null hypothesis is true, and the true percentage of published results that are Type I errors. In symbols, it's the difference between $P[\text{Significance}|H_0]$ and $P[H_0|\text{Significance}]$. The latter reflects the probability of the null hypothesis given the data, which requires Bayesian analysis, including a prior probability for the hypothesis, to understand. It's the same as the difference between the percentage of healthy patients who test positive for a disease and the percentage of those testing positive who are actually healthy. To calculate the probability we actually want, we need to include the base rate—that is, the prior probability.

Ioannidis had done his example calculations by considering an idealized situation where all the possible genetic associations one could study were grouped into two discrete categories: those that were *real* associations, with some assumed magnitude of effect, say, on the incidence rate of the disease, and those that were *not real*, where the effect was 0. This makes the inference similar to our base rate neglect problems, where a condition such as a disease is either *present* or *not present*.

However, for many types of problems, there is more of a *spectrum* of possibilities. Between any two variables, there might be a tiny association, still real but not practically meaningful in any sense. Thinking in the Fisherian way about making inferences concerning a population, for example, it would be very unlikely for any given pair of variables to have exactly 0 association with each other if the whole population was considered.[18] Say, as in our example in the last chapter, we were doing a study of the relationship between political party affiliation and household income, and we're interested in rejecting the null hypothesis that these are independent, meaning, in particular, that the proportion of people in different income brackets doesn't vary according to their political affiliation. If we were able to survey literally *everyone*, we'd almost certainly find the proportions weren't exactly equal because a difference of even one person in any income bracket would break the equality. Since the only *randomness* in this kind of surveying problem comes from the sampling process, we could say with near certainty that if our sample size was large enough, we'd find some statistically significant effect, and we'd be correct in rejecting the null hypothesis!

Berkson originally theorized this problem back in 1938. He said it should be genuinely troubling to statisticians, since it meant that rejecting a null hypothesis based on a sample was, in principle, meaningless: "I suppose it would be agreed by statisticians that a large sample is always better than a small sample. If, then, we know in advance the P that will result from a large sample, there would seem to be no use in doing it on a smaller one. But since the result of the former test is known, it is no test at all."[19] In other words, collecting data samples is supposed to lead us toward the conclusion we *would* reach if, ideally, we had access to the whole population. But without doing any research at all, we know from the beginning that certain kinds of null hypotheses, when applied to the whole population, are almost surely false. So what's the point of doing the sample?

Bakan described the same phenomenon in 1966 and demonstrated it empirically by running statistical tests on data gathered from 60,000 people. He showed that no matter how he divided up his subjects—those east of the Mississippi versus those west of it, those in the north versus those in the south, those in Maine versus those in the rest of the United States, etc.—the tests for differences between the two groups always came out to be significant, with small p-values. In 1968, David Lykken of the University of Minnesota called this the "ambient noise level of correlations."[20] He and Meehl demonstrated it with an analysis of 57,000 questionnaires that had been filled out by Minnesota high school students. The survey included a wide range of questions about the students' families, leisure activities, attitudes toward school, extracurricular organizations, etc. The two found that, of the 105 possible cross-tabulations of variables, every single association was statistically significant, and 101 (96 percent) of them had p-values less than 0.000001. So, for example, birth order (oldest, youngest, middle, only child) was significantly associated with religious views and also with family attitudes toward college, interest in cooking, membership in farm youth clubs, occupational plans after school, and so on.

Meehl called this the "crud factor," meaning the general observation that "in psychology and sociology everything correlates with everything."[21]

But as Meehl emphasized, these were not simply results obtained purely by chance: "These relationships are not, I repeat, Type I errors. They are facts about the world, and with $N = 57,000$ they are pretty stable. Some are theoretically easy to explain, others more difficult, others completely baffling. The 'easy' ones have multiple explanations, sometimes competing, usually not. Drawing theories from a pot and associating them whimsically with variable pairs would yield an impressive batch of H_0-refuting 'confirmations.'"[22] That is, any one of these 105 findings could, according to standard practice, be wrapped in a theory and published in a journal.

As larger samples become easier to collect, these kinds of small-effect results can be expected more and more. For example, a 2013 study on more than 19,000 participants showed that people who met their spouses online tended to have higher reported rates of marital satisfaction than those who met in person, with a tiny p-value of 0.001. It sounds like an impressive, and very topical, result until you see that the observed difference was minuscule: an average "happiness score" of 5.64 versus 5.48 on a 7-point scale—that is, less than a 3 percent relative improvement.[23]

The relationship between sample size and statistical significance was a regular point of confusion and criticism early on, as it is to this day. Bakan described how journal editors in the 1960s frequently got mixed up when judging the quality of research papers:

> The author knows of instances in which editors of very reputable psychological journals have rejected papers in which the p values and n's were small on the grounds that there were not enough observations, clearly demonstrating that the same mode of thought is operating in them. Indeed, rejecting the null hypothesis with a small n is indicative of a strong deviation from null in the population, the mathematics of the test of significance having already taken into account the smallness of the sample. Increasing the n increases the probability of rejecting the null hypothesis; and in these studies rejected for small sample size, that task has already been accomplished. These editors are, of course, in some sense the ultimate "teachers" of the profession; and they have been teaching something which is patently wrong![24]

That is, low p-values seemed like a good thing, and large sample sizes seemed like a good thing, so to be *really* important, a result should contain *both*. But this meant by necessity that the size of the claimed effect was likely to be tiny because it otherwise wouldn't have taken so large a sample to find it at a given level of significance. So whatever theory the statistical analysis was meant to support in the first place could be entirely inconsequential.

If the arbiters of publishable research cared only about studies that rejected the null hypothesis with high power and a low chance of being a false positive, then a simple procedure to get published would be to always reject the null hypothesis! Since the hypothesis is always false anyway, you could claim you'd never commit a Type I error (rejecting the null even though it's true), and since you were always rejecting the null, there would be no chance of a Type II error (accepting the null even though it's false).

All of this confusion serves to underscore the point that hypothesis testing is meaningless without alternatives against which to test. When the hypothesis that a population correlation is exactly 0 or that a population proportion is exactly 1/2 is tested against its simple negation—that the correlation is not 0 or that the proportion is not 1/2—the null hypothesis will always lose if the amount of data is large enough. But that's no surprise because these hypotheses should have basically 0 prior probability anyway. Instead, we need to give hypotheses a fighting chance by stating them such that their prior probability is not 0 or, even better, to treat hypotheses on a continuum and assign prior and posterior probability distributions.

The real problem, though, is that the mere fact a result is statistically significant holds no content at all about the *size* of the effect. With a large enough sample size, any minuscule effect (of which there is almost always some) would be detectable, but that shouldn't be taken as validation of any *theory* of why the effect should be present. That is, as first articulated by psychologist Edwin Boring in 1919,[25] a scientific hypothesis is never just a statistical hypothesis—that two statistics in the population are different from each other, that two variables are correlated, that a treatment has some nonzero effect—but also an attempt at explaining why, by how much, and why it matters. Forgetting this is what Stephen Ziliak and Deirdre McCloskey in *The Cult of Statistical Significance* (2008) called "the Error of the Third Kind." As they put it, "Statistical significance is not a scientific test. It is a philosophical, qualitative test. It does not ask how much. It asks 'whether.' Existence, the question of whether, is interesting. But it is not scientific."[26]

Furthermore, statistical significance at best addresses only one kind of error that can beset an experiment—namely, sampling error. The weird correlations that show up in large-sample experiments make clear that other types of systematic error, like selection bias or the presence of confounding variables, are often at work. Understanding and controlling for those factors requires more careful thinking than just turning the handle on a mechanical process.

So the common practice, in force from about 1930 to the present, of judging whether research findings were publication worthy based only on statistical significance created the possibility that two kinds of bad scientific research could enter the literature. One was a simple Type I error, where, by a fluke of random sampling, data was obtained that passed a threshold of significance despite there being no real effect present; this could be expected more often as researchers were sifting through many possible associations until they found one that worked, per Ioannidis's dire predictions. The other possibility was a Type III error, where the effect was real in a statistical sense but did not actually support the scientific theory it was supposed to, perhaps because the sample was so large that the procedure found a tiny effect of little to no scientific value. It could be that another factor, something specific to that experiment and not thought of by the researcher, could explain the finding in a way that made it of no practical use to anyone else. Older and younger siblings in Minnesota high schools in 1966 might have had genuinely different feelings about college when asked a certain way, but that's only scientifically meaningful if the result generalizes beyond that one particular time and place.

A way to ferret out both types of error, which started to become more fashionable in the mid-2000s, was to test whether results were *replicable*. If the effect claimed by a given study was real, in the sense of actually having approximately the magnitude the study claimed and not just being the result of trying hypotheses until one met the threshold of significance by chance, then a high-powered repeat of the experiment should probably find it again. And if a small effect was actually supportive of some scientific theory and not just a result specific to a particular population or the product of some unexplained systematic bias the researcher had failed to account for, a study conducted somewhere else by someone else on some other experimental subjects should probably find it too.

Replication had long been the cornerstone of scientific truth and an integral part of the scientific method. So the question was essentially, Were all these new statistical results science or just noise?

In 2005, Ioannidis reviewed 49 medical studies conducted from 1990 to 2003, 45 of which had claimed a therapy was effective. In total, these studies had been cited over 1,000 times in the research literature. He found 7 of the 45 (16 percent) were later contradicted by subsequent studies, meaning no significant effect was found; another 7 had claimed stronger effects than subsequent studies found; 20 of them (44 percent) were essentially replicated; and the remainder had gone largely unchallenged.[27] This was somewhat disturbing evidence, but it was still possible the problem was limited to these kinds of studies, characterized, as he had predicted, by having many possible associations to consider.

EROTIC IMAGES AND ESP

Something happened in 2011, though, to open up the eyes of scientists to the possibility that the replication problem was much bigger and closer to home than just a few old medical studies. In January of that year, Daryl Bem, a social psychologist and professor emeritus at Cornell University, published a study in the prestigious *Journal of Personality and Social Psychology* claiming to demonstrate the existence of ESP ability. The paper, titled "Feeling the Future: Experimental Evidence for Anomalous Retroactive Influences on Cognition and Affect," contained a statistical analysis of the results of nine different experiments that were conducted over a 10-year period and included more than 1,000 college student participants.[28]

The experiments all involved some kind of precognition, the most famous being a test in which subjects tried to guess ahead of time where an image would appear on a computer screen, either on the left or on the right, with the true locations chosen at random. The images were classified into different categories according to a scale of "valence and arousal," including some neutral images and some erotic ones. What Bem found, in a study of 50 female students and 50 male students over a large number of sessions, was that their overall success rate in guessing the location of the erotic images was 53.1 percent, a sizable deviation from chance with an associated p-value of 0.01. The success rate for nonerotic images was not distinguishable from chance. So Bem concluded that something about the arousing imagery stimulated the participants' ability to predict their locations, maybe as the evolutionary adaptation of a "precognitive ability to anticipate sexual opportunities."[29]

Bem's description of his choice of a random number generator (RNG) and the implications this would have for the conclusion the participants were *clairvoyant* or merely *psychokinetic* makes the paper worth reading on its own:

> In the experiment just reported, for example, there are several possible interpretations of the significant correspondence between the participants' left/right responses and the computer's left/right placements of the erotic target pictures:
>
> 1. Precognition or retroactive influence: The participant is, in fact, accessing information yet to be determined in the future, implying that the direction of the causal arrow has been reversed.
> 2. Clairvoyance/remote viewing: The participant is accessing already determined information in real time, information that is stored in the computer.

3. Psychokinesis: The participant is actually influencing the RNG's placements of the targets.

4. Artifactual correlation: The output from the RNG is inadequately randomized, containing patterns that fortuitously match participants' response biases. This produces a spurious correlation between the participant's guesses and the computer's placements of the target picture.[30]

The relationship between parapsychology and statistics goes back a long time. In chapter 2, we met Samuel Soal, who was among the first to try using the new techniques of statistics to evaluate paranormal claims in the 1930s and '40s. Fisher himself wrote a short note in 1929 titled "The Statistical Method in Psychical Research," in which he commented on an ongoing discussion of card-guessing experiments; his advice was that any statistically significant deviation above chance could be taken as evidence for ESP (committing the same fallacy of misinterpreting statistical significance that would plague generations of researchers since).[31]

By the late 1970s and early '80s, parapsychology was experiencing a faddish popularity, and the number of such experiments had substantially increased, but the field was suffering from its own crisis of replication. Isolated experiments had seemed to demonstrate phenomena such as ESP, but the evidence never seemed to withstand scrutiny or manifest itself under repeated trials. At a conference held by the Parapsychology Foundation in San Antonio in 1983 on "The Repeatability Problem in Parapsychology," members of the community discussed research strategies for making their results more convincing, including performing studies with larger numbers of subjects and combining studies via meta-analysis. Among these methodological strategies, they also emphasized reaching out to *skeptics* in academia as potential collaborators to add legitimacy to any positive results they found.

Bem's story bears some superficial similarities to Soal's. Like Soal, he had impeccable academic credentials, receiving a doctorate in social psychology from the University of Michigan after studying physics at the graduate level at MIT. Bem held positions at Carnegie Mellon, Stanford, and Harvard Universities before becoming a professor at Cornell in 1978. Also like Soal, his earliest encounters with parapsychology were spent debunking claims of ESP abilities. Bem had experience as a stage magician and liked to perform occasional feats of mentalism for his students as a lesson about how easily one can be fooled by observation.[32] Because of Bem's academic background and his experience debunking phonies, parapsychologists who considered themselves serious scientists reached out to him.

Bem was contacted by an ESP researcher named Charles Honorton, the director of the Psychophysical Research Laboratories at the Forrestal Research Center in Princeton, New Jersey. Honorton specialized in what are called *ganzfeld* experiments (German for "whole field"). In these experiments, a subject, the *receiver*, is placed in a state of sensory deprivation—sitting in a comfortable chair in a dimly lit room, eyes covered, with headphones playing white noise— while a *sender* tries to transmit a randomly chosen image to them telepathically. Honorton had theorized that depriving a person of other sensory stimuli would enhance their ESP abilities, or *psi*, and he claimed to have experimental evidence showing a significant deviation from chance.

Bem reviewed Honorton's experiments in 1983 and, to his surprise, found everything seemed to check out. He wasn't able to dismiss the evidence by any of his conventional explanations, so he became a believer in ESP. He had a history of being an iconoclast, unafraid to take on traditionally accepted psychological dogma. The skeptical attitude he once had concerning paranormal research, then, had the potential to be transformed, in light of Honorton's results, into a skeptical attitude concerning the conventional wisdom *against* parapsychology.

The two coauthored a paper in 1994 titled "Does Psi Exist? Replicable Evidence for an Anomalous Process of Information Transfer," published in the *Psychological Bulletin*.[33] The thing that stands out the most about this paper, and others like it in the parapsychology community, is that it does not read like typical pseudoscience. It begins with a remarkably circumspect acknowledgment of the disciplinary headwinds any such studies are likely to face and the reasonable conservatism academic psychologists generally feel regarding phenomena they can't explain. What follows is a discussion of all the standard statistical concepts: significance, power, effect sizes, confidence intervals, multiple comparisons, the kinds of selective reporting bias Ioannidis had described for medical studies, and why the small nature of the effects they were studying necessitates larger sample sizes and meta-analyses.

They also examined flaws in previous psi studies and offered alternative hypotheses for the effects, including *sensory leakage*, meaning the incidental transmission of information to the subject by the experimenter, perhaps unknown to either of them. Their analysis concluded that the ganzfeld experiments (or what they called *autoganzfeld*, with the images for transmission being randomly chosen by a computer) had controlled for all the relevant variables, and in 329 sessions involving subjects guessing images out of a set of four, they achieved a hit rate of 32 percent versus the chance expectation of 25 percent, with an associated p-value of 0.002. That is, according to their statistical methodology, which appeared to be sound, they had observed a sizable deviation and rejected the null hypothesis at a very low significance level.

Bem wasn't the only person with a legitimate academic background making this kind of claim. For example, Jessica Utts, a professor of statistics at University of California, Davis (and now at UC Irvine), who later became the president of the American Statistical Association, conducted a similar survey of the existing research around that time and concluded there was positive evidence for the existence of psi. Her paper, "Replication and Meta-Analysis in Parapsychology," published in the journal *Statistical Science*,[34] contained, among other things, a sharp criticism of the nature of replication when it came to statistical methods in psychological studies. Bem and Honorton had also directed criticism that way. After expressing a hope their studies would provoke others to take up the task of trying to replicate their results, they cautioned against overinterpreting any *negative* results as proof against their theory given that a false negative could easily happen for an effect this small: "Would-be replicators also need to be reminded of the power requirements for replicating small effects. Although many academic psychologists do not believe in psi, many apparently do believe in miracles when it comes to replication."[35] Reminiscent of Soal's analysis of his card-guessing experiments (only this time without the fraud, hopefully), these papers, including Bem's 2011 results, seemed in every way like the kinds of analyses that *would* establish psi as a real phenomenon if such a thing were ever going to happen.

Alternatively, they could be read as a devastating critique of research methods in psychology. Their critique was twofold: (1) Given that Bem and others had followed all the same experimental protocols and used all the same statistical methods as other psychological studies, the fact they reached an unreasonable conclusion (psi is real) should call into question potentially *all* psychological research. (2) If the institution of academic psychology tried to dismiss this psi research because it had not been sufficiently replicated, how confident could they really be in the established results they had never tried in earnest to replicate? Lee Ross, a Stanford University professor of social psychology and one of Bem's peer reviewers, put it this way: "The level of proof here was ordinary. I mean that positively as well as negatively. I mean it was exactly the kind of conventional psychology analysis that [one often sees], with the same failings and concerns that most research has."[36]

Even before it was published, "Feeling the Future" ignited a firestorm of controversy. Some academics dismissed it as nonsense. Ray Hyman, emeritus professor of psychology at the University of Oregon and a longtime critic of

ESP research, referred to the paper as "an embarrassment for the entire field."[37] Cornell, meanwhile, issued a press release describing the results as the capstone of Bem's distinguished career.[38] The general public was more willing to accept these results than academic psychologists were; surveys had shown that something like 60 percent of Americans already believed in the existence of psychics. The possibility that science had finally *proven* the existence of paranormal phenomena, combined with the salacious idea that college students looking at porn were especially paranormal, naturally captured the public's imagination. The paper was covered prominently in the *New York Times* and *Wired*, and Bem was invited to appear on the *Colbert Report*.

THE MATH

Eric-Jan Wagenmakers of the University of Amsterdam's Department of Psychology saw a preprint of Bem's paper while at a conference in Berlin and set to work writing a rebuttal with three of his colleagues. In their article, which was published alongside Bem's in the *Journal of Personality and Social Psychology*,[39] they made three major arguments about Bem's paper:

1. Given a very low prior probability for the existence of ESP, even strong evidence in favor of it should leave the reader with a very low posterior probability.
2. Bem's evidence was not really all that strong but rather provided almost as much support for the null hypothesis as the alternatives given the size of the effect he was claiming.
3. Bem had perhaps blurred the lines between *exploratory* and *confirmatory* analysis, allowing him to cherry-pick possible relationships from among his data.

Argument (1) was a case for Bayesian inference. Wagenmakers et al. referenced the idea originally attributed to Pierre-Simon Laplace that "extraordinary claims require extraordinary evidence." They argued that the existence of psi would, indeed, constitute an extraordinary claim. For one thing, we have no working model of how ESP could function in a way consistent with known laws of physics. Second, we have ample evidence no one in the world has the ability to predict the future of random outcomes above chance expectation given all the world's casinos have not been bankrupted by any gamblers with ESP. (What could stimulate a "precognitive ability to anticipate sexual opportunities" more than the prospect of becoming an instant millionaire?)

Bem had reported a success rate of 53.1 percent for his subjects when trying to guess the location of the images versus a chance expectation of 50 percent. In total, this was measured over 1,560 guesses, split into sessions of 12 or 18 across 100 different subjects. Ignoring for the moment the effect of testing the same person multiple times and just treating each image as an independent guess, under the chance hypothesis the total probability of attaining 53.1 percent correct—that is, 828/1,560—would be

$$P[828 \text{ correct out of } 1,560 \mid \text{hit rate} = 0.5] = \binom{1,560}{828}(0.5)^{828}(0.5)^{7,320} \approx 0.11\%$$

The corresponding probability, assuming a hit rate of 53.1 percent, would be

$$P[828 \text{ correct out of } 1,560 \mid \text{hit rate} = 0.531] = \binom{1,560}{828}(0.531)^{828}(0.469)^{7,320} \approx 2.02\%$$

The second probability is approximately 19 times greater than the first one. But this would still leave someone with even a modestly skeptical prior pretty unconvinced. If, say, a person assigned a 1-in-1,000 probability to the theory of mild ESP in the context of erotic images, their Bayesian inference would involve comparing the pathway probabilities: (0.999) · (0.0011) versus (0.001) · (0.0202). The first pathway makes up over 98 percent of the total, so the skeptic would still be about 98 percent sure the data had just happened by chance, even though it was a statistically significant deviation. Wagenmakers et al. made the point, in agreement with the criticism of NHST we've seen many times, that to take a statistically significant result as indication the null hypothesis is false with high probability is to commit a logical fallacy.

Their point (2) was a little more subtle but also pertained to the Bayesian inference for or against the hypothesis given the data. The question, which, as we have seen, often goes unanswered in NHST, is this: If the null hypothesis *isn't* true, then what could better explain the data? Specifically, how much support does the data actually give any of these alternatives? If it turns out the only other theory is one making the data almost as unlikely as the null hypothesis does, then observing that data is really no evidence against the null at all, even though the orthodox significance test says to reject it.

For example, if the possible hit rate of 53.1 percent was the only alternative hypothesis under consideration, then the data would surely count as *some* evidence and move the probability needle at least a little in that direction.

Even if, as our back-of-the-envelope estimate showed, it increased a skeptic's probability assignment only from 1-in-1,000 to 1-in-50, that was still progress.

But Wagenmakers et al. argued a hit rate of 53.1 percent is not the only alternative that could explain the data, even granting ESP were real! That is, the effect of psi could result in a general improvement from 50 percent to *50.01 percent*, say, and the success rate shown in the data could be a combination of this effect *and* chance. To make a baseball analogy, imagine a struggling hitter is told by a teammate that using a pair of lucky batting gloves could raise their average from .200 to .210. Then, while borrowing the gloves for a week, the hitter goes on a hot streak and sees their average over that period spike to .350. Even if our priors didn't rule out the possibility of lucky batting gloves altogether, how much credit should be given to the gloves and how much to chance?

The reason this matters comes back to the idea of a test's *power*. Any test with a given finite sample size will be able to reliably distinguish only between there being no effect and there being some effect of a particular magnitude. If the truth of the matter is the null hypothesis is false but only by a very tiny degree—as we said before is generally the case when discussing large population parameters—then a test without enough power to detect the difference will often fail to reject the null hypothesis even though it's false; that is, we will commit a Type II error.

However, another insidious consequence of an underpowered test is that even when the significance test *does* say to reject the null hypothesis at some significance level, this may not count for much in favor of the alternative. We can see how this would play out in the inference table (table 6.3), assuming any prior probability p for the null hypothesis and $(1 - p)$ for the alternative. Because our test is not really able to distinguish between these probabilities, how we assign them to the null and alternative hypotheses doesn't really change at all, even though we "reject" the null! This is, once again, a by-product of the

TABLE 6.3 **Inference for an underpowered test**

Hypothesis H	Prior probability $P[H \mid X]$	Sampling probability $P[D \mid H \text{ and } X]$	Pathway probability $P[H \mid X]P[D \mid H \text{ and } X]$	Relative proportion $P[H \mid D \text{ and } X]$
Null hypothesis	p	α	$p\alpha$	$\approx p$
Almost the null hypothesis	$1 - p$	$\approx \alpha$	$\approx (1 - p)\alpha$	$\approx 1 - p$

confusing language of significance testing when it runs afoul of the correct Bayesian inference.

Another side effect of low power is that, assuming the study did find a statistically significant effect, its estimated *size* is necessarily many times greater than the true effect size. For example, in our urn-testing experiment in the last chapter, the null hypothesis is that the ratio of white:total pebbles in the urn is exactly 0.5, and we said that, with a sample of size $n = 32$, we will reject the null only if the sample ratio falls into the 5 percent rejection region, which we found to be

$$R \leq 0.3125 \text{ or } R \geq 0.6875$$

If the truth of the matter is that the urn ratio is actually 0.5000001, though, the true probability of this happening is effectively 5 percent because the true alternative hypothesis is so close to the null. But then assume this actually happens and we find ourselves in one of these 5 percent of cases—say we get $R = 0.6875$. Then our estimated "effect" is (0.6875 − 0.5), or 0.1875, which is **1,875,000** times greater than the true effect size of 0.0000001.

Bem chose his sample sizes so he would have about an 80 percent power to detect an effect of the size equal to the one he found. Somewhat conveniently, he claimed this effect size was right in the middle of the range "reported in the psi literature."[40] The critics applied a different idea, arguing that even under the assumption some deviation from chance was really possible, the *prior probability distribution* for the effect size should concentrate most of the probability toward small effects. This was based on a general derivation when little is known about the size of a possible new effect.[41] After including all the possible small effects with their associated probabilities, they arrived at a total probability for the data under the alternatives only 1.64 times greater than the probability for the data under the null, rather than the factor of 19 Bem had claimed.[42] In other words, the data did not really represent strong evidence against the null hypothesis at all.

This calculation was damning criticism of Bem's results even if one did not grant the skeptical prior probabilities Wagenmakers et al. assumed in argument (1). What they computed in point (2) is something called a *Bayes factor*, which gives a sort of compromise between frequentist methods and Bayesian ones. Many researchers, educated in statistical orthodoxy and still clinging to some idea of objectivity, are made uncomfortable by the important role of the prior probabilities for the hypothesis and alternatives in the full Bayesian inference.

Why is it appropriate to bring one's own skepticism about the hypothesis into the analysis of a particular study? What if that skepticism were turned around and someone claimed a *high* prior probability for the hypothesis that ESP is real, resulting in an even higher posterior probability? If priors were just about feelings, how could one person's feelings about their own theory be judged inferior or superior to another's? (As we have argued, prior probabilities are *not* about feelings but about *information*; however, this view is not yet the most common one.)

The question answered by the Bayes factor is this: Given *any* starting point regarding the relative prior probabilities of the null and alternative hypotheses, how much support does the data lend just by itself? Mathematically, this can be found by comparing the prior and posterior probabilities for the null (H_0) *and* alternative (H_1) hypotheses. Since Bayes' theorem tells us the posterior probability of each is proportional to its pathway probability to the data, D, if we divide one by the other, we will have

$$\frac{P[H_0 \mid DX]}{P[H_1 \mid DX]} = \frac{P[H_0 \mid X]}{P[H_1 \mid X]} \cdot \frac{P[D \mid H_0 X]}{P[D \mid H_1 X]}$$

The second fraction on the right-hand side is the Bayes factor (denoted BF_{01} when calculated the way we have here). It measures how much the ratio of the probability of the null hypothesis to the probability of the alternative hypothesis changes after observing the data. Thus, it's agnostic to what the actual prior probabilities originally were. If the Bayes factor is 10, someone who thinks the null is initially 1/10 as likely as the alternative will assign them equal posterior probabilities; someone who initially considers them equally likely will think the null is now 10 times likelier than the alternative, etc. Of course, as in the analysis of Bem's results, the calculation may still be a *little* Bayesian because the "alternative" may actually represent a range of alternatives, weighted according to some prior probability distribution. But at least that distribution of the likely effect *size* could be argued separately from any dogmatism about whether ESP was or wasn't a real phenomenon *at all*.

Harold Jeffreys first proposed the idea of Bayes factors in his *Theory of Probability*. A value of 1 means no change in the relative likelihood of the null and alternative hypotheses. From there, the lower the Bayes factor, the greater the support for the scientific theory. Jeffreys interpreted it as "decisive evidence for H_1" if the factor was less than 1/100 and "decisive evidence for H_0" if the factor was greater than 100. Factors between 1/3 and 3 were "not worth more than a bare mention."[43]

TABLE 6.4 **Classification of Bayes factors by Wagenmakers et al.**

Bayes factor	Interpretation
<1/100	Extreme evidence for H_1
1/100–1/30	Very strong evidence for H_1
1/30–1/10	Strong evidence for H_1
1/10–1/3	Substantial evidence for H_1
1/3–1	Anecdotal evidence for H_1
1	No evidence
1–3	Anecdotal evidence for H_0
3–10	Substantial evidence for H_0
10–30	Strong evidence for H_0
30–100	Very strong evidence for H_0
>100	Extreme evidence for H_0

Wagenmakers et al. softened the language a bit and classified Bayes factors by ranges, as shown in table 6.4.[44] They found a Bayes factor of 0.61 for Bem's erotic image results, suggesting this was "anecdotal" evidence that gave slightly more support in favor of the alternative than the null. For the other eight studies in Bem's paper, they computed Bayes factors in a similar way and found they were mostly anecdotal, either for the null or against it, with three of the studies actually providing "substantial" support for the *null* hypothesis. Only one of the nine gave substantial support for the alternative hypothesis. So, taking into account the likely effect sizes, in no way were Bem's results conclusive evidence for the psi hypothesis, even for someone who came to the analysis without any strong preconceptions for or against the idea.

Argument (3) claimed Bem had not begun his studies with a clear hypothesis in mind but instead allowed the data to guide him to whatever theory it might *accidentally* support. Evidence for this was mostly found in the specific ways Bem had phrased his hypotheses—in particular, in the fact that only erotic images could stimulate a significant psi ability in the subjects. If that claim had been part of Bem's theory from the beginning, why include the other images at all? Why not just use the ones with the greatest chance of producing an effect when demonstrating the very existence of the effect would have been enough?

Even though the paper was written to suggest all the effects found in the analysis were what Bem had expected from the beginning, there was no way to know for sure, but there was plenty of evidence he had kept slicing up the data until some statistics had crossed the threshold of significance just by chance.

THE AFTERMATH

All told, the assessment of Bem's results by Wagenmakers et al. painted a pretty unimpressive picture. The data Bem had gathered amounted to weak support for his own theory, likely tailored after the fact to make the data support it even more than was really true, and the theory itself was so implausible it shouldn't have been taken seriously even if the evidence *had* been very strong. But the larger point, as the critics said, was not so much the fact this paper was bunk but rather what it represented as an indictment of statistical methods in psychology. Everyone seemed to agree Bem had followed all the rules for legitimate research. So what if he had let his data guide him toward a more specific hypothesis about his theory? That was pretty common practice in the social sciences—and often inevitable given how messy the data collection process is and how difficult and expensive it can be to gather new data to test a particular narrow version of a hypothesis. Finding experimental participants and testing them is hard work, so it makes sense to try many different variations of the experiment at the same time and to record as much useful information as you can to sift through later. As Bem himself once wrote:

> To compensate for this remoteness from our participants, let us at least become intimately familiar with the record of their behavior: the data. Examine them from every angle. Analyze the sexes separately. Make up new composite indexes. If a datum suggests a new hypothesis, try to find further evidence for it elsewhere in the data. If you see dim traces of interesting patterns, try to reorganize the data to bring them into bolder relief. If there are participants you don't like, or trials, observers, or interviewers who gave you anomalous results, place them aside temporarily and see if any coherent patterns emerge. Go on a fishing expedition for something—anything—interesting.[45]

After a researcher has gone on such a fishing expedition, requiring that they then set up a separate study to prove the fish was real seemed unduly harsh. What if the data were *observational*, say, consisting of records of people's behavior during the last financial crisis? Did one have to engineer an all-new financial crisis to truly test the theory? And Wagenmakers's points about the prior and posterior probabilities for Bem's hypothesis seemed entirely out of bounds. No previous research work had ever involved prior probabilities. Frequentist methods had been the industry standard for as long as anyone could remember, and all of a sudden people were supposed to just switch to Bayesianism? Wasn't this all a little *subjective*?

Yes and yes. Social science had gotten by long enough using mechanical methods of inference such as NHST, and the claim had always been that no serious damage had been done. Here was actual damage. Bem's paper was a wake-up call that it was time to rethink those methods, become more strict about the design of experiments and treatment of data, and learn how to interpret evidence in terms of what it meant for the hypothesis in question and its alternatives. The rebuttal by Wagenmakers et al. concluded:

> Bem played by the implicit rules that guide academic publishing. In fact, Bem presented many more studies than would usually be required. It would therefore be mistaken to interpret our assessment of the Bem experiments as an attack on research of unlikely phenomena; instead, our assessment suggests that something is deeply wrong with the way experimental psychologists design their studies and report their statistical results. It is a disturbing thought that many experimental findings, proudly and confidently reported in the literature as real, might in fact be based on statistical tests that are explorative and biased (see also Ioannidis, 2005). We hope the Bem article will become a signpost for change, a writing on the wall: Psychologists must change the way they analyze their data.[46]

The funny thing was that maybe Bem agreed. In the opening of their 1994 paper, Bem and Honorton indirectly referenced Laplace, saying "We psychologists are probably more skeptical about psi for several reasons. First, we believe that *extraordinary claims require extraordinary proof.*"[47] And the closing paragraphs practically goaded reviewers into using Bayesian methods:

> More generally, we have learned that our colleagues' tolerance for *any* kind of theorizing about psi is strongly determined by the degree to which they have been convinced by the data that psi has been demonstrated. We have further learned that their diverse reactions to the data themselves are strongly determined by their a priori beliefs about and attitudes toward a number of quite general issues, some scientific, some not. In fact, several statisticians believe

that the traditional hypothesis-testing methods used in the behavioral sciences should be abandoned in favor of Bayesian analyses, which take into account a person's a priori beliefs about the phenomenon under investigation. . . .

In the final analysis, however, we suspect that both one's Bayesian a prioris and one's reactions to the data are ultimately determined by whether one was more severely punished in childhood for Type I or Type II errors.[48]

In 1996, a hoax perpetrated by physics professor Alan Sokal on the philosophy journal *Social Text* had demonstrated the occasional intellectual laziness and gullibility of journal editors. He successfully submitted the nonsense article "Transgressing the Boundaries: Towards a Transformative Hermeneutics of Quantum Gravity," which was laden with scientific jargon and unsupported conclusions he knew the editors would find amenable, to the point where they wouldn't bother scrutinizing the rigor of his arguments.[49] He revealed as much a few months later, and the hoax caused quite a stir, provoking spirited arguments about academic authority and research methods in philosophy.

Could Bem, a former magician with a history of using deception to teach his audience lessons in credulousness, have been attempting a similar hoax for social science? Ray Hyman said about Bem, "He's got a great sense of humor. I wouldn't rule out that this is an elaborate joke."[50] The theory was that Bem had never *really* believed in ESP at all but had used parapsychological studies as a way to call attention to flaws in more traditional research. The statistician Andrew Gelman at Columbia University said he heard rumors to this effect from a friend who knew Bem, and he suggested the fact that Bem's results had only *just barely* met the thresholds for significance was further evidence in support of the theory.[51] Why rush to publish marginal results when he could have taken time to perform a larger study, more narrowly designed to discern a larger effect?

If it was a hoax, though, it would have to rank among the most elaborate of all time, requiring a multiple-decades-long commitment to the bit that was Andy Kaufman-esque. It would also torch Bem's own academic legacy assuming the result of all the heightened attention to research methods was that *all* previously established results in the field were called into question, including Bem's. What Bayes factor would we give this hypothesis over the alternative: that he genuinely believed in ESP? We'll likely never know the truth of his intentions, but, regardless, Bem's paper did provoke a wave of introspection among psychologists, especially regarding the need for replication.

To take the provocation a step further, Wharton business school professor Joseph Simmons, together with his colleague Uri Simonsohn and Leif Nelson from the University of California, Berkeley, published an article in 2011 to show how flexibility in data manipulation and the reporting of methods could easily lead to a false positive result. They illustrated the technique by using statistical trickery to make it appear students who listened to the Beatles song "When I'm Sixty-Four" became, on average, a year and a half younger afterward than those who listened to a control song.[52]

It happened in the same year that the field of social psychology attracted even more public attention to itself for all the wrong reasons. In September 2011, Tilburg University in the Netherlands announced it was suspending Professor Diederik Stapel from his position as dean of the School of Social and Behavioral Sciences over allegations he had fraudulently manipulated data in several studies. A group of three researchers under his supervision had blown the whistle on Stapel after observing instances of questionable activity, including the possibility he had completely fabricated much of his data. Stapel later admitted it and claimed full responsibility for his actions. He had been a rising star of social psychology, and the fallout from the revelations was massive. Some 58 publications he had contributed to had to be retracted, representing years of ruined work for his unwitting coauthors.

With the spotlight shining brightly on the discipline of psychology, in November 2011 University of Virginia professor Brian Nosek organized a collaboration among 270 people to attempt to replicate 100 psychology studies whose results had been published in prominent journals in 2008, a project that would take nearly four years to complete. In January 2013, Nosek and Jeffrey Spies founded the Center for Open Science (COS), a nonprofit organization with a mission to increase the transparency and integrity of data-driven scientific research. Funded by an initial grant of $5 million, the COS gives researchers a way to connect with others who are interested in improving scientific practice as well as tools to help manage and share data and methods. For example, one of the research practices they have promoted is the preregistration of methods, meaning that the ways data in a particular study will be analyzed are all recorded ahead of the actual data collection. This could help scientists from even *unintentionally* being led to change what tests they run based on what relationships emerge in the data—what Gelman called the "garden of forking paths" problem.[53]

To model what they considered good scientific practice, Nosek and his coauthor Matt Motyl attempted to replicate a study of their own.[54] Their study, which involved whether people with more extreme political views were able to perceive shades of the color gray less accurately than those with moderate views, had been inspired by similar recent studies on *embodiment* that investigated the surprising ways body and mind were linked. They initially tested their theory of color perception with a large group of nearly 2,000 participants and found a significant effect with a p-value of 0.01; people with extreme political views saw the world in black and white, literally. It was academically sexy and media friendly, the kind of result that was sure to land in a high-impact journal and generate ample PR buzz.

Before publishing, though, they decided to perform a replication study with another 1,300 participants and preregistered methods. They were very confident the results would replicate; according to their analysis, the replication had a 99.5 percent chance of reaching significance with a p-value of less than 0.05. Instead, the attempted replication came out with a p-value of 0.59, and the effect had disappeared. Had they not conducted the second experiment, the two could very easily have gotten their paper published while legitimately claiming that they had followed all the standard rules of academia.

In August 2015, Nosek and his team of 270 collaborators released the results of their psychology replication project. Of the 100 papers they studied, 97 had originally claimed to discover a significant effect. The replication studies used the original materials when possible and large enough sample sizes that they would have high power (at least 80 percent) to detect the effects that were claimed. The experimental protocols were all reviewed and approved by the original authors. They found they were able to replicate only 35 of these 97 results (36 percent), which they defined as achieving a statistically significant effect in the same direction as the original.[55] Of those effects they *did* replicate, they found the average size of the effect to be about half the original.

Three years later Nosek and another group of collaborators released the results of a similar project to replicate experiments in social science. The 21 studies they considered had all been published in the prestigious journals *Science* and *Nature* between 2010 and 2015. Once again, the team contacted the original authors whenever possible to obtain materials and to get their approval for the replication efforts. The replications were high-powered, with sample sizes on average five times greater than the samples used in the original experiments. They found they were able to successfully replicate only 13 out of the 21 results (62 percent), with average effect sizes of the successful replications being only about 75 percent of the original size.[56]

Wagenmakers, along with another group of colleagues, computed Bayes factors for the null versus the alternative hypotheses in each study, as he had done for Bem's paper, and showed that 8 out of 21 attempted replications had data that gave more support to the null hypothesis than the alternative.[57] Remember, this means not only is the effect not statistically significant, but also the data actually should make the null hypothesis *more* likely regardless of one's prior probabilities. Of the remaining 13 studies, 4 gave only moderate support to the alternative hypothesis.

A representative example of the studies included in the replication project is "Analytic Thinking Promotes Religious Disbelief" by Will Gervais and Ara Norenzayan of the University of British Columbia, originally published in *Science* in 2012.[58] The study concerned the relationship between various cognitive processes and religious sentiment. The authors' hypothesis was that if a person was stimulated into a mode of analytical thinking, their self-reported level of disbelief in religion would go up. In one experiment, subjects had to reason their way through a few math problems and then answer survey questions about their religious faith and their belief in God, angels, and the devil. Another experiment to test their theory used *visual priming* to stimulate analytic thinking. Subjects were randomly presented with an image of either Rodin's sculpture *The Thinker* or Myron's *Discobolus*, the control image. Previous pilot studies had shown that viewing *The Thinker* was significantly associated with improved performance on a test of logical reasoning ability, so the authors hypothesized it could stimulate the same kind of thinking they expected would interfere with religious belief.

They assigned 26 subjects to view *The Thinker* and 31 to view the control image and then asked the participants to rate their belief in God on a scale from 0 to 100. The result was a significantly lower belief in God for those who had viewed *The Thinker*; those subjects reported a mean God-belief score of 41.42 versus the control group's 61.55, with an associated p-value of 0.029. The authors concluded: "In sum, a novel visual prime that triggers analytic thinking also encouraged disbelief in God."[59] The suggestion, hinted at but not explicitly claimed by the authors, that religion is incompatible with analytic thought made the paper the predictable subject of controversy. Atheists hailed it as scientific proof that religion was irrational; religious people were understandably offended at the suggestion that the source of their faith was a lack of reasoning ability.

The replication study used 531 participants (compared to the original's 57), with 262 viewing *The Thinker* and 269 viewing *Discobolus*. With a sample so large, the replicators could expect to find an effect even half the size of the one reported in the original study with a power of 90 percent. Instead, they found no significant difference in belief in God between the two groups. In fact, the mean God-belief score in the *Thinker* group was slightly *higher* (62.94) than in the control group (60.38).[60] The Bayes factor computed for the null hypothesis (no difference) versus the alternative hypothesis (some difference, with a given distribution of effect sizes) was 17.78, indicating strong support in favor of the *null*.[61]

Looking back on the experience, Gervais said, "In hindsight, our study was outright silly. When we asked them a single question on whether they believe in God, it was a really tiny sample size, and barely significant . . . I'd like to think it wouldn't get published today."[62] He and Norenzayan issued a comment gracefully withdrawing support for their original finding.[63] Again, the original study was in no way egregious or misrepresentative of contemporary methods. It was published in a very prestigious journal and has, up to this date, been cited 494 times in other research literature.

While Nosek and his collaborators were working on their replication efforts, people in other disciplines began to think more critically about replicating results within their own fields as well. In 2012, Glenn Begley, a biotech consultant working at Amgen, and Lee Ellis, at the University of Texas, published a paper summarizing their finding that out of 53 preclinical cancer studies, only 6 (11 percent) could be replicated.[64] Ioannidis and Nosek teamed up in 2013 with five coauthors from the Universities of Oxford and Bristol to examine the issues caused by low statistical power in neuroscience research. By comparing sample sizes reported in 730 individual studies with the effect sizes confirmed in 49 meta-analyses of those studies, they estimated the median statistical power of neuroscience studies to be 21 percent.[65] This meant neuroscience studies were generally far too small to reliably detect the effects they were designed to study, and even statistically significant results couldn't be taken as strong evidence against the null hypothesis, as we saw earlier.

Even more disturbingly, they found animal model studies (think mice in mazes) could have a median power as low as *18 percent* and neuroimaging (MRI) studies of structural abnormalities in the brain had a median power of *8 percent*. They discussed the ethical implications of the inefficiency this entailed:

> There is ongoing debate regarding the appropriate balance to strike between using as few animals as possible in experiments and the need to obtain robust, reliable findings. We argue that it is important to appreciate the waste associated with an underpowered study—even a study that achieves only 80 percent power still presents a 20 percent possibility that the animals have been sacrificed without the study detecting the underlying true effect. If the average power in

neuroscience animal model studies is between 20–30 percent, as we observed in our analysis above, the ethical implications are clear. Low power therefore has an ethical dimension—unreliable research is inefficient and wasteful. This applies to both human and animal research.[66]

In 2014, Ioannidis and Steven Goodman started the Meta-Research Innovation Center at Stanford (METRICS), a research center within the Stanford University School of Medicine that aims to improve reproducibility by studying how science is practiced and published and by developing better ways for the scientific community to operate.

A study released in June 2015 in the journal *PLOS Biology* analyzed the economic costs of irreproducible results in life sciences—specifically, those for preclinical trials for new drugs. Lead author Leonard Freedman, a scientist with experience in pharma research, and two coauthors surveyed the previous literature on the rates of irreproducibility (higher rates being worse) in published preclinical trials and estimated the average rate to be about 50 percent.[67] They identified several key factors contributing to the failure of the studies to replicate, including contaminated lab materials, poor experimental design, and incorrect methods of data analysis. Considering the size of the pharmaceutical industry, they estimated the money wasted on replicating faulty trials to be $28 billion per year in the United States alone.

A replication project for economics research conducted by a group of 18 collaborators found they were able to replicate 11 out of 18 experiments (61 percent) published in the *American Economic Review* and the *Quarterly Journal of Economics* in 2011–2014. The replication studies all had power 90 percent to detect effects of the original size. On average, the effects they did find to be significant were 66 percent of the originals.[68] A 2017 study conducted by Ioannidis suggested "the majority of the average effects in the empirical economics literature are exaggerated by a factor of at least 2 and at least one-third are exaggerated by a factor of 4 or more."[69]

Another similar project for cancer research, the Reproducibility Project: Cancer Biology, is now underway, coordinated by the COS and Science Exchange. The teams of researchers are investigating 54 high-impact cancer biology studies published between 2010 and 2012. Initial results are roughly consistent with those of the other replication projects so far.[70]

Thanks to these replication projects and others like them, we can begin to see the scale of the crisis. So far all indications are that it is massive.

THE SIGNIFICANCE OF INSIGNIFICANCE

Do the anti-inflammatory drugs known as selective COX-2 inhibitors (like the drug Celebrex) cause an increased risk of heart problems? If you skimmed the medical literature with a significance testing mindset, you might be led to believe there is considerable doubt on the subject. For instance, a trial with 5,500 participants conducted in 1999–2000 on the drug Vioxx found no statistically significant difference in the number of heart attacks between the treatment group and the control group, who had taken naproxen.[71] A more recent study in 2013 examined the effects of COX-2 inhibitors (among other anti-inflammatory drugs) on atrial fibrillation, a type of abnormal heart rhythm, and concluded the "use of selective COX-2 inhibitors was not significantly related to atrial fibrillation occurrence, except in patients with chronic kidney or pulmonary disease."[72]

The latter finding was seemingly at odds with an earlier study in 2011 that found COX-2 inhibitors *were* significantly associated with an increased risk of atrial fibrillation or flutter, to the tune of about a 27 percent relative increase over the control.[73] So which was it, significant or insignificant? The authors of the 2013 study explained the discrepancy by saying the two had been measuring slightly different outcomes; by including atrial flutter as well as fibrillation, the earlier study had cast too wide a net. Focusing on atrial fibrillation led the later authors to a more specific finding, one they claimed was "unique."[74]

But the authors of the 2011 study (including Kenneth Rothman, who, as we saw, tried to ban *p*-values from the *American Journal of Public Health*) disagreed. Not only were there medical reasons to doubt the explained difference, they said, but also the two studies had reported basically the same thing![75] The 2013 study had actually found a 20 percent increase in the relative risk of atrial fibrillation. It's just that because the sample was smaller, the result didn't cross the threshold of statistical significance. The 95 percent confidence interval for the increase in risk spanned from −3 percent to +48 percent, containing the value 0, so the data was, by a slim margin, not quite enough to reject the null hypothesis of no difference.

A similar dynamic was at work in the earlier 1999 Vioxx study, although with some additional data-fudging. The report explained that five patients in the treatment group had suffered heart attacks, compared with only one in the control group, but that difference implied a *p*-value of more than 0.2, so it did not count as statistically significant.[76] Later it came to light that actually three more people in the treatment group also had heart attacks, including a 73-year-old woman who died. A scientist at Merck, the drug company running the trial, initially judged that she had probably died of a heart attack but was overruled by a senior executive who classified the cause of death as unknown, "so we don't raise concerns."[77] Vioxx was voluntarily taken off the market in 2004, but not before it was linked to approximately 140,000 incidences of heart disease just in the United States.[78]

In terms of statistics, though, what was really at stake was whether an *insignificant* finding—more precisely, the lack of a statistically significant result that would be cause to reject the null hypothesis—could be taken to have any meaning. That is, is insignificance significant? By the standards of orthodox theory, the answer is a little controversial. Fisher was very clear that the null hypothesis is never confirmed, only rejected, but he did also say that significance tests could tell a scientist what to ignore.[79] Neyman and Pearson explicitly described the possibility of both Type I and Type II errors, implying there must be some cost to a *non*rejection of the null hypothesis if it is, in fact, false. Untwisting the multiple negations in that sentence, it was kosher, in the Neyman-Pearson framework anyway, for a scientist to make decisions based on a failure to find a significant difference, to the same extent they could based on having found one. So they—and the drug companies and the government regulators—could ignore a 20 percent increase in atrial fibrillation risk or five times more heart attacks (once they threw out the other three, that is) because it hadn't met the threshold for significance at the 5 percent level. That was simply balancing one type of error against another. As Ziliak and McCloskey described in their analysis of the Vioxx case, "Statistical significance, as the authors of the Vioxx study were well aware, is used as an on-off switch for establishing scientific credibility. No significance, no risk to the heart. That appears to have been their logic."[80]

These incidents, and many others like them, have led the scientific community to finally say enough is enough. In March 2019, an article in *Nature* cosigned by more than 800 research scientists called for an end to the concept of statistical significance altogether. In the authors' words, "Let's be clear about what must stop: we should never conclude there is 'no difference' or 'no association' just because a *P* value is larger than a threshold such as 0.05 or, equivalently, because a confidence interval includes zero. Neither should we conclude that two studies conflict because one had a statistically significant result and the other did not. These errors waste research efforts and misinform policy decisions."[81] It was, by their estimates, an incredibly widespread problem. A tally of 791 articles in five journals showed that roughly half had wrongfully interpreted a lack of significance as confirmation of the null.

Prominent statisticians have likewise agreed that it's time to move on. In 2016, the American Statistical Association (ASA) had issued a statement on the use of p-values, in which it cautioned against their many forms of misuse.[82] In March 2019, the editors of the *American Statistician*, including Ron Wasserstein, the executive director of the ASA, published a special issue of the journal titled "Statistical Inference in the 21st Century: A World Beyond p < 0.05" with even stronger words of warning. In the introduction to the issue, the editors wrote, "The *ASA Statement on P-Values and Statistical Significance* stopped just short of recommending that declarations of 'statistical significance' be abandoned. We take that step here. We conclude, based on our review of the articles in this special issue and the broader literature, that it is time to stop using the term 'statistically significant' entirely. Nor should variants such as 'significantly different,' '$p < 0.05$,' and 'nonsignificant' survive, whether expressed in words, by asterisks in a table, or in some other way."[83]

The editors appeared ready to back off from p-values as meaningful altogether: "For example, no p-value can reveal the plausibility, presence, truth, or importance of an association or effect. Therefore, a label of statistical significance does not mean or imply that an association or effect is highly probable, real, true, or important. Nor does a label of statistical nonsignificance lead to the association or effect being improbable, absent, false, or unimportant."[84] For now, p-values remain in standard usage, but the prevailing sentiment that seems to be emerging is a holistic view of statistical methods, of which p-values are merely a part. As I'll argue in the next chapter, this is a move in the right direction, just an incrementally small one that's unlikely to fix anything unless it leads to more radical upheaval.

It's also a move that's about 100 years too late. In particular, it's an inescapable fact that significance or lack thereof *has been* the standard for publication since the time of Fisher. Even if "proving the null" was never condoned, it was the practical effect of selecting only the significant results for further inquiry. Outside the few studies, like the COX-2 inhibitor drug trials, that claimed to draw meaningful conclusions from a lack of significance, there are many more that were simply thrown away because their authors considered them unimportant. Looking back on Meehl's "crud factor" examples in another way, the null hypothesis is, in nearly every circumstance, almost certainly false. Many of the alternatives, meaning the nonnull theories, are true in a way that's uninteresting because they applied only to that one experimental time and place, but *many more might still be relevant.* What this means potentially is that there's an altogether unseen crisis of replication hiding in the shadows: *the failure of failed experiments to fail again.* The (bad) replication crisis we know about is of Type I; there might be a wholly different (good) crisis of Type II. There might be a rich vein of results just waiting to be mined.

These events were foreseeable as soon as the concept of significance was born. (Just ask any of the many authors who foresaw them.) Significance testing was always based on a classification of results into significant/insignificant without regard to effect size or importance; no attempts to rehabilitate it now can change that fundamental aspect nor repair the damage significance testing has already caused. This yes/no binary has well and truly mixed things up. To have any hope of unmixing them, we need to completely reimagine what statistical inference is all about.

WHY IT'S A CRISIS

The evidence is piling up that the areas of science in which NHST has long been the standard are now experiencing a widespread failure of replication. This is not a coincidence. Psychology has produced the most visible examples, owing mostly to the media friendliness of its studies and the attention-grabbing headlines when those studies fail to replicate. But the crisis is by no means confined to psychology, as the various projects have abundantly shown, and the costs of failed replications in other research fields may be much higher. Whether a claim like "Biomechanical remodeling of the microenvironment by stromal caveolin-1 favors tumor invasion and metastasis"[85] is reproducible may not get as much media attention as whether a study claiming you can force yourself to be happy by holding a pen between your teeth[86] can be replicated, but it is of arguably greater consequence.

Summarizing the results of the replication efforts we've seen just in this chapter (table 6.5), no discipline stands out as having an especially great track record.

These failures of replication are compounded by the fact that effect sizes even among the findings that do replicate are generally found to be substantially smaller than what the original studies claimed. This is to some extent the fault of a lack of power, since a low-powered study necessarily reports a larger effect than is really there, as we saw with the urn example. However, boosting the power by making sample sizes larger is not necessarily a cure-all either because it invites the variety of *Type III* errors we saw previously, where a statistical effect may be real in the orthodox sense but lacking in scientific value and reproducibility.

TABLE 6.5 Evidence of replication problems in different areas of science

Field	Estimated replication success rate
Medicine	59% ($n = 34$)
Psychology	36% ($n = 97$)
Social science	62% ($n = 21$)
Preclinical cancer studies	11% ($n = 53$)
Preclinical pharmacology studies	$< 50\%$ ($n > 100$)
Economics	61% ($n = 18$)

The replication crisis wasn't inevitable, even given the problems with statistical methods, and there may still be non-Bayesian ways to lessen its severity. For example, Ioannidis's paper in 2005 gave a Bayesian argument that if the prior probability for a claimed effect was low, then even after having passed a test for statistical significance, the posterior probability should still be low. He predicted this situation would manifest in research areas where many possible effects *could* be studied and there was no way from the start to know which of them would turn out to be real; in effect, this was saying the real-world frequencies would match the probability assignments. But nothing forced researchers to search for hypotheses this way. If the effects people were examining generally had a *high* rate of being true because of some other selection criterion, then we should expect most or almost all of published research findings to be true, naturally.

In particular, then, we might expect the replication problem to be *worse* than we currently think it is if all the results published outside the top tier of prestigious journals are included, since the high-profile studies are generally backed by some research theory that should raise their prior probabilities of being true. Of course, this is somewhat balanced out by a bias toward unexpected results, even in top-tier journals. The institutions of science and the journalists that cover them are not generally interested in studies that simply confirm something we already believe to be true. It's so much more interesting when the hypothesis is *surprising*.

Indeed, there is still no way to be *sure* these replication failures are actually a problem at all. No study ever has 100 percent power, and so many of the replication results are surely Type II errors—failing to reject the null when it's false.

It could be that, through some combination of not repeating the exact conditions of the original experiment correctly and just having bad luck, the replication studies failed to validate effects that are genuinely true. (That is certainly the position taken by many of the authors of the original studies.) But since replication is ultimately supposed to be an integral part of the scientific method and the way pseudoscience like parapsychology gets dismissed by the scientific community, all these findings should be, at the very least, troubling. What does it mean for a scientific theory if one study finds a statistically significant effect and a replication study finds a statistically significant effect in the opposite direction?

Eventually replication efforts will sort the real results from the others. But at what cost? Replication studies are time consuming, expensive, and generally at odds with an academic community that incentivizes new and exciting results. According to the 2016 survey in *Nature*, of the more than 1,500 scientists surveyed across multiple disciplines, only 24 percent had published a replication study of any kind during their career, and only 13 percent had published a failed replication.[87] From the perspective of an already resource-strapped scientist, this makes total sense. How many professors have ever gotten tenure based on attempts to replicate someone else's work?

The real problem, though, is that, for decades, the method for deciding which findings got published has made *no distinction* between those that should have a high posterior likelihood of being true and those that should not. This is where Bayesian analysis could have helped prevent much of the replication crisis. Frequentist methods gave cover for research theories that should never have been taken seriously. By lumping together all claims, including ridiculous ones like Bem's psi hypothesis, that pass a test for significance based *only* on how likely the hypothesis makes the observed data, the orthodox statistical methods made it possible for many false positives to contaminate the published results. Bayesian analysis would have required the truly surprising claims to be substantiated by truly impressive data and, even then, would have cautioned us to look for alternative explanations, as we saw with the Soal-Goldney experiments in chapter 2.

Many will object: How am I supposed to know how high to set the bar? If the burden of proof changes from theory to theory, how will I know how to evaluate any given one? The answer: You do it by examining the claim within the broader context of what you know about the world. That is, in probability terms you need the prior probability $P[H|X]$ for the hypothesis H, and that depends on X, all the information you carry with you about your past observations and the theories you believe about how things work. It makes your life history, wisdom, and experience as a human being relevant. In short, it makes statistics a tool of human understanding, not a mysterious oracle issuing cryptic rulings about significance. Sometimes you will be wrong, but that's life.

To return to Gervais and Norenzayan's *Thinker* study, for example, with a Bayesian mindset we would need to assess the likelihood, before considering the data, that a brief encounter with art could have such an effect on people's religious beliefs. Past experience should make us pretty skeptical, especially given the size of the claimed effect: about a 33 percent reduction in average belief in God after looking at a statue for 15 seconds. If art could have such an influence, we'd find that any trip to a museum would send us careening between belief and nonbelief. Or if *The Thinker* somehow wielded a unique atheistic power, its unveiling in Paris in 1904 should have corresponded with a mass exodus from organized religion. Instead, we experience our own religious beliefs, and those of our society, as relatively stable through time. Maybe we're not so dogmatic as to rule out the *Thinker* → atheism hypothesis altogether, but a prior probability of 1 in 1,000, somewhere between the chances of being dealt a full house and four-of-a-kind in poker, could be around the right order of magnitude. The data, which the authors claimed was unlikely to have arisen by chance, would need to be that much more unlikely to shake us of our skepticism. According to the study, the data was about 12 times more probable under an assumption of an effect of the observed magnitude than it would have been under an assumption of pure chance. Putting this claim into Bayes' theorem with our prior probability assignment, we'd end up saying the probability for the effect based on this experiment was 0.012, or about 1 in 83, a mildly interesting blip but almost certainly not worth publishing.

Among the many causes of the replication crisis is the problem of multiple comparisons—the ability of one or more researchers to sort through the possible associations present in the data until one pops out as being significant just by chance—but Bayesian analysis has a built-in safeguard: the *prior probability*. As we saw with "The Sure-Thing Hypothesis" example in the last chapter, we may pick a single test to apply to our data from among a huge number of possible tests and still end up with a reasonable inference so long as we honestly account for the prior probability associated with that test. That is, when we make our statistical hypothesis more specific (a sequence of 60,000 rolls of a die is predetermined to be *exactly this one*; college students have ESP but only in *this particular way*), we must lower the prior probability given to that hypothesis and set a higher evidentiary standard. Whether one scientist is checking multiple hypotheses to find one that works with a given data set or multiple scientists are all separately trying different ideas until one of them gets lucky makes no difference.

Bayesian analysis also takes care of the "crud factor," the tiny background correlations that tend to exist between any two measured variables and can show up as statistically significant in large enough samples. Instead of just classifying results as significant or insignificant, the Bayesian posterior distribution always properly reports the likely size of any claimed effect or association. Thus, a researcher can gauge at a glance if they've found something truly noteworthy or just an ephemeral quirk of the data, likely to vanish when the experiment is repeated. This also cuts off the converse problem of concluding an effect must not exist because the data didn't show statistical significance. Bayesian analysis does not chase after the question of "whether" but stays grounded in "how much" and "how likely."

It seems now we are in the situation predicted by Meehl in 1967 and for the reason he described: there are researchers in various fields who have accumulated publications but left behind little or nothing of any value because *their methods of analysis never made logical sense in the first place.* The fundamental claim of orthodox statistics is that observing unlikely data under an assumed hypothesis is necessarily a reason to doubt the hypothesis. As people have pointed out countless times over the years, that claim was never logically true, but this seems to be a lesson we are destined to learn and relearn many times. As we have seen, the fallacy was present in the very first statistical inference, Bernoulli's analysis of his urn-drawing examples, and it continues to plague science to this day.

Bernoulli's Fallacy is buried deep in modern scientific practice. It was planted there by the early frequentist statisticians, who were especially motivated to think of statistics as a completely objective discipline free from interpretation or prior judgment. They were able to get away with that mistaken idea because the problems they faced, such as estimating a completely unknown mean value in a population, made it such that the prior information played a relatively minor role, so there was no harm in ignoring it. What the examples shown here demonstrate is that, for many problems of modern science, prior information is often *essential*—to drastically elevate the standards for claims violating our prior understanding of the world (like Bem's ESP experiments); to tell us the effect we're looking for is likely close to nil, so we need high-powered studies if we want to detect it with any likelihood (as in the neuroscience studies considered by Ioannidis and others); to tell us another study has already established a decent estimate for our effect, so we can't take an insignificant result as proof of the null (as in the COX-2 inhibitor trials); or to remind us we are searching for a needle in a haystack, so what we are most often going to find is hay (as in the data mining that takes place in genetics research and elsewhere). Marginal improvements can be made through other changes in methodology, but if statistical techniques continue to ignore the necessary ingredient of prior information, we can expect replication problems to continue.

Since the frequentist methods such as NHST grew out of an incomplete understanding of probability, it's no surprise we find them inadequate to handle all the complex problems of scientific research. To begin to fix modern science, then, we must pull these ideas out by the root and replace them with methods that follow a coherent logic of probabilistic inference.

In the next chapter, we'll give specific directions for how to do just that. The replication *crisis*, from the Greek *krinein* meaning "to decide," offers us a unique chance to decide whether scientists of the future will adopt better research practices or continue to repeat the mistakes of the past. The opportunity cuts both ways. The replication crisis may finally provide sufficient motivation for people to care enough about the logic of statistics that they listen to what the critics have been yelling about for almost a century now, but science will reap the benefits as well. The version of statistics that gets reborn from the ashes of frequentism may help scientists avoid future replication problems, and it might even do something much more important: assist them in understanding the subjects they are truly passionate about. For too long, statistical tests have been largely meaningless hurdles a researcher has had to surmount on the way to publication or, worse, means to add a patina of legitimacy to shoddy experiments that have contributed little of value to the world. If we want statistics to be more than just an empty formality, we must make it a genuinely useful tool with which scientists can better scrutinize their data and that of their peers.

7

THE WAY OUT

I cannot conceal here that I foresee many obstacles in special applica-
tions of these rules that can often lead to shameful mistakes if caution
is not observed.

~Jacob Bernoulli

In this chapter, we'll recommend specific steps that should be taken in order to rid ourselves of Bernoulli's Fallacy once and for all. These are addressed to a general "you," representing the community of scientists, statisticians, psychologists, lawyers, doctors, engineers, economists, management professionals, and anyone else doing any kind of reasoning with probabilities, but it could just as well be read as "us." These things apply to all of us, and no one of them will be easy given the long history of resistance to this way of thinking about probability and statistics. The benefit will be that we will right the ship of statistics and provide the next generation with a way of doing probabilistic inference that handles uncertainty in a logical way. Critics have been complaining about the weather long enough. Now it's time to do something about it.

ABANDON THE FREQUENTIST INTERPRETATION AND ITS LANGUAGE

The orthodox statistical methods were forced to be what they are because they rely on the frequentist interpretation of probability. If all probability can mean is long-run frequency in a series of trials, then the only component of a typical experiment that can have a probability given to it is the data, since this is the only thing that could be said to vary over repeated trials. Basing all probabilistic inferences on this sampling probability of the data therefore creates the awkwardness of the standard methods of statistical inference and makes them subject to Bernoulli's Fallacy.

The idea that probability means frequency, then, is ultimately too narrow to be of any practical use. While it carries a certain appeal because of the empirical connection to what we know to be the right answers for simple questions related to dice or coins, we need probability to answer many more questions than just those. We need a way to describe probabilistic *inference* about past events or possible hypotheses concerning the state of the world, and for this, the frequency view is inapplicable. This broader understanding of probability has been around for a very long time—since before there was such a thing as mathematical probability.

Furthermore, as we saw in chapter 1, the frequentist view isn't really all that convincing even for the problems for which it's designed. The frequentist answer to common probability questions, such as the probability of a coin flip resulting in heads, requires us to stipulate that the answer 1/2 *would* be obtained as the long-run frequency if the sequence of coin flips were performed endlessly, in such a way that some of the conditions are held constant, such as the shape and physical condition of the coin, while others are allowed to vary, such as the exact initial conditions of each flip. No real physical experiment could ever confirm this, and if an experiment were conducted and the long-run frequency appeared to tend toward some number other than 1/2, the frequentist response would be that something was physically wrong with the coin or with the way it was being flipped. So not only is the frequentist answer *not* empirical, but also it's so dogmatic that it would overrule any empirical evidence against it. These issues become even more pronounced when considering the probability of very unlikely events, such as someone winning the lottery or being guilty of murder.

Adapting the frequentist view from simple games of chance to real-world probabilities also carries with it the reference class problem (which characteristics do we take as fixed, and which are allowed to vary?), as was already apparent to Jacob Bernoulli when he considered estimating mortality. This issue makes the supposedly objective frequentist probabilities just as subject to interpretation as any Bayesian prior. Consider, for example, the probabilities in the Sally Clark case, which we encountered in chapter 2. Dr. Roy Meadow came up with the figure of 1 in 73 million for the probability of two infants dying of SIDS by applying adjustments to the observed frequency of SIDS (about 1 in 1,300) to account for what was known about the Clark family: they were nonsmokers with steady jobs, and Sally was over the age of 26. How could he know he had adjusted for all the right factors? Why not include the fact that she and her husband were both solicitors? The more specific information about the Clarks he included, the less available data he would have to go on, until his sample size was reduced to 1. He also assumed pairs of SIDS deaths in a family would be independent, so their probabilities should get multiplied together. This was obviously a faulty assumption, but given the lack of data on such rare events, wouldn't any correction for their dependency be somewhat subjective?

Defining probabilities in terms of populations, as Ronald Fisher did, doesn't help much either. As we saw in chapter 4, the suggestion that all probability is about sampling from a population requires one to imagine that population to be infinitely large and to have the characteristics it's supposed to have in order to make the probability calculations come out correctly. Results like Pierre-Simon Laplace's theorem of the convergence of the sum of any large number of independent variables to a normal distribution would require one to somehow think of those sums as belonging to some population of theoretical sums. Why bother with such a bizarre and artificial construction? Furthermore, how can one ever test counterfactual hypotheses about a population empirically? That is, to say that a sampling probability *would* emerge as the frequency if the population had a property we know it *not* to have requires an imaginary series of samples from an imaginary population. In what sense is this empirical?

Fisher and the other early frequentist statisticians defined probability in terms of sampling from a population because that was usually what they were doing. But physicists, astronomers, chemists, doctors, psychologists, and anyone else not concerned with sampling from a population need a different way of describing probabilities altogether. As Jerzy Neyman himself put it, "The trouble is that what we [statisticians] call modern statistics was developed under strong pressure on the part of biologists. As a result, there is practically nothing done by us which is directly applicable to problems of astronomy."[1]

The better, more complete interpretation of probability is that it measures the *plausibility* of a proposition given some assumed *information*. This extends the notion of deductive reasoning—in which a proposition is derivable as a logical consequence of a set of premises—to situations of incomplete information, where the proposition is made more or less plausible, depending on what is assumed to be known. Deductive reasoning, in this framework, is probability with 1s and 0s. Or viewed another way, probability is deductive reasoning with uncertainty.

Since all probability assignments are therefore dependent on assumed information, we need to adopt a notation for writing probabilities that requires the information to be accounted for. That is, instead of writing $P[A]$ for the probability of some proposition A, we should always write $P[A|X]$ and be forced to say what information X the probability depends on. This may seem like a small change of notation, but it implies a world of difference. In the standard textbooks, this would be called a *conditional probability*, but we need to stop thinking of this as a separate *kind* of probability. All probability is conditional. By using this notation, we are constantly reminded that probability is a function of our knowledge or lack thereof and not an inherently measurable quantity of a physical system. Frequencies are measurable; probabilities are not. Trying to measure probabilities would be like solving an equation for y without knowing what the equation is.

Abandoning the frequentist interpretation of probability doesn't mean that frequencies are irrelevant, though. Instead, they can be a valuable part of the background information on which a probability is conditioned. We may know, for example, that of the last 1,000 times we performed some action, 600 of them produced a desired outcome, and the other 400 produced a different one. Whether that means we assign the probability 0.6 to the desired outcome happening again, though, depends on what else we know about these outcomes. And whether our assigned probabilities actually match the observed frequencies on an ongoing basis may hinge on whether our background assumptions are *correct*. Probability as logic gives us a general framework for using this information to guide us toward a more accurate understanding of the world, whereas probability as frequency gives us only a prediction that may or may not come true.

The other thing we must rid ourselves of is the language of orthodox statistics that it inherited from eugenics. As we saw in chapter 4, the principal founders of what we consider statistical orthodoxy were motivated by a particular set of questions they needed to understand in order to advance their eugenicist agenda. Francis Galton's attempts to understand to what extent human characteristics would be passed from one generation to the next and what characteristics of desirable or undesirable people generally went together led him to the concepts of *correlation* and *regression*. Even though it was Francis Edgeworth who coined the term, Karl Pearson's attempts to discriminate between different subpopulations led to the growth of *significance testing*. It was originally applied to find a significant difference between a population distribution and the normal curve, which Pearson took to suggest the presence of more than one race. Later the tool was applied to detecting the *significant differences* between the racial subgroups themselves, but in both cases, the mere existence of a difference was all that mattered for purposes of discrimination, with no consideration for how big or meaningful the difference was. Fisher's attempts to answer similar questions of variable dependencies led to *linear discriminant analysis* and *analysis of variance*. All of these ideas were developed with the ultimate goal of purifying the human race in the background, and the language of eugenics lingers in all of them still.

So, in addition to giving up the idea of probability as sampling from a population, we should rid ourselves of language that describes members of that population as deviating from or varying around some mean and that asks whether two subpopulations have significant differences and whether we might determine relationships by regression or analysis of variance. In any real problem of probabilistic inference, the things we care about may not be *variable* at all but simply quantities we don't yet fully understand. To speak of our uncertainty as having a standard deviation or variance could lead to nonsense or worse. These terms are misnomers, and they enable destructive statistical thinking because of the wide range of statistical problems that have nothing to do with populations or heredity—but that's not all. They also carry an ethical instruction, calling on us to hunt down deviance and punish impurity, affirming a particular kind of ableist and (in the words of bell hooks) imperialist, white-supremacist, capitalist, patriarchal[2] cultural violence. Surely, as in a linear regression, we can make the model assumption $y = ax + b$ without having to call it the equivalent of *linear mongrelization*.

Following are a few possible replacements for the standard terminology:

- Random variable → Unknown
- Standard deviation → Uncertainty or, expressed via its inverse, precision
- Variance or covariance → Second central moment
- Linear regression → Linear modeling
- Significant difference → N/A (instead, report a probability distribution showing the likely size of the difference)

We must likewise get rid of the statistical terminology surrounding estimators, with all the built-in normative advertisements for one kind of estimator or another. The properties of being *unbiased, consistent,* or *efficient,* and so on, which characterize the sampling distribution of an estimator over many repetitions of the experiment, *aren't meaningful* if what we care about is drawing inferences from a single set of observations. As we saw in chapter 5, the Bayesian methods of inference don't depend on estimators having any of these properties—or on estimators at all. The model that connects hypothetical parameter values to data probabilities gives us everything we need to know.

Nothing ever forces us to make a single estimate of an unknown parameter, and if, for whatever reason, it becomes useful to do so, the posterior probability distribution for the parameter gives us many possibilities from which to choose. We may similarly discard such concepts as *sufficient* and *ancillary* statistics. Those things, if they exist, will be found automatically in the course of doing the inference, and this may happen without our even noticing. Sufficient statistics were useful in the orthodox framework only because they helped reassure the user that no more information was available in the data beyond their chosen estimator; since we know in Bayesian analysis that we're *always* using *all* the available information, we have no need for such reassurances.

One of the more insidious consequences of treating probabilities as though they are real, measurable quantities (outside of quantum theory, arguably) is that it tempts us into thinking that our current uncertainty about something is *final*. That is, it stops us from investigating to try to learn more. As Edwin Jaynes wrote:

> Indeed, there is no such thing as a "stochastic process" in the sense that the individual events have no specific causes. One who views human diseases or machine failures as "stochastic processes", as described in some orthodox text-books, would be led thereby to think that in gathering statistics about them he is measuring the one controlling factor—the physically real "propensity" of a person to get a disease or a machine to fail—and that is the end of it.[3]

Instead of merely concentrating on trying to discern some signal from an inherently noisy set of observations, we should focus more on trying to understand and eliminate the noise. The attention to detail and the creativity this requires has always been what sets great experimental scientists apart. When Louis Pasteur attempted to disprove the theory of spontaneous generation, he was confronted with the fact that early experiments did, indeed, seem to show that life could arise spontaneously under some conditions but not others. Had he simply stopped there, we might today have a stochastic theory of the probability of spontaneous generation; instead, he performed ingenious experiments to determine what circumstances could lead to the presence of life in a growth medium, confirming in each case that it was only through some *contamination* by outside dust that microscopic living organisms were introduced into the medium. It is because of these experiments that we now have the germ theory of disease. One such high-quality experiment is worth a hundred statistical tests rejecting the null hypothesis at whatever level of significance. Once we jettison the bureaucracy of frequentist statistics, we can spend more time doing actual science.

The logical version of probability should be taught as early as possible to prime students to think of statistics in terms of information and plausibility before they get to college. Those who go on to become research scientists will naturally apply Bayesian reasoning to their data, and ultimately this will become the standard expected by journals. The education and publication feedback loop in academia will likely always exist, but it has the potential to be a virtuous rather than a vicious cycle.

Getting rid of the useless concepts (significance testing, estimators, sufficient and ancillary statistics, stochastic processes) will amount to cutting out probably 90 percent of the standard statistics curriculum. It might even mean giving up on statistics as a separate academic discipline altogether, but that's alright. Probability as a topic should rightfully split time between its parents, math and philosophy, the way logic does. Bayesian statistical inference contains exactly one theorem of importance anyway, and its practical techniques can be taught in a single semester-long course in applied math. There needn't be a whole university department dedicated to it, any more than there needs to be a department of the quadratic formula.

DON'T FEAR THE PRIOR

That is, do Bayesian inference. The reason so many people over the years have hesitated to use the Bayesian method is that it requires an assignment of prior probabilities. This has bothered practitioners from Bayes onward because it opens up the possibility that any conclusions we draw from the inference could be heavily dependent on this choice, which feels subjective and arbitrary. As we discussed in the example of the Bertrand paradoxes, there are also theoretical problems with assigning a uniform prior probability distribution for continuous parameters, since this distribution is not invariant with respect to a change of variables (for example, area versus side length as the "unknown" measurement of the size of a square). This was exactly the argument that Fisher repeatedly used to dismiss Bayesian methods as invalid.[4]

Get over it.

First of all, the answer is usually that if the choice of prior probabilities feels arbitrary, it is only because we have not properly specified all the prior information we are bringing to a problem. For example, Jaynes's answer to one of the

Bertrand paradox problems exploited an unspoken assumption that the problem is invariant with respect to location. As is always the case, our probability assignments must respect the invariances that are present in our background information. So it may be that those invariances exist but have not been properly articulated.

Jaynes, in his theory of *transformation groups*, investigated how different "sorts" of ignorance for different types of parameters (scale and location parameters, specifically) would lead to different probability assignments.[5] If changing the location of a problem gives us no useful information about the value of a location parameter, then we must assign it a uniform probability distribution, etc. He also presented a general argument that, in the absence of one of these obvious transformations, we should assign the distribution that maximizes *entropy*, which is simply a way of quantifying the lack of information available to us. (Both of these topics are beyond the reach of this book, but with a little practice, the techniques are not difficult to learn.)

Second, it is often the case that the choice of prior probabilities doesn't really affect the final answer in any significant way. For example, in our examples in chapter 5, to do the Bayesian inferences we occasionally assumed a *uniform* prior for a variable that could have an unbounded range of values, such as the position of our lost spinning robot. Technically this is improper because there is no uniform probability distribution on the whole real line.[6] However, whether we use that prior or the uniform distribution on the interval [−10,000,000, 10,000,000] makes no discernible difference in the final analysis because the data gives us so much information that this feature of the prior distribution is washed away. By letting the absence of a precisely specified probability distribution keep us from attempting the calculation, we are letting the perfect be the enemy of the good enough. A few concrete calculations would show that it often simply doesn't matter, and if it does matter, we may *learn something* about which form of prior information gives us reasonable conclusions about the problem at hand or to which of our assumptions our conclusions are particularly sensitive.

Third, if we want to avoid the charge of being biased in favor of our conclusions, we can start the inference with a very weak prior to express skepticism. We could even *crowdsource* and *preregister* these priors, the same way that many researchers are now preregistering their methods, if we want to avoid the possibility of changing the priors to give us the answers we want. Conversely, a partisan who began an inference with an unreasonably high prior probability in order to reach the conclusion they wanted would at least be required to reveal their bias! It would make for an altogether unimpressive paper if the author started with, say, a 99 percent probability for an effect and then demonstrated results that raised the probability to 99.9 percent. Bayesian reasoning requires us to put all our cards on the table.

Finally, those who defend frequentist statistics have no right to call anything arbitrary. The users of orthodox statistical methods claim to draw inferences only from the objective facts, meaning the sampling probabilities, but they're able to do so only because they make arbitrary decisions about the reference class, arbitrarily assign probabilities to outcomes that have not been observed, and arbitrarily ignore the important role of prior probabilities and alternative hypotheses, as we have shown many times. It's that way of thinking that led to Daryl Bem's proof of ESP being published in a leading journal of psychology. The prior probability plays the same role in inference that the premise plays in logical deduction, and the sampling probability alone is simply not enough information on which to base an inference. Any attempts to do so are illogical. Bayes' theorem is a theorem, as mathematically valid as the Pythagorean theorem, and it makes about as much sense to do statistical inference without the former as it does to make architectural drawings without the latter.

Various compromises do exist to help ease people into the Bayesian way of thinking. In chapter 6, we described the use of Bayes factors, which express the support given by the data alone in favor of a given hypothesis irrespective of our prior probability assignment for that hypothesis. These can provide a friendly middle ground because they allow different readers of a paper with different perspectives on the reasonableness of the hypothesis to all get some meaningful information from the analysis. Ultimately whether a paper is worth publishing should depend on the strength of its conclusions, though, not just its methods, so it will be somewhat up to consensus opinion whether a certain amount of support for an unreasonable hypothesis is deemed important or interesting.

As we saw in Wagenmakers's analysis of Bem's results and the various replication efforts since, there may still be some leftover Bayesian analysis needed to come up with the Bayes factors, anyway, in the form of prior probability distributions for the effect *size*, assuming the effect is real. For whatever reason, this seems to be easier for many people to swallow. Another kind of compromise that we saw in chapter 2 is *prior elicitation*: turning the Bayesian problem around to ask what prior probability would be *necessary* in order for a given amount of evidence to be persuasive. Again, this contains exactly the same informational content as the Bayesian inference but can sometimes give a more palatable presentation of the results.

But as we saw in "The Problem of Divided Data" example in chapter 5, the other key difference is that Bayesian inference is never over! In the Bayesian paradigm, we never claim a final answer. That is, we may begin a problem with one assumed state of uncertainty about, say, a parameter value and end the problem with another state of uncertainty, but we never claim to definitively reject any particular hypothesis about the value so long as it's consistent with the end state of our knowledge. This provides a much richer method of analysis than significance testing, which reduces the entire procedure down to a binary yes/no decision about rejecting the hypothesis.

IGNORE THE DATA YOU DIDN'T GET; FOCUS ON THE HYPOTHESES YOU DIDN'T ASSUME

That is, stop using null-hypothesis significance testing (p-values). As we saw in the last chapter, there is a long history of criticism of NHST, specifically concerning the fact that p-values are often misinterpreted by the researchers who use them. In 2012, Charles Lambdin compiled a list of 12 common misconceptions about p-values:

1. A p value is the probability the results will replicate if the study is conducted again (false).
2. We should have more confidence in p values obtained with larger Ns than smaller Ns (this is not only false but backwards).
3. A p value is a measure of the degree of confidence in the obtained result (false).
4. A p value automates the process of making an inductive inference (false, you still have to do that yourself—and most don't bother).
5. Significance testing lends objectivity to the inferential process (it really doesn't).
6. A p value is an inference from population parameters to our research hypothesis (false, it is only an inference from sample statistics to population parameters).
7. A p value is a measure of the confidence we should have in the veracity of our research hypothesis (false).
8. A p value tells you something about the members of your sample (no it doesn't).
9. A p value is a measure of the validity of the inductions made based on the results (false).
10. A p value is the probability the null is true (or false) given the data (it is not).
11. A p value is the probability the alternative hypothesis is true (or false; this is false).
12. A p value is the probability that the results obtained occurred due to chance (very popular but nevertheless false).[7]

By 2014, the situation had gotten so bad, especially in light of the ongoing crisis of replication, that the new editor of the journal *Basic and Applied Social Psychology (BASP)*, David Trafimow, published an editorial stating that the journal would no longer require submissions to use inferential statistics—specifically, NHST. The editorial stated bluntly: "The null hypothesis significance testing procedure has been shown to be logically invalid and to provide little information about the actual likelihood of either the null or experimental hypothesis."[8] The following year the journal banned NHST altogether.[9] In order to publish in the journal now, authors must remove all references to significant differences, *p*-values, and the tests that generate them.

Even in the *Nature* article announcing that *BASP* had banned *p*-values because they were so often misinterpreted, the author misinterpreted *p*-values, defining them by stating "The closer to zero the *P* value gets, the greater the chance the null hypothesis is false."[10] The journal subsequently issued this correction: "*P* values do not give the probability that a null hypothesis is false, they give the probability of obtaining data at least as extreme as those observed, if the null hypothesis was true. It is by convention that smaller *P* values are interpreted as stronger evidence that the null hypothesis is false." The fact that this is done "by convention" and not in reference to any actual probabilities is exactly the problem.

In 2015, the American Statistical Association (ASA) convened a two-day meeting of experts from various backgrounds in statistics and scientific research to come up with an official stance on *p*-values and statistical significance. After multiple revisions over the following months, they issued a statement with six core principles, one of which did not mention *p*-values at all and four of which concerned what *p*-values are *not*:

1. *P*-values can indicate how incompatible the data are with a specified statistical model.
2. *P*-values do not measure the probability that the studied hypothesis is true, or the probability that the data were produced by random chance alone.
3. Scientific conclusions and business or policy decisions should not be based only on whether a *p*-value passes a specific threshold.
4. Proper inference requires full reporting and transparency.
5. A *p*-value, or statistical significance, does not measure the size of an effect or the importance of a result.
6. By itself, a *p*-value does not provide a good measure of evidence regarding a model or hypothesis.[11]

The one positive statement about p-values—that they can indicate how incompatible the data are with a specified model—is incomplete to the point of being misleading. There is no sense in which data can be said to be incompatible with a given model unless the *alternatives* to the model are specified. Without referencing these alternatives, the orthodox p-value argument essentially claims something like "This observation and the ones more extreme than it are very improbable under the null hypothesis; therefore, something is wrong with the null hypothesis." Careful frequentists will, of course, stop short of saying the hypothesis itself is unlikely, mostly because that claim has no meaning in frequentist thought, but still, according to the procedure, we have reason to doubt the hypothesis or even reject it.

As we have seen, this is the same mistake of thinking at the heart of base rate neglect, or the prosecutor's fallacy, and many other examples we have considered. While it is true that the fact the data is unlikely under a given hypothesis has the *potential* to give us reason to doubt the hypothesis, we can do so only if we have *another* hypothesis at the ready, along with its prior probability and the probability it assigns to the data. Trying to do inference without these other ingredients is what makes significance testing such an awkward and illogical enterprise. A cup of sugar has the potential, in the presence of other ingredients, to form a cake, but no amount of creativity can turn it into a cake on its own.

As we discussed in chapter 5, the definition of the p-value contains a tacit admission of its own inadequacy in the sense that the value isn't simply equal to the probability of the data under the null hypothesis. Instead, it requires us to imagine what other, more extreme data we *could* have gotten. The choice of what counts as more extreme often reveals the alternative hypotheses we're implicitly testing against; tail regions of the probability distribution are usually chosen because under a different hypothesis they'd be less "tail-like."

Statistical inference using p-values only sort of works under special circumstances for certain kinds of problems and falls apart for other common problems, as we saw in chapter 5. In an article in the *American Statistician* accompanying the ASA's statement on p-values, Ronald Wasserstein (ASA's executive director) and Nicole Lazar (editor in chief of the journal) gave the context explaining why the statement was deemed necessary but also added that these problems had been understood for a long time:

> Let us be clear. Nothing in the ASA statement is new. Statisticians and others have been sounding the alarm about these matters for decades, to little avail. We hoped that a statement from the world's largest professional association of

statisticians would open a fresh discussion and draw renewed and vigorous attention to changing the practice of science with regards to the use of statistical inference.[12]

But the ASA's "clarifying" comments and its implicit blessing to keep using p-values so long as their meaning is properly understood and they are supported by other modes of analysis all miss the point. *P*-values were taught to begin with only *because* of their potential to be misinterpreted. They seemed to give the right answers for the survey sampling problems Pearson and Fisher considered because they overlapped with the Bayesian methods for those problems, as we have seen. The conditions that made this overlap possible were (1) a simple, clearly defined set of alternative hypotheses and (2) weak prior information showing no preference for one hypothesis over another, neither of which can be expected to hold true in general. The fundamental logic of significance testing was always flawed from the beginning. We should no more be teaching p-values in statistics courses than we should be teaching phrenology in medical schools.

To combat the problem of replication—in particular, the high incidence of false positive results that seem to be made possible by the mechanical application of the "$p < 0.05$" threshold—many people have suggested lowering that threshold to something more difficult. For example, a 2017 paper in *Nature Human Behavior* coauthored by 72 scientists and statisticians advocated changing the standard from $p < 0.05$ to $p < \mathbf{0.005}$.[13]

In spirit, this would actually move significance testing closer to how Fisher intended it to be used. In an oft-cited passage from *The Design of Experiments*, he once tried to draw a distinction between a single experiment producing a significant result and an experimental *procedure*:

> In order to assert that a natural phenomenon is experimentally demonstrable we need, not an isolated record, but a reliable method of procedure. In relation to the test of significance, we may say that a phenomenon is experimentally demonstrable when we know how to conduct an experiment which will rarely fail to give us a statistically significant result.[14]

Of course, the distinction is illusory, since an experiment repeated many times is functionally the same as a larger experiment with more trials. The assumption of independence between trials under the null hypothesis means that the data probabilities are simply multiplied together, so how we choose to group the samples for the purpose of performing that multiplication is arbitrary.

If, for example, significance of an experimental result was defined by a certain test statistic S falling in the rightmost 5 percent of its probability distribution under H_0, then insisting that it do so twice in a row in two subsequent repetitions of the experiment is mathematically equivalent to insisting that the pair (S_1, S_2) fall in the "upper-rightmost corner" of its distribution, an event that under H_0 has probability $(0.05)^2 = 0.0025$. Fisher's "rarely fail" criterion, if it were ever specified with precision, could easily be translated into a single test for a similarly defined joint statistic, with a correspondingly tiny threshold for significance.

While raising the bar for what counts as significant would certainly help reduce the number of false positives (at the expense of more false negatives), it would not fix the underlying issue that significance testing makes no sense.

Another popular remedy is to report confidence intervals for any estimated effects instead of just single yes/no decisions about whether the effect exists at some level of significance. As we saw in chapter 5, confidence intervals do provide more information than single tests of significance, but they're all based on the same underlying idea. A confidence interval reports all the hypotheses that would not have been rejected at the chosen level of significance, so the logic for each one of them is the same as the single null-hypothesis significance test. Whether the null hypothesis is rejected at the 5 percent level is equivalent to the "null effect" value being contained in the 95 percent confidence interval, and so on.

All of these non-Bayesian approaches suffer from the same inherent problem, which is that they attempt to do inference armed only with the sampling probability for the data; that is, they commit Bernoulli's Fallacy. Significance tests suffer from a particular pathology in that they base their inferences not only on the observed data itself but also on the *possible* data, of an even more extreme character, that *could* have been observed. As we saw with "The Problem of Optional Stopping," "The Lucky Experimenter," and "The Malfunctioning Digital Scale" examples in chapter 5, different ideas of how the experiment could have gone—and what counts as more extreme—can lead to very different inferences.

The Bayesian method does not suffer from these illusions because it is oriented in the correct logical direction: based on a given observation, we assign probabilities to the various hypotheses we care about. What matters are the alternative *hypotheses*, not the possibly alternative *data*. Like Sherlock Holmes solving a murder mystery, we're concerned only with making inferences based on the facts of the case as given, not with the hypothetical crimes a suspect could have committed but did not.

GET USED TO APPROXIMATE ANSWERS

Probability has presented computational difficulties since its earliest days. It's almost certain that the development of theoretical probability was delayed significantly because all of the interesting nontrivial examples are hard to compute. We saw in chapter 3 how even a simple binomial distribution problem could involve computations that were very laborious to do by hand. This is why Abraham de Moivre's discovery of the normal approximation to the binomial was so important.

Bernoulli had approached his original problem by means of a more computationally friendly lower bound, but it came at the cost of making his method impractical. At the very end of *Ars Conjectandi*, he included an example of his method to compute a sufficient sample size so that a sample from an urn with a ratio of white:total pebbles of 3:5 would have a probability at least 1,000 times greater of being inside the interval [29/50, 31/50] than outside it. That is, he found the sample size n necessary for the sample ratio to be within 1/50 of its true value with a "moral certainty" defined by a probability of 1,000/1,001. The sample size he came up with was 25,500. Stephen Stigler suggested that a possible reason this example came at the very end of the book (with some perfunctory closing remarks likely added later by Bernoulli's nephew) was that Bernoulli saw that large value for n and simply quit. At the time, a sample of size 25,500 for any real-world experiment was ridiculously implausible; that was greater than the population of Basel, Switzerland, where Bernoulli lived.[15]

Even in the time of Pearson and Fisher, the computations required for probability and statistics problems could still be onerous. As we saw in chapter 4, part of Fisher's genius was his ability to compute various quantities—such as the sampling distribution of the sample correlation coefficient or Gosset's t-statistic—*analytically*, meaning he could derive exact formulas for the answers. But he was perhaps trapped in a bubble of thinking about only those questions because they were the ones he could answer, and the problems he was concerned with gave him the reassurance that they were enough. For example, Fisher's usual assumption was that a population was normally distributed, which allowed him to use all the pleasant mathematical properties of the exponential function embedded in the normal distribution. As we described in our "Finding a Lost Spinning Robot" example, had some important statistic been assumed to have a Cauchy distribution instead, none of those mathematical niceties would be present. So Fisher's fondness for concepts like sufficient statistics might have been shaken if these problems, for which no sufficient statistic exists, had actually been of relevance to him. Conversely, if he had cared less about being able to compute sufficient statistics, he might have explored the world of possibilities, including Cauchy-distributed data.

The affinity felt by the early statisticians for the normal distribution is still present in the way that data sets are usually discussed in terms of a few summary statistics: the mean, the standard deviation, and, if the data contains pairs of values, their correlation coefficient. This makes sense if the data can be assumed to be normally distributed, but for other distributions, they might not tell the whole story. For example, in 1973 the English statistician Francis Anscombe constructed an example of four different data sets, now called *Anscombe's quartet*, all having the same, or very nearly the same, summary statistics for the two variables x and y.[16] That is, the four data sets shared these characteristics: sample mean of x, sample mean of y, sample variance of x, sample variance of y, and sample correlation between x and y. Based just on that summary, we might expect them to look roughly the same too. Yet a glance at the graphs of the data sets in figure 7.1 reveals some obvious differences.

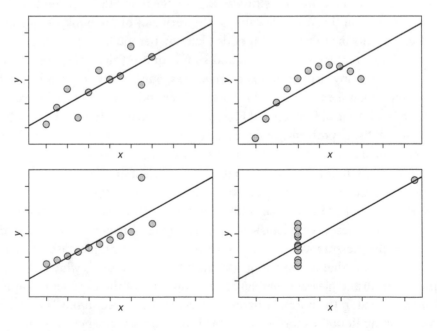

FIGURE 7.1

Anscombe's quartet.

Source: Wikimedia.

The top-left data set looks like what we'd expect from a pair of normally distributed variables with some strong correlation, a linear relationship with some dispersion around the line of best fit. The others clearly have different kinds of relationships: perfectly nonlinear (top right), perfectly linear with a single outlier (bottom left), and no variation in x at all apart from a single outlier (bottom right). Anscombe's point was that graphs can be an important analytical tool to discern patterns in the data and that we should not be afraid to trust our eyes. Joseph Berkson called this the "interocular traumatic test"; you know what the data means when the conclusion hits you right between the eyes.[17] But a graph is necessarily an approximation. Working with nonnormal data may frequently require other kinds of analysis. And in many cases, if we try to model the nonnormality of the data, we may be able to carry out the calculations only approximately because no exact answer exists.

The same phenomenon often gets in the way of doing Bayesian analysis analytically. Depending on the combination of our assignment of the prior distribution and the sampling probability for the data, we may be left with calculations that involve some horrendous sums or integrals that we can't calculate exactly. For example, this would not be a completely unusual situation facing the modern Bayesian scientist: Suppose we're trying to model a linear relationship between two variables x and y, $y = ax + b$, with an unknown slope a and intercept b, and we assume that every measurement of y comes with some error term ε, to which we assign a normal distribution with an uncertainty σ. Maybe, though, this is a new technique of measurement, and we don't know the typical size of the error, so we assign σ itself a probability distribution; according to Jaynes's guidance concerning scale parameters, we would likely set this distribution to be proportional to $1/\sigma$. Furthermore, we might have restrictions on a and b; suppose we suspect b is likely close to 0, according to a normal distribution with uncertainty 1, and we know a must be positive, so we assign it an exponential distribution proportional to e^{-a}. Imagine what we really care about, for our scientific theory, is whether a is greater than 1 and b is positive. Putting all these assumptions together would then require us to compute an integral of pathway probabilities of the following kind:

$$\int_1^\infty e^{-a} \int_0^\infty e^{-b^2/2} \int_0^\infty \frac{1}{\sigma} \prod_{i=1}^n e^{-\left(y_i - (ax_i + b)\right)^2/(2\sigma^2)}\, d\sigma\ db\ da$$

Analytically that expression is nothing short of disgusting. But the good news is that we don't *have to* evaluate every expression analytically if a numerical approximation will suffice. Modern computational techniques make this kind of thing a snap. If we need to compute a tough integral, we can either divide up the parameter space into some number of grid points and use them up to approximate the integral, or if that proves inadequate, we can use Markov Chain Monte Carlo methods to simulate random variables with the appropriate distributions. The statistical language R has a wide variety of user-friendly statistical techniques already implemented, and the powerful Stan programming language implements numerically efficient ways of approximating the kinds of calculations that are needed for Bayesian inference problems.

In the world of Bayesian inference, we may take any given constant, like the scale of our measurement error in the above example, and decide to treat it as an unknown with a probability distribution. That distribution may, in turn, depend on unknown parameters, to which we can assign probability distributions, etc. Some of those parameters may be what are called *nuisance parameters* because we don't have a way of observing them directly—only their second- or third-order effects on the data we *can* observe. But these aren't such a nuisance for us at all because we can always sum (or integrate) over their probability distributions and condition them out.

So, among other things, Bayesian inference is *more fun* than frequentist inference because we are free to play around with different model structures to our heart's delight, once we rid ourselves of the burden of needing every question to have an exact analytical answer. Of course, for any given amount of data, the strength of the inferences we're able to draw about a model will decrease with the model's complexity. But our inferences will contain built-in warnings in the form of the widths of the posterior distributions. That is, if we try to fit a model with too many parameters to a relatively small data set, we will find that the posterior distributions indicate a great amount of uncertainty, as well they should. This is the means by which Bayesian inference protects us from the danger of *overfitting*, which frequentist methods typically do not. A frequentist model inference typically returns a single estimated value for any parameter, with no built-in warnings about the uncertainty of that estimate, unless it falls into one of the cases where an analytic distribution (and hence a p-value) exists.

Bernoulli was stymied in his analysis by computational problems. Fisher sought out questions he could answer analytically, but they were the wrong questions. If we're willing to live with approximations, we can ask the right ones.

GIVE UP ON OBJECTIVITY; TRY FOR VALIDITY INSTEAD

Finally, we must get over the obsession with objectivity that has plagued statisticians since the earliest days of statistics. This obsession has fluctuated over the years but reached its zenith during the late 19th and early 20th centuries when the first modern statisticians—namely, Galton, Pearson, and Fisher—were setting the terms of what statistics would mean for the next 100 years. As we have seen, they were operating under particularly contentious circumstances. The project begun by Adolphe Quetelet of importing probabilistic methods into the social sciences had not yet been completely accepted as valid, the theory of evolution was under attack from all sides and needed theoretical support, and the statisticians were trying to advance a controversial agenda of eugenics that could not allow for the appearance of subjectivity if it was going to be accepted as scientific truth.

In *Social Sciences as Sorcery*, Stanislav Andreski argued that, in their search for objectivity, social science researchers had settled for a cheap version of it, which allowed them to hide behind mechanical methods to "churn out the tedious door-to-door surveys which pass for sociology."[18] Instead, he wrote, we should concern ourselves with the *moral* objectivity we need to simultaneously live in the world and study it:

> The ideal of objectivity is much more complex and elusive than the pedlars of methodological gimmicks would have us believe; and . . . it requires much more than an adherence to the technical rules of verification, or recourse to recondite unemotive terminology: namely, a moral commitment to justice—the will to be fair to people and institutions, to avoid the temptations of wishful and venomous thinking, and the courage to resist threats and enticements.[19]

Even if we express our theories in the most precise technical language and back them with the most exact measurements, we cannot escape the fact that all science is a human enterprise and is therefore subject to human desire, prejudice, consensus, and interpretation. What we can do is try (and most often fail) to be honest about the factors that influence us and avoid serving any unjust masters who would push us toward whatever research conclusions suit them best. The eugenics movement, for example, should be understood as a cautionary tale about the dangers of failing to do this introspection while attempting to sail under the flag of objectivity. In other words, we should try to be objective—not in the impossible sense that Galton, Pearson, and Fisher claimed granted them unquestionable authority but in the way they failed to be when they let the political interests of the ruling class dictate the outcome of their research before it began.

The rallying cry of frequentist statistics has always been that we should let the data speak for itself. As Fisher wrote in 1932, "Conclusions can be drawn from data alone . . . if the questions we ask seem to require knowledge prior to these, it is because . . . we have been asking somewhat the wrong questions."[20] But the harsh reality is that the data does not speak for itself and never has. The history of science consists entirely of examples when a given observation could have been interpreted multiple ways, and it was up to the scientist and the larger community to determine which of those explanations would be accepted. Not even astronomy, the supposedly exact science where probability first started to show its usefulness outside of gambling questions, has been immune. For example, when astronomers of the 1800s first noticed that Uranus was deviating from its predicted orbit, what they were presented with, logically speaking, was a contradiction between their observations and their assumptions:

1. There are seven planets in the Solar system.
2. The recorded positions of Uranus are reasonably accurate.
3. Newton's laws of gravitation are correct.

They took this to mean there must be another planet out there, rejecting assumption (1), and, indeed, in September 1846 Johann Gottfried Galle found Neptune in the night sky almost exactly where Urban Le Verrier predicted it had to be.

Why didn't they give up on Newton's laws instead? The answer is that they had a higher *prior probability* for the correctness of the Newtonian theory than they did for the nonexistence of an eighth planet. They could just as well have concluded that *both* Newton's laws and the hypothesis of seven planets were correct and that their observations of the position of Uranus had simply been in error. However, they had evidence against this in the form of assigned probability distributions for the errors of their astronomical observations. We believe their inferences were correct, but in no way did the data speak for itself.

Prior probabilities, and the subjectivity they represent, are the reason we believe in things like the existence of Neptune and we don't believe in results like the existence of cold fusion, demonstrated empirically by Martin Fleischmann and Stanley Pons in 1989, or Daryl Bem's proof of the existence of psi, or any number of other bogus scientific-looking findings. This is a good thing!

The scientific community has been acting subjectively this entire time by accepting some of these results but not others. Until we come to terms with the reality of what scientific inference is, though, we will always be tempted to rely on frequentist methods of inference because they *promise* objectivity, and we will have no language with which to distinguish the bogus experiments that appear to pass all the necessary tests from the legitimate ones.

In his 1981 article titled "Son of Seven Sexes: The Social Destruction of a Physical Phenomenon"[21] and, in greater detail, in his 1998 book with Trevor Pinch titled *The Golem: What You Should Know About Science*, the British sociologist of science Harry Collins argued that scientific experiments are subject to "experimenter's regress," meaning a logical loop between experimental evidence and theory.[22] The results of experiments, particularly surprising or controversial ones, can be trusted only if the experiments are known to be sound; however, as is often the case, an experiment is *known* to be sound only if it produces the results we expect. So it would seem that no experiment can ever convince us of something surprising. This situation was anticipated by the ancient Greek philosopher Sextus Empiricus. In a skepticism of induction that predated David Hume's by 1,500 years, he wrote: "If they shall judge the intellects by the senses, and the senses by the intellect, this involves circular reasoning inasmuch as it is required that the intellects should be judged first in order that the sense may be judged, and the senses be first scrutinized in order that the intellects may be tested [hence] we possess no means by which to judge objects."[23] In other words, you can't know what's there until you observe it, but you can't trust your observations unless they confirm what's actually there.

If we insist on perfect objectivity, this loop poses an enormous scientific problem. However, as Thomas Bayes attempted to demonstrate in his response to Hume that spawned an entirely new mode of probabilistic inference, if we are willing to settle for *probable* knowledge instead, then we may have a way out of this trap. More and more accumulated evidence can make us more certain of the truth of our theories—but only within the subjective confines of the assumptions we make about the world. But this is a practical necessity anyway, so let's get on with it. In fact, this way of establishing credibility and authority by degrees is much more the norm than deductive proof. We accept authoritative judgments all the time in our daily lives at every level from restaurant recommendations to national politics, not because those judgments have been conclusively proven to be correct but because the statements accord with our prior understanding of the world and because the authority figures have shown themselves to be credible, sound, and trustworthy—in other words, *probable*—through previous experience. Scientific authority is no different.

The true logical tools of science have never been Aristotle's *syllogisms* but rather his *enthymemes*: partial syllogisms with missing premises and logical leaps, based on probable truths and consensus opinions. Assembling those pieces correctly is the essence of mathematical probability theory. We can never hope to determine whether our assumptions are *true* (that is, until we have other possible assumptions to compare them against, relying on even further assumptions), but we may have a chance of determining whether we have processed their consequences in a way that is *valid*.

Bernoulli's mathematical treatment of probabilistic inference was a gift to the world, but unfortunately it also contained the seeds of a great fallacy that has grown to such proportions that it now threatens the integrity of diverse fields from science to medicine to legal justice. The problem was that he failed to appreciate the inherent subjectivity of probability assignments and claimed an objective inference based purely on data was possible. This was a dangerous idea then, and it continues to be dangerous to this day. Objective probability appeared to work at the time only because the simple examples like tossing dice and drawing from an urn seemed to lend themselves to this way of thinking. While making intuitive sense, these examples had the unfortunate consequence of muddling the definition of probability right from the start, as the properties of those physical systems make the probabilities nearly a fact of nature, leading one reasonably to conclude that probability is a *quality of objects* rather than of our *perception and knowledge of those objects*. They reduced what should have been a philosophical question to a counting problem about enumerating a set of outcomes, with answers that would seem to be confirmed if we actually started doing experiments and tallying up results.

That version of probability immediately proves itself inadequate when we try to apply it to our daily lives. Probability now saturates the conversations we have about the world around us, from politics to economics to sports. We find ourselves speaking the language of probability to each other on a regular basis, using words like *odds* and *likelihood* but without a common understanding of what we mean. Part of the reason that people respond so viscerally to probability questions is that we can sense this inner tension. We all seem to recognize that these tools are important, even essential, to conducting our lives, but none of us (myself included) are perfectly adept at using them.

Understanding the idea of probability as logic gives us some insight into the question of why people react so strongly to probability arguments.

Once we accept that probability has something to do with our inner state of uncertainty about a proposition given a set of assumptions, we are led to the somewhat destabilizing conclusion that, when we make judgments of a probabilistic nature, *our assumptions matter*. Since people will usually not share the same prior information, it may be that others would, when faced with the same hypothetical situation, reach entirely different conclusions. A quantification of uncertainty requires us to first admit *that we are uncertain*, and that realization may occasionally say more about us than we are ready to hear. It's one thing to be uncertain about the result of a coin toss—but something else entirely to be uncertain about whether a debt will be repaid or whether a person convicted of murder is actually innocent. The tools of probability can assist us with determining the consequences of a set of assumptions, but they cannot justify those assumptions for us.

Probability is at once close to us and yet always receding, just out of reach. It has a beautifully rich mathematical theory— given elegant formalism in the work of such luminaries as Laplace, Carl Friedrich Gauss, and Andreï Kolmogorov—but that theory is mediated through common words like *average* and *chance*, and it promises to assist us with solving our everyday problems. As such, it is the most active fault line there is between theory and application. It is, at its best, a lens through which fuzzy ideas come more into focus, but too often it can be a tool of manipulation and specious argumentation instead.

Probability opens doors for us and then trips us up. With a handful of simple concepts and definitions, it brings a whole world of problems more under our control but also makes us vulnerable to fallacies we otherwise might never have considered. It is perhaps the defining logic of our time, an era when more and more of our decision-making is based on data and statistical analysis. The Age of Information is often also the Age of Incomplete and Contradictory Information, and the various information sources we depend on are frequently misleading or full of holes.

Probability is ultimately nothing more than a codification of our ability to reason with less than perfect information, as we are constantly required to do. Data is wonderful to have, but data can also be manipulated and distorted to tell a story that isn't true or to suggest certainty where really there is none. Probability can be the antidote. It reminds us—or perhaps provides the language with which we remind *ourselves*—of the boundaries of our understanding, and it keeps us conscious of the fact that abundant data is not the same thing as knowledge. The era of Big Data must also therefore be an era of Big Probability.

But first we need to understand what we're doing.

NOTES

INTRODUCTION

1. Edwin T. Jaynes, *Probability Theory: The Logic of Science*, ed. G. Larry Bretthorst (Cambridge: Cambridge University Press, 2003), 80.
2. Jaynes, *Probability Theory*, 281.
3. Jaynes, *Probability Theory*, xix.
4. Jaynes, *Probability Theory*, 31–32.
5. Dennis Newman, "The History of Statistics in the 17th and 18th Centuries, Against the Changing Background of Intellectual, Scientific and Religious Thought: Lectures by Karl Pearson, 1921–1933," *Journal of the Royal Statistical Society: Series A (General)* 143, no. 1 (1980): 78–79.
6. Fritz Strack, Leonard L. Martin, and Sabine Stepper, "Inhibiting and Facilitating Conditions of the Human Smile: A Nonobtrusive Test of the Facial Feedback Hypothesis," *Journal of Personality and Social Psychology* 54, no. 5 (1988): 768.
7. Eric-Jan Wagenmakers et al., "Registered Replication Report: Strack, Martin, & Stepper (1988)," *Perspectives on Psychological Science* 11, no. 6 (2016): 917–928.
8. John A. Bargh, Mark Chen, and Lara Burrows, "Automaticity of Social Behavior: Direct Effects of Trait Construct and Stereotype Activation on Action," *Journal of Personality and Social Psychology* 71, no. 2 (1996): 230.
9. Stéphane Doyen et al., "Behavioral Priming: It's All in the Mind, but Whose Mind?," *PLOS One* 7, no. 1 (2012).
10. Dana R. Carney, Amy J. C. Cuddy, and Andy J. Yap, "Power Posing: Brief Nonverbal Displays Affect Neuroendocrine Levels and Risk Tolerance," *Psychological Science* 21, no. 10 (2010): 1363–1368.
11. Eva Ranehill et al., "Assessing the Robustness of Power Posing: No Effect on Hormones and Risk Tolerance in a Large Sample of Men and Women," *Psychological Science* 26, no. 5 (2015): 653–656.
12. Chandra Sripada, Daniel Kessler, and John Jonides, "Methylphenidate Blocks Effort-Induced Depletion of Regulatory Control in Healthy Volunteers," *Psychological Science* 25, no. 6 (2014): 1227–1234.
13. Martin S. Hagger et al., "A Multilab Preregistered Replication of the Ego-Depletion Effect," *Perspectives on Psychological Science* 11, no. 4 (2016): 546–573.
14. Monya Baker, "Is There a Reproducibility Crisis?," *Nature* 533 (2016): 452–454.

1. WHAT IS PROBABILITY?

1. John Edwin Sandys, ed., *The Rhetoric of Aristotle* (Cambridge: Cambridge University Press, 1909), 139.
2. Edward P. J. Corbett, W. Rhys Roberts, and Ingram Bywater, *The Rhetoric and the Poetics of Aristotle* (New York: Modern Library, 1984), 140.
3. Corbett et al., *The Rhetoric and the Poetics of Aristotle*, 22.
4. Marcus Tullius Cicero, *De Natura Deorum. I.* (Bryn Mawr, Pa.: Thomas Library, Bryn Mawr College, 1997), 5.
5. Jacob Bernoulli, *On the Law of Large Numbers*, trans. Oscar Sheynin (Berlin 2005), 10, http://www.sheynin.de/download/bernoulli.pdf.
6. Bernoulli, *On the Law of Large Numbers*, 10.
7. Bernoulli, *On the Law of Large Numbers*, 20.
8. Bernoulli, *On the Law of Large Numbers*, 19.
9. John Venn, *The Logic of Chance: An Essay on the Foundations and Province of the Theory of Probability, with Especial Reference to Its Logical Bearings and Its Application to Moral and Social Science, and to Statistics*, 3rd ed. (London: Macmillan, 1888), vi.
10. Venn, *The Logic of Chance*, 97–98.
11. Pierre-Simon Laplace, *Essai philosophique sur les probabilités* (Paris: Bachelier, 1825), 74.
12. Venn, *The Logic of Chance*, 96.
13. Ronald A. Fisher, *Statistical Methods and Scientific Inference* (Edinburgh: Oliver and Boyd, 1956), 25.
14. Venn, *The Logic of Chance*, 14.
15. Actually, at all but a finite number of points, but whatever.
16. Venn, *The Logic of Chance*, 225.
17. Stephen M. Stigler, "The True Title of Bayes's Essay," *Statistical Science* 28, no. 3 (2013): 283–288.
18. Thomas Bayes, "LII. An Essay Towards Solving a Problem in the Doctrine of Chances. By the Late Rev. Mr. Bayes, F.R.S. Communicated by Mr. Price, in a Letter to John Canton, A.M.F.R.S," *Philosophical Transactions of the Royal Society of London* 53 (1763): 370–418.
19. Pierre-Simon Laplace, *Théorie analytique des Probabilités*, 2 vols. (Paris: Courcier Imprimeur, 1812).
20. Laplace, *Essai philosophique sur les probabilités*, 7.
21. Venn, *The Logic of Chance*, 208.
22. Laplace, *Théorie analytique des Probabilités*.
23. J. M. Keynes, *A Treatise on Probability* (London: Macmillan, 1962).
24. Frank P. Ramsey, "Truth and Probability," in Horacio Arló-Costa, Victor F. Hendricks, and Johan van Benthem, eds., *Readings in Formal Epistemology* (Cham, Switz.: Springer, 2016), 21–45.
25. Harold Jeffreys, *Theory of Probability*, 3rd ed. (Oxford: Clarendon Press, 1961), 401.
26. Benjamin Yandell, *The Honors Class: Hilbert's Problems and Their Solvers* (Boca Raton, FL: CRC Press, 2001).
27. Andreï Nikolaevich Kolmogorov, *Foundations of the Theory of Probability*, trans. Nathan Morrison, with an added bibliography by Albert T. Bharucha-Reid, 2nd English ed. (Mineola, NY: Dover, 2018).
28. Lewis Campbell and William Garnett. *The Life of James Clerk Maxwell: With a Selection from His Correspondence and Occasional Writings and a Sketch of His Contributions to Science* (New York: Macmillan, 1882), 143.

29. Keynes, *A Treatise on Probability*, 10.
30. Harold Jeffreys, "Probability, Statistics, and the Theory of Errors," *Proceedings of the Royal Society of London. Series A, Containing Papers of a Mathematical and Physical Character* 140, no. 842 (1933): 527–528.
31. Ronald A. Fisher, "Probability Likelihood and Quantity of Information in the Logic of Uncertain Inference," *Proceedings of the Royal Society of London. Series A, Containing Papers of a Mathematical and Physical Character* 146, no. 856 (1934): 3.
32. Edwin T. Jaynes, *Probability Theory: The Logic of Science*, ed. G. Larry Bretthorst (Cambridge: Cambridge University Press, 2003).
33. Jaynes, *Probability Theory*, 74.
34. Jaynes, *Probability Theory*, 52.
35. Jaynes, *Probability Theory*, 293.
36. Jaynes, *Probability Theory*, 82–83.
37. Jaynes, *Probability Theory*, 411.
38. Jaynes, *Probability Theory*, 89.
39. Martin Gardner, *The 2nd Scientific American Book of Mathematical Puzzles and Diversions* (New York: Simon and Schuster, 1961), 51.
40. Assuming each child to have exactly one of two genders, already only an approximation of the truth.
41. Craig R. Fox and Jonathan Levav, "Partition-Edit-Count: Naive Extensional Reasoning in Judgment of Conditional Probability," *Journal of Experimental Psychology: General* 133, no. 4 (2004): 626.
42. Marilyn vos Savant, "Ask Marilyn," *Parade*, September 9, 1990.
43. Marilyn vos Savant, *The Power of Logical Thinking: Easy Lessons in the Art of Reasoning . . . and Hard Facts About Its Absence in Our Lives* (New York: Macmillan, 1997), 7.
44. vos Savant, *The Power of Logical Thinking*, 7.
45. John Tierney, "Behind Monty Hall's Doors: Puzzle, Debate, and Answer?" *New York Times*, July 21, 1991.
46. vos Savant, *The Power of Logical Thinking*, 9.
47. vos Savant, *The Power of Logical Thinking*, 10.
48. Tierney, "Behind Monty Hall's Doors."
49. Gardner, *The 2nd Scientific American Book of Mathematical Puzzles and Diversions*, 220–232.

2. THE TITULAR FALLACY

1. The number had been increased from 40 to 41 after the election of 1229 resulted in a tie and was settled, you guessed it, by lot.
2. Miranda Mowbray and Dieter Gollmann, "Electing the Doge of Venice: Analysis of a 13th Century Protocol," in *20th IEEE Computer Security Foundations Symposium* (Los Alamitos, CA: IEEE, 2007), 295–310.
3. Jacob Bernoulli, *On the Law of Large Numbers*, trans. Oscar Sheynin (Berlin, 2005), 19, http://www.sheynin.de/download/bernoulli.pdf.
4. Bernoulli, *On the Law of Large Numbers*, 19.

5.   Bernoulli, *On the Law of Large Numbers*, 28.

6.   For the numbers we've chosen here, the answer turns out to be about $n = 4,046$. Get an urn and try it out!

7.   The smallest sample that works for the numbers given here is around $n = 16,700$, but Bernoulli's answer would have been much larger due to his upper-bound argument.

8.   We use the labels *sampling* and *inferential* in context to refer, respectively, to the probabilities associated to the experimental outcome (the *sample*) and to those associated to an experimental hypothesis about which we wish to draw inferences. This is not to say these are different "types" of probability, though. As discussed in the previous chapter, there is only one type of probability. One person's sampling probability may be another's inferential one, depending on the situation.

9.   What makes this even more frustrating is that we *can* integrate the first statement over any probability distribution for F and conclude "$P[S$ is close to $F]$ is high." Yet this is still not the inferential probability statement we want.

10. Ronald A. Fisher, "Inverse Probability," *Mathematical Proceedings of the Cambridge Philosophical Society*, vol. 26, no. 4 (Cambridge: Cambridge University Press, 1930): 528–535.

11. Technical details: if our prior probability distribution for the urn fraction was uniform on $[0, 1]$, then after observing 8 green and 22 red, our posterior probabilities would have a Beta(9, 23) distribution, for which the probability of being between 3/30 and 13/30 is about 0.963.

12. Brian Everitt and Anders Skrondal, *The Cambridge Dictionary of Statistics* (Cambridge: Cambridge University Press, 2002).

13. S. G. Soal, "Experimental Evidence for Extra-Sensory Perception," *Nature* 185 (1960): 950–951.

14. Charles Edward Mark Hansel, *The Search for Psychic Power: ESP and Parapsychology Revisited* (Buffalo, NY: Prometheus Books, 1989), 106.

15. Hansel, *The Search for Psychic Power*, 106.

16. David Hume, *An Enquiry Concerning Human Understanding: A Critical Edition*, ed. Tom L. Beauchamp (Oxford: Oxford University Press, 2000), 87.

17. S. G. Soal and Frederick Bateman, *Modern Experiments in Telepathy* (New Haven, CT: Yale University Press, 1954), 203.

18. Hansel, *The Search for Psychic Power*.

19. Tom Stoppard, *Rosencrantz and Guildenstern Are Dead*, (New York: Grove, 1967), 16.

20. Richard P. Feynman, Robert B. Leighton, and Matthew Sands, *Six Easy Pieces: Essentials of Physics Explained by Its Most Brilliant Teacher* (Reading, MA: Addison-Wesley, 1995), xxi.

21. Ronald A. Fisher, *Statistical Methods and Scientific Inference* (Edinburgh: Oliver and Boyd, 1956), 39.

22. Jordan Ellenberg, *How Not to Be Wrong: The Power of Mathematical Thinking* (New York: Penguin, 2015).

23. David M. Eddy, "Probabilistic Reasoning in Clinical Medicine: Problems and Opportunities," in *Judgment Under Uncertainty: Heuristics and Biases* (Cambridge: Cambridge University Press, 1982), 249–267.

24. Leila Schneps and Coralie Colmez, *Math on Trial: How Numbers Get Used and Abused in the Courtroom* (New York: Basic Books, 2013).

25. Roy Meadow, *ABC of Child Abuse*, ed. Roy Meadow, 3rd ed. (London: BMJ Publishing Group, 1997), 29.

26. Royal Statistical Society, "Royal Statistical Society Concerned by Issues Raised in Sally Clark Case," news release, October 23, 2001.

27. Ray Hill, "Multiple Sudden Infant Deaths—Coincidence or Beyond Coincidence?," *Paediatric and Perinatal Epidemiology* 18, no. 5 (2004): 320–326.

28. Hill, "Multiple Sudden Infant Deaths."

29. R v. Clark, [2003] EWCA Crim 1020.

30. People v. Collins, 68 Cal. 2d 319, 438 P.2d 33, 66 Cal. Rptr. 497 (1968), 9.

31. "Statistiek in het strafproces," *NOVA/Den Haag Vandaag*, Petra Greeven and Marcel Hammink (Hilversum, Neth.: VARA/NOS, November 4, 2003).

32. Mark Buchanan, "Statistics: Conviction by Numbers," *Nature* 445, no. 7125 (2007): 255.

33. Ronald A. Fisher, *The Design of Experiments*, (London: Oliver and Boyd, 1935), 11.

34. Ronald A. Fisher, "On the Mathematical Foundations of Theoretical Statistics," *Philosophical Transactions of the Royal Society of London. Series A, Containing Papers of a Mathematical or Physical Character* 222 (1922): 311.

35. Fisher, *Statistical Methods and Scientific Inference*, 41–42.

3. ADOLPHE QUETELET'S BELL CURVE BRIDGE

1. Pierre-Simon Laplace, *Théorie analytique des Probabilités*, 2 vols. (Paris: Courcier Imprimeur, 1812), i.

2. Pierre-Simon Laplace, "Mémoire sur les probabilités," *Mémoires de l'Académie Royale des sciences de Paris* 1778 (1781), trans. Richard J. Pulskamp: 227.

3. Stephen M. Stigler, *The History of Statistics: The Measurement of Uncertainty Before 1900* (Cambridge, MA: Belknap Press, 1986), 135.

4. Stigler, *The History of Statistics*, 165.

5. Stigler, *The History of Statistics*, 170.

6. Silvan S. Schweber, "The Origin of the *Origin* Revisited," *Journal of the History of Biology* 10, no. 2 (1977): 229–316.

7. Stigler, *The History of Statistics*, 172.

8. Stigler, *The History of Statistics*, 171.

9. It being a continuous probability distribution, the way to get probabilities from the normal curve is by integrating—that is, finding the area under the curve. So the probability of a variable with the normal distribution being, say, between a and b is the integral from a to b of $f(x)\, dx$.

10. Here is a proof that can fit in an endnote: The derivative with respect to x of the sum of the squared errors is $2 \cdot (x - 10) + 2 \cdot (x - 12) + 2 \cdot (x - 17) = 2 \cdot [3x - (10 + 12 + 17)]$. Setting this equal to 0 and solving yields $x = (10 + 12 + 17)/3$. QED.

11. Stigler, *The History of Statistics*, 13.

12. Stigler, *The History of Statistics*, 145.

13. Stigler, *The History of Statistics*, 214.

14. Stigler, *The History of Statistics*, 214.

15. Stigler, *The History of Statistics*, 214.

16. Stigler, *The History of Statistics*, 203.

17. Stigler, *The History of Statistics*, 217.

18. William Spottiswoode, "On Typical Mountain Ranges: An Application of the Calculus of Probabilities to Physical Geography," *Journal of the Royal Geographical Society of London* 31 (1861): 149–154.

19. Siméon Denis Poisson, *Recherches sur la probabilité des jugements en matière criminelle et en matière civile* (Paris: Bachelier, 1837).
20. Stigler, *The History of Statistics*, 194.
21. Stigler, *The History of Statistics*, 194–195.
22. Stigler, *The History of Statistics*, 194.
23. John Stuart Mill, *A System of Logic, Ratiocinative and Inductive, Being a Connected View of the Principles of Evidence and the Methods of Scientific Investigation* (London: John W. Parker, 1843), 2:81.
24. Robert Leslie Ellis, *On the Foundations of the Theory of Probabilities* (London: John W. Parker, 1843).
25. Antoine Augustin Cournot, *Exposition de la théorie des chances et des probabilités* (Paris: L. Hachette, 1843), ii.
26. Jakob Friedrich Fries, *Versuch einer Kritik der Principien der Wahrscheinlichkeitsrechnung* (Braunschweig: F. Vieweg u. sohn, 1842), v–vi.
27. Robert Leslie Ellis, "Remarks on an Alleged Proof of the 'Method of Least Squares,' Contained in a Late Number of the Edinburgh Review," *London, Edinburgh, and Dublin Philosophical Magazine and Journal of Science*, 3rd ser., 37, no. 251 (1850): 321–328.
28. William Fishburn Donkin, "On Certain Questions Relating to the Theory of Probabilities," *London, Edinburgh, and Dublin Philosophical Magazine and Journal of* Science, 4th ser., 1, no. 5 (1851): 353–368.
29. Boole's argument was against Bayes's uniform prior for a continuous variable by means of two *related* discrete distributions: the number of successes out of n trials and the sequence of successes and failures out of n for any n. Our example captures the same essential idea with less overhead.
30. George Boole, *An Investigation of the Laws of Thought: On Which Are Founded the Mathematical Theories of Logic and Probabilities* (London: Walton and Maberly, 1854), 370.
31. Edwin T. Jaynes, "The Well-Posed Problem," *Foundations of Physics* 3, no. 4 (1973): 477–492.

4. THE FREQUENTIST JIHAD

1. Karl Pearson, *The Life, Letters and Labours of Francis Galton* (Cambridge: Cambridge University Press, 1924), 1: 229.
2. Francis Galton and Charles Darwin shared a grandfather but not a grandmother, so some sources refer to Darwin as Galton's half-cousin. Since it seems an unnecessarily awkward and genealogically precise term, I'll just refer to them as cousins.
3. Francis Galton, letter to the editor, *The Times* (UK), June 5, 1873.
4. Francis Galton, *Memories of My Life* (New York: Dutton, 1909).
5. Francis Galton, "Hereditary Talent and Character," *Macmillan's Magazine* 12, no. 157–166 (1865): 318–327.
6. Francis Galton, *Hereditary Genius: An Inquiry Into Its Laws and Consequences*, 2nd ed. (London: Macmillan, 1892), 342.
7. Galton, *Hereditary Genius*, xi.
8. Francis Galton, "Statistics by Intercomparison, with Remarks on the Law of Frequency of Error," *London, Edinburgh, and Dublin Philosophical Magazine and Journal of Science* 49, no. 322 (1875): 33–46.

9. Galton, *Hereditary Genius*, 31–32.

10. Francis Galton, *Natural Inheritance* (London: Macmillan, 1889), 66.

11. Galton, *Natural Inheritance*, 62.

12. Galton, "Statistics by Intercomparison," 45.

13. Here is a short Laplacean proof: For any random variable X, we define the characteristic function $\varphi_X(t)$ as the expected value $E[\exp(itX)]$, from which it follows that if X and Y are independent, then $\varphi_{X+Y}(t) = \varphi_X(t)\varphi_Y(t)$. If (and only if) X is normally distributed, then $\varphi_X(t)$ is of the form $\exp(it\mu - \sigma^2 t^2/2)$. Multiplying two such functions produces a function of the same form. QED.

14. Essentially the game Plinko from *The Price Is Right*.

15. Francis Galton, "Regression Towards Mediocrity in Hereditary Stature," *Journal of the Anthropological Institute of Great Britain and Ireland* 15 (1886): 252–253.

16. The keen-eyed reader will notice we said in the previous chapter that linear regression was descended from Laplace and Gauss's method of least squares. Mechanically Gauss's and Galton's methods are, in fact, exactly the same, but the contextual meanings of all the component parts are so different that it wasn't until Pearson and Fisher clarified the relationship many years later that this fact was fully understood.

17. Stephen M. Stigler, *The History of Statistics: The Measurement of Uncertainty Before 1900* (Cambridge, MA: Belknap Press, 1986), 304.

18. Today this term is an offensive racial slur, considered hate speech since the days of South Africa's apartheid government, but at the time of Pearson's writing, it did not have such a pejorative meaning and simply referred to a Black person in Africa—about on par with the historical usage of the word *negro*. It is noteworthy, though, that of all the many examples he *could* have picked, Pearson chose one involving races and disease resistance; see subsequent discussion of Pearson and racism.

19. Karl Pearson, *The Grammar of Science*, 2nd ed. (London: Adam and Charles Black, 1900), 407.

20. Egon S. Pearson, *Karl Pearson: An Appreciation of Some Aspects of His Life and Work* (Cambridge: Cambridge University Press, 1938), 19.

21. Karl Pearson, *National Life from the Standpoint of Science: An Address Delivered at Newcastle November 19, 1900* (London: Adam and Charles Black, 1901), 43–44.

22. K. Pearson, *National Life from the Standpoint of Science*, 17–18.

23. K. Pearson, *National Life from the Standpoint of Science*, 19–20.

24. K. Pearson, *National Life from the Standpoint of Science*, 23.

25. K. Pearson, *The Grammar of Science*, 369.

26. Karl Pearson, "III. Contributions to the Mathematical Theory of Evolution," *Philosophical Transactions of the Royal Society of London A* 185 (1894): 72.

27. K. Pearson, *National Life from the Standpoint of Science*, 48.

28. Karl Pearson and Margaret Moul, "The Problem of Alien Immigration Into Great Britain, Illustrated by an Examination of Russian and Polish Jewish Children: Part II," *Annals of Eugenics* 2, no. 1–2 (1927): 125.

29. K. Pearson and Moul, "The Problem of Alien Immigration Into Great Britain: Part II," 126–127.

30. Karl Pearson and Margaret Moul, "The Problem of Alien Immigration Into Great Britain, Illustrated by an Examination of Russian and Polish Jewish Children: Part I," *Annals of Eugenics* 1, no. 1 (1925): 8.

31. K. Pearson and Moul, "The Problem of Alien Immigration Into Great Britain: Part I," 49.

32. K. Pearson and Moul, "The Problem of Alien Immigration Into Great Britain: Part I," 43.

33. K. Pearson, *The Life, Letters and Labours of Francis Galton*, 1:vii.

34. Donald MacKenzie, "Statistical Theory and Social Interests: A Case-Study," *Social Studies of Science* 8, no. 1 (1978): 35–83.

35. George Smith and L. G. Wickham Legg, *The Dictionary of National Biography, 1931–1940: With an Index Covering the Years 1901–1940 in One Alphabetical Series* (London: Oxford University Press, 1949), 683.

36. Ronald A. Fisher, "The Bearing of Genetics on Theories of Evolution," *Science Progress in the Twentieth Century (1919–1933)* 27, no. 106 (1932): 275.

37. Ronald A. Fisher, "Student," *Annals of Eugenics* 9, no. 1 (1939): 1–9.

38. Also a parapsychologist and later a president of the Society for Psychical Research, two years after Samuel Soal.

39. Egon S. Pearson, "Studies in the History of Probability and Statistics. XX: Some Early Correspondence Between W. S. Gosset, R. A. Fisher and Karl Pearson, with Notes and Comments," *Biometrika* 55, no. 3 (1968): 446.

40. E. Pearson, "Studies in the History of Probability and Statistics," 446.

41. An old joke in the math community: A physicist and a mathematician are attending a lecture about geometry in nine-dimensional space. The physicist, baffled and struggling to keep up, sees the mathematician smiling and nodding along to the lecture. Later the physicist asks the mathematician, "How were you able to visualize anything in nine dimensions?" The mathematician says, "Easy! I just imagined it in n dimensions and then let n equal nine."

42. E. Pearson, "Studies in the History of Probability and Statistics," 449.

43. E. Pearson, "Studies in the History of Probability and Statistics," 451.

44. H. E. Soper et al., "On the Distribution of the Correlation Coefficient in Small Samples. Appendix II to the Papers of 'Student' and R. A. Fisher," *Biometrika* 11, no. 4 (1917): 328–413.

45. Ronald A. Fisher, *Statistical Methods for Research Workers*. Biological Monographs and Manuals No. 3 (Edinburgh: Oliver and Boyd, 1925), vii.

46. Stephen T. Ziliak and Deirdre Nansen McCloskey, *Cult of Statistical Significance: How the Standard Error Costs Us Jobs, Justice, and Lives* (Ann Arbor: University of Michigan Press, 2010).

47. Ronald A. Fisher, *The Genetical Theory of Natural Selection* (Oxford: Clarendon Press, 1930), 35.

48. Fisher, *The Genetical Theory of Natural Selection*, 36–37.

49. "Who Is the Greatest Biologist of All Time?," Edge, accessed April 28, 2020, https://www.edge.org/conversation/who-is-the-greatest-biologist-of-all-time.

50. William R. Inge, "Some Moral Aspects of Eugenics," *Eugenics Review* 1 (1909–1910): 30.

51. Stephen Stigler, "Fisher in 1921," *Statistical Science* 20, no. 1 (2005): 33.

52. Ronald A. Fisher, "Some Hopes of a Eugenist," *Eugenics Review* 5, no. 4 (1914): 309.

53. Fisher, *The Genetical Theory of Natural Selection*, 226.

54. Fisher, *The Genetical Theory of Natural Selection*, 227.

55. Fisher, *The Genetical Theory of Natural Selection*, 182.

56. Adolf Hitler and Ralph Manheim, *Mein Kampf* (Boston: Houghton Mifflin, 1971), 439–440.

57. Edwin Black, *War Against the Weak: Eugenics and America's Campaign to Create a Master Race* (New York: Four Walls Eight Windows, 2003), 275–276.

58. Black, *War Against the Weak*, 259.

59. "Report of Committee for Legalizing Eugenic Sterilization," *Postgraduate Medical Journal* 6, no. 61 (1930): 13.

60. Pauline Mazumdar, *Eugenics, Human Genetics and Human Failings* (Florence: Routledge, 1992), 203–204.

61. F. William Engdahl, *Seeds of Destruction: The Hidden Agenda of Genetic Manipulation* (Montreal: Global Research, 2007), 81.

62. Sheila Faith Weiss, "After the Fall: Political Whitewashing, Professional Posturing, and Personal Refashioning in the Postwar Career of Otmar Freiherr Von Verschuer," *Isis* 101, no. 4 (2010): 745.

63. UNESCO, *The Race Question in Modern Science: Race and Science* (New York: Columbia University Press, 1961), 497.

64. UNESCO, *The Race Question in Modern Science*, 497.

65. Ronald A. Fisher, *Natural Selection, Heredity, and Eugenics: Including Selected Correspondence of R. A. Fisher with Leonard Darwin and Others*, ed. J. H. Bennett (Oxford: Clarendon Press, 1983), 191–192.

66. Constance Reid, *Neyman–from Life* (New York: Springer-Verlag, 1982), 126.

67. Ronald A. Fisher, *Statistical Methods and Scientific Inference* (Edinburgh: Oliver and Boyd, 1956), 2–3.

68. E. Pearson, "Studies in the History of Probability and Statistics," 456.

69. E. Pearson, *Karl Pearson*, 53.

70. Jerzy Neyman, *Lectures and Conferences on Mathematical Statistics and Probability* (Washington, DC: Graduate School, U.S. Department of Agriculture, 1952), 210.

71. Fisher, *Statistical Methods and Scientific Inference*, 76.

72. Fisher, *Statistical Methods and Scientific Inference*, 7.

73. Ronald A. Fisher, "Statistical Methods and Scientific Induction," *Journal of the Royal Statistical Society: Series B (Methodological)* 17, no. 1 (1955): 70.

74. Oliver H. Lowry et al., "Protein Measurement with the Folin Phenol Reagent," *Journal of Biological Chemistry* 193 (1951): 265–275.

75. Fred Hoyle, *Mathematics of Evolution* (Memphis, TN: Acorn Enterprises, 1999), 5.

76. Reid, *Neyman–from Life*, 82–85.

77. Student, "Probable Error of a Correlation Coefficient," *Biometrika* 6, no. 2–3 (1908): 302–303.

78. K. Pearson, *The Grammar of Science*, 146.

79. Fisher, *Statistical Methods for Research Workers*, 10.

80. Ronald A. Fisher, "On the Mathematical Foundations of Theoretical Statistics," *Philosophical Transactions of the Royal Society of London. Series A, Containing Papers of a Mathematical or Physical Character* 222 (1922): 326.

81. Fisher, "On the Mathematical Foundations of Theoretical Statistics," 312.

82. Fisher, *Statistical Methods for Research Workers*, 2–3.

83. Fisher, *Statistical Methods and Scientific Inference*, 18–19.

84. Sandy L. Zabell, "R. A. Fisher on the History of Inverse Probability," *Statistical Science* 4, no. 3 (1989): 247–256.

85. George Boole, "XII. On the Theory of Probabilities," *Philosophical Transactions of the Royal Society of London* 152 (1862): 228.

86. Zabell, "R.A. Fisher on the History of Inverse Probability," 247–256.

87. Ronald A. Fisher, "Inverse Probability," *Mathematical Proceedings of the Cambridge Philosophical Society* 26, no. 4 (1930): 531.

88. Leonard J. Savage, "On Rereading R. A. Fisher," *Annals of Statistics* 4, no. 3 (1976): 473.

89. Fisher, *Statistical Methods and Scientific Inference*, 43.

90. J. H. Bennett, *Statistical Inference and Analysis: Selected Correspondence of R. A. Fisher* (Oxford: Oxford University Press, 1990), 61.

91. Ronald A. Fisher, "Probability Likelihood and Quantity of Information in the Logic of Uncertain Inference," *Proceedings of the Royal Society of London. Series A, Containing Papers of a Mathematical and Physical Character* 146, no. 856 (1934): 4.

92. Fisher, *Statistical Methods and Scientific Inference*, 25.

93. Bernard J. Norton, "Karl Pearson and Statistics: The Social Origins of Scientific Innovation," *Social Studies of Science* 8, no. 1 (1978): 21.

94. For example, Thomas Huxley (1825–1895), an early advocate for Darwin's theory and the first to suggest that man had evolved from apes, did a great deal of comparative study of "biometric" factors such as skull and jawbone sizes between Europeans and Africans in an attempt to prove that Africans were evolutionarily "closer" to apes.

95. Karl Pearson, "On the Inheritance of the Mental and Moral Characters in Man: II," *Biometrika* 3 (1904): 156.

96. K. Pearson, "On the Inheritance of the Mental and Moral Characters in Man: II," 160.

97. Arthur Henry Lane, *The Alien Menace: A Statement of the Case*, 5th ed. (London: Boswell, 1934), 50.

98. K. Pearson and Moul, "The Problem of Alien Immigration Into Great Britain: Part I," 8.

99. Ian Hacking, "Karl Pearson's History of Statistics," *British Journal for the Philosophy of Science* 32, no. 2 (1981): 177–183.

100. Joan Fisher Box, *R. A. Fisher, the Life of a Scientist* (New York: Wiley, 1978), 2. Italics in original.

101. Ronald A. Fisher, "Dangers of Cigarette-Smoking," *British Medical Journal* 2, no. 5039 (1957): 297.

102. K. Pearson, *The Grammar of Science*, 468.

103. Galton, "Regression Towards Mediocrity in Hereditary Stature," 247.

104. Graham J. Baker, "Christianity and Eugenics: The Place of Religion in the British Eugenics Education Society and the American Eugenics Society, c. 1907–1940," *Social History of Medicine* 27, no. 2 (2014): 281–302.

105. James Hamilton Francis Peile, "Eugenics and the Church," *Eugenics Review* 1, no. 3 (1909): 163.

106. K. Pearson, *National Life from the Standpoint of Science*, 26.

107. Pius XI, "Casti Connubii," *Acta Apostolicae Sedis* 22 (1930): 539–592.

108. Jayne Woodhouse, "Eugenics and the Feeble-Minded: The Parliamentary Debates of 1912–14," *History of Education* 11, no. 2 (1982): 133.

109. Gilbert Keith Chesterton, *Eugenics and Other Evils* (London: Cassell, 1922).

110. Lancelot Thomas Hogben, *Principles of Evolutionary Biology* (Cape Town: Juta, 1927), 100.

111. Lancelot Thomas Hogben, *Genetic Principles in Medicine and Social Science* (New York: Knopf, 1932), 210.

112. "Social Biology," *New Statesman and Nation*, December 26, 1931, 816–817.

113. Francis Galton, *Probability, the Foundation of Eugenics: The Herbert Spencer Lecture Delivered on June 5, 1907* (Oxford: Clarendon Press, 1907), 29–30.

5. THE QUOTE-UNQUOTE LOGIC OF ORTHODOX STATISTICS

1. "Which Statistics Test Should I Use?," Social Science Statistics, accessed April 28, 2020, https://www.socscistatistics.com/tests/what_stats_test_wizard.aspx.
2. Edwin T. Jaynes, *Probability Theory: The Logic of Science*, ed. G. Larry Bretthorst (Cambridge: Cambridge University Press, 2003), 492.
3. Jaynes, *Probability Theory*, 511–514.
4. Jaynes, *Probability Theory*, 514.
5. Harold Jeffreys, "Probability, Statistics, and the Theory of Errors," *Proceedings of the Royal Society of London. Series A, Containing Papers of a Mathematical and Physical Character* 140, no. 842 (1933): 529.
6. Harold Jeffreys, "On the Theory of Errors and Least Squares," *Proceedings of the Royal Society of London. Series A, Containing Papers of a Mathematical and Physical Character* 138, no. 834 (1932): 48–55.
7. Ronald A. Fisher, "The Concepts of Inverse Probability and Fiducial Probability Referring to Unknown Parameters," *Proceedings of the Royal Society of London. Series A, Containing Papers of a Mathematical and Physical Character* 139, no. 838 (1933): 343–348.
8. Harold Jeffreys, "Probability and Scientific Method," *Proceedings of the Royal Society of London. Series A, Containing Papers of a Mathematical and Physical Character* 146, no. 856 (1934): 12.
9. Ronald A. Fisher, "Probability Likelihood and Quantity of Information in the Logic of Uncertain Inference," *Proceedings of the Royal Society of London. Series A, Containing Papers of a Mathematical and Physical Character* 146, no. 856 (1934): 4.
10. Jeffreys, "Probability, Statistics, and the Theory of Errors," 532.
11. Fisher, "Probability Likelihood and Quantity of Information," 3.
12. Ronald A. Fisher, *Statistical Methods and Scientific Inference* (Edinburgh: Oliver and Boyd, 1956), 8.
13. Fisher, *Statistical Methods and Scientific Inference*, 25.
14. Ronald A. Fisher, *Statistical Methods for Research Workers*, 5th ed. Biological Monographs and Manuals No. 5 (Edinburgh: Oliver and Boyd, 1932), 10–11.
15. Fisher, *Statistical Methods for Research Workers*, 22–23.
16. Sandy L. Zabell, "R. A. Fisher and Fiducial Argument," *Statistical Science* 7, no. 3 (1992): 369–387.
17. Ronald A. Fisher, "Inverse Probability," *Mathematical Proceedings of the Cambridge Philosophical Society*, vol. 26, no. 4 (1930): 533.
18. Fisher, "Inverse Probability," 532.
19. If we had at least two samples, instead of using σ we could use the sample standard deviation and have a better pivotal quantity that allows fiducial inference via Student's t-distribution.
20. M. S. Bartlett, "The Information Available in Small Samples," *Mathematical Proceedings of the Cambridge Philosophical Society* 32, no. 4 (1936): 560–566.
21. Fisher, "The Concepts of Inverse Probability and Fiducial Probability," 348.
22. Ronald A. Fisher, "A Note on Fiducial Inference," *Annals of Mathematical Statistics* 10, no. 4 (1939): 386.
23. Ronald A Fisher, *Contributions to Mathematical Statistics* (New York: Wiley, 1950), 35.173a.
24. Ronald A. Fisher, "Statistical Methods and Scientific Induction," *Journal of the Royal Statistical Society: Series B (Methodological)* 17, no. 1 (1955): 75.

25. Zabell, "R. A. Fisher and Fiducial Argument," 380.

26. Ronald A. Fisher, *The Design of Experiments*, 6th ed. (London: Oliver and Boyd, 1951), 195 –196.

27. Ronald A. Fisher, *The Design of Experiments*, 9th ed. (London: Oliver and Boyd, 1971), 198.

28. Ronald A. Fisher, "On the 'Probable Error' of a Coefficient of Correlation Deduced from a Small Sample," *Metron* 1 (1921): 25.

29. Fisher, *Contributions to Mathematical Statistics*, 1.2b.

30. Fisher, *Statistical Methods and Scientific Inference*, 1st ed. (1956), 77.

31. Fisher, *Statistical Methods and Scientific Inference*, 3rd ed. (1973), 35.

32. Leonard J. Savage, "The Foundations of Statistics Reconsidered," in *Proceedings of the Fourth Berkeley Symposium on Mathematical Statistics and Probability, Volume 1: Contributions to the Theory of Statistics* (Berkeley, CA: University of California Press, 1961), 578.

33. Robert J. Buehler and Alan P. Feddersen, "Note on a Conditional Property of Student's *t*," *Annals of Mathematical Statistics* 34, no. 3 (1963): 1098–1100.

34. Savage, "Discussion," 926.

35. Some people have continued trying to rescue Fisher's idea of fiducial inference. For one example, see Teddy Seidenfeld, "R. A. Fisher's Fiducial Argument and Bayes' Theorem," *Statistical Science* 7, no. 3 (1992): 358–368.

36. Ronald A. Fisher, "The Statistical Method in Psychical Research," *Proceedings of the Society for Psychical Research* 39 (1929): 191.

37. This isn't technically quite right; it's only the MVUE if the samples are not allowed to repeat. To keep the formulas simple, we've allowed for repetition, but for large N, the difference is minimal.

38. Endnote-sized derivation: The maximum is $\leq m$ if and only if all k samples are $\leq m$, which happens with probability $(m/n)^k$. Taking $P[\text{Max} \leq m] - P[\text{Max} \leq (m-1)]$ gives $P[\text{Max} = m]$.

39. For larger sample sizes, the canonical advice would be to use Pearson's chi-squared test, but given the numbers here, that test is considered inappropriate.

40. One of the proposed approaches is to rank all the possible tables according to their probabilities and define the *tail* to consist of all those tables starting from the least probable and working up to some threshold of improbability. However, as we know from the "Malfunctioning Digital Scale" example, this is not generally a good procedure because the least likely data possibilities may include some events that we wouldn't necessarily take as evidence *against the hypothesis* at all. Improbable data is not, in and of itself, necessarily interesting. How can we be sure that the same phenomenon isn't happening here?

41. Fisher, *Statistical Methods for Research Workers*, 6–7.

42. Jaynes, *Probability Theory*, 253.

43. Technically the rank-order statistics are minimally sufficient, meaning we can forget about the exact sequence of observations, but that's it.

44. Stephen M. Stigler, "Studies in the History of Probability and Statistics. XXXIII Cauchy and the Witch of Agnesi: An Historical Note on the Cauchy Distribution," *Biometrika* 61, no. 2 (1974): 377.

45. Endnote-sized proof: The probability of the result of the tangent of a uniformly distributed angle being less than any value x is given by the inverse tangent of x; taking the derivative then gives the Cauchy distribution as the probability density function. QED.

46. Thomas J. Rothenberg, Franklin M. Fisher, and Christian Bernhard Tilanus, "A Note on Estimation from a Cauchy Sample," *Journal of the American Statistical Association* 59, 306 (1964): 460–463.

47. Jaynes, *Probability Theory*, 500.

6. THE REPLICATION CRISIS/OPPORTUNITY

1. Joseph Berkson, "Tests of Significance Considered as Evidence," *Journal of the American Statistical Association* 37, no. 219 (1942): 326.

2. William W. Rozeboom, "The Fallacy of the Null-Hypothesis Significance Test," *Psychological Bulletin* 57, no. 5 (1960): 422.

3. David Bakan, "The Test of Significance in Psychological Research," *Psychological Bulletin* 66, no. 6 (1966): 428.

4. Paul E. Meehl, "Theory-Testing in Psychology and Physics: A Methodological Paradox," *Philosophy of Science* 34, no. 2 (1967): 114.

5. Paul E. Meehl, "Theoretical Risks and Tabular Asterisks: Sir Karl, Sir Ronald, and the Slow Progress of Soft Psychology," *Journal of Consulting and Clinical Psychology* 46, no. 4 (1978): 817.

6. Stanislav Andreski, *Social Sciences as Sorcery* (New York: St. Martin's Press, 1972), 114–115.

7. John P. Campbell, "Some Remarks from the Outgoing Editor," *Journal of Applied Psychology* 67, no. 6 (1982): 698.

8. Jacob Cohen, "The Earth Is Round ($p < .05$)," *American Psychologist* 49, no. 12 (1994): 997.

9. Ruma Falk and Charles W. Greenbaum, "Significance Tests Die Hard: The Amazing Persistence of a Probabilistic Misconception," *Theory and Psychology* 5, no. 1 (1995): 78.

10. Persi Diaconis and David Freedman, "The Persistence of Cognitive Illusions," *Behavioral and Brain Sciences* 4, no. 3 (1981): 333–334.

11. Gerd Gigerenzer, "The Superego, the Ego, and the Id in Statistical Reasoning," in *A Handbook for Data Analysis in the Behavioral Sciences: Methodological Issues*, ed. G. Keren and C. Lewis (Hillsdale, NJ: Erlbaum, 1993), 330.

12. Charles Lambdin, "Significance Tests as Sorcery: Science Is Empirical—Significance Tests Are Not," *Theory and Psychology* 22, no. 1 (2012): 67–90.

13. Ronald L. Wasserstein and Nicole Lazar, "The ASA Statement on p-Values: Context, Process, and Purpose," *American Statistician* 70, no. 2 (2016): 129.

14. Robert A. J. Matthews, "Why *Should* Clinicians Care About Bayesian Methods?," *Journal of Statistical Planning and Inference* 94, no. 1 (2001): 44.

15. Roger Peng, "The Reproducibility Crisis in Science: A Statistical Counterattack," *Significance* 12, no. 3 (2015): 30–32.

16. Helen M. Colhoun, Paul M. McKeigue, and George Davey Smith, "Problems of Reporting Genetic Associations with Complex Outcomes," *The Lancet* 361, no. 9360 (2003): 868.

17. John P. A. Ioannidis, "Why Most Published Research Findings Are False," *PLOS Medicine* 2, no. 8 (2005): e124.

18. Fisher himself, of course, never ran into this problem because his populations were always "hypothetically infinite."

19. Joseph Berkson, "Some Difficulties of Interpretation Encountered in the Application of the Chi-Square Test," *Journal of the American Statistical Association* 33, no. 203 (1938): 527.

20. David T. Lykken, "Statistical Significance in Psychological Research," *Psychological Bulletin* 70, no. 3 (1968): 154.

21. Paul E. Meehl, "Why Summaries of Research on Psychological Theories Are Often Uninterpretable," *Psychological Reports* 66, no. 1 (1990): 204.

22. Meehl, "Why Summaries of Research on Psychological Theories Are Often Uninterpretable," 206.

23. John T. Cacioppo et al., "Marital Satisfaction and Break-Ups Differ Across On-Line and Off-Line Meeting Venues," *Proceedings of the National Academy of Sciences* 110, no. 25 (2013): 10135–10140.
24. Bakan, "The Test of Significance in Psychological Research," 430.
25. Edwin G. Boring, "Mathematical Vs. Scientific Significance," *Psychological Bulletin* 16, no. 10 (1919): 335–338.
26. Stephen T. Ziliak and Deirdre Nansen McCloskey. *The Cult of Statistical Significance: How the Standard Error Costs Us Jobs, Justice, and Lives* (Ann Arbor: University of Michigan Press, 2008), 4–5.
27. John P. A. Ioannidis, "Contradicted and Initially Stronger Effects in Highly Cited Clinical Research," *JAMA* 294, no. 2 (2005): 218–228.
28. Daryl J. Bem, "Feeling the Future: Experimental Evidence for Anomalous Retroactive Influences on Cognition and Affect," *Journal of Personality and Social Psychology* 100, no. 3 (2011): 407–425.
29. Bem, "Feeling the Future," 422.
30. Bem, "Feeling the Future," 410.
31. Ronald A. Fisher, "The Statistical Method in Psychical Research," *Proceedings of the Society for Psychical Research* 39 (1929), 189–192.
32. Daniel Engber, "Daryl Bem Proved ESP Is Real Which Means Science Is Broken," *Slate*, May 17, 2017, accessed April 28, 2020, https://slate.com/health-and-science/2017/06/daryl-bem-proved -esp-is-real-showed-science-is-broken.html
33. Daryl J. Bem and Charles Honorton, "Does Psi Exist? Replicable Evidence for an Anomalous Process of Information Transfer," *Psychological Bulletin* 115, no. 1 (1994): 4–18.
34. Jessica Utts, "Replication and Meta-Analysis in Parapsychology," *Statistical Science* 6, no. 4 (1991): 363–378.
35. Bem and Honorton, "Does Psi Exist?,"13.
36. Engber, "Daryl Bem Proved ESP Is Real."
37. Benedict Carey, "Journal's Paper on ESP Expected to Prompt Outrage," *New York Times*, January 5, 2011.
38. George Lowery, "Study Showing That Humans Have Some Psychic Powers Caps Daryl Bem's Career," *Cornell Chronicle*, December 6, 2010.
39. Eric-Jan Wagenmakers et al., "Why Psychologists Must Change the Way They Analyze Their Data: The Case of Psi: Comment on Bem (2011)," *Journal of Personality and Social Psychology* 100, no. 3 (2011): 426–432.
40. Bem, "Feeling the Future," 409.
41. Mithat Gönen et al., "The Bayesian Two-Sample t Test," *American Statistician* 59, no. 3 (2005): 252–257.
42. Wagenmakers et al., "Why Psychologists Must Change the Way They Analyze Their Data," 430.
43. Harold Jeffreys, *Theory of Probability*, 3rd ed. (Oxford: Oxford University Press, 1961), 432.
44. Wagenmakers et al., "Why Psychologists Must Change the Way They Analyze Their Data," 429.
45. Daryl J. Bem, "Writing the Empirical Journal Article," in *The Compleat Academic: A Practical Guide for the Beginning Social Scientist*, ed. Mark P. Zanna and John M. Darley (Hove, East Sussex, UK: Psychology Press, 1987), 172.
46. Wagenmakers et al., "Why Psychologists Must Change the Way They Analyze Their Data," 431.
47. Bem and Honorton, "Does Psi Exist?," 4 (emphasis added).
48. Bem and Honorton, "Does Psi Exist?," 16.

49. Alan D. Sokal, "Transgressing the Boundaries: Towards a Transformative Hermeneutics of Quantum Gravity," *Social Text* 46/47 (1996): 217–252.
50. Carey, "Journal's Paper on ESP Expected to Prompt Outrage."
51. "A New Bem Theory," Statistical Modeling, Causal Inference, and Social Science, accessed April 28, 2020, https://andrewgelman.com/2013/08/25/a-new-bem-theory.
52. Joseph P. Simmons, Leif D. Nelson, and Uri Simonsohn, "False-Positive Psychology: Undisclosed Flexibility in Data Collection and Analysis Allows Presenting Anything as Significant," *Psychological Science* 22, no. 11 (2011): 1359–1366.
53. Andrew Gelman, and Eric Loken, "The Statistical Crisis in Science," *American Scientist* 102, no. 6 (2014): 460–465.
54. Brian A. Nosek, Jeffrey R. Spies, and Matt Motyl, "Scientific Utopia: II. Restructuring Incentives and Practices to Promote Truth Over Publishability," *Perspectives on Psychological Science* 7, no. 6 (2012): 615–631.
55. Open Science Collaboration, "Estimating the Reproducibility of Psychological Science," *Science* 349, no. 6251 (2015): aac4716.
56. Colin F. Camerer et al., "Evaluating the Replicability of Social Science Experiments in *Nature* and *Science* Between 2010 and 2015," *Nature Human Behaviour* 2, no. 9 (2018): 637–644.
57. Eric-Jan Wagenmakers et al., "Supplement: Bayesian Analyses for 'Evaluating Replicability of Social Science Experiments in Nature and Science,' " accessed April 28, 2020, https://osf.io/nsxgj/
58. Will M. Gervais and Ara Norenzayan, "Analytic Thinking Promotes Religious Disbelief," *Science* 336, no. 6080 (2012): 493–496.
59. Gervais and Norenzayan, "Analytic Thinking Promotes Religious Disbelief," 495.
60. Nick Buttrick et al., "Replication of Analytic Thinking Promotes Religious Disbelief," Center for Open Science, accessed April 28, 2020, https://osf.io/r4dvc/.
61. Wagenmakers et al., "Supplement: Bayesian Analyses," 10–11.
62. Brian Resnick, "More Social Science Studies Just Failed to Replicate. Here's Why This Is Good," *Vox*, August 27, 2018, accessed April 28, 2020, https://www.vox.com/science-and-health/2018/8/27/17761466/psychology-replication-crisis-nature-social-science
63. Will M. Gervais and Ara Norenzayan, "Author Comment," Center for Open Science, accessed April 28, 2020, https://osf.io/q64td/.
64. C. Glenn Begley and Lee M. Ellis, "Raise Standards for Preclinical Cancer Research," *Nature* 483, no. 7391 (2012): 532.
65. Katherine S. Button et al., "Power Failure: Why Small Sample Size Undermines the Reliability of Neuroscience," *Nature Reviews Neuroscience* 14, no. 5 (2013): 365–376.
66. Button et al., "Power Failure," 372.
67. Leonard P. Freedman, Iain M. Cockburn, and Timothy S. Simcoe, "The Economics of Reproducibility in Preclinical Research," *PLOS Biology* 13, no. 6 (2015): e1002165.
68. C. F. Camerer et al., "Evaluating Replicability of Laboratory Experiments in Economics," *Science* 351, no. 6280 (2016): 1433–1436.
69. John P. A. Ioannidis, T. D. Stanley, and Hristos Doucouliagos, "The Power of Bias in Economics Research," *Economic Journal* 127, no. 605 (2017): F250.
70. "Reproducibility Project: Cancer Biology," ed. Roger J. Davis et al., eLife, accessed April 28, 2020, https://elifesciences.org/collections/9b1e83d1/reproducibility-project-cancer-biology/.

71. Jeffrey R. Lisse et al., "Gastrointestinal Tolerability and Effectiveness of Rofecoxib Versus Naproxen in the Treatment of Osteoarthritis: A Randomized, Controlled Trial," *Annals of Internal Medicine* 139, no. 7 (2003): 539–546.

72. Tze-Fan Chao et al., "The Association Between the Use of Non-steroidal Anti-Inflammatory Drugs and Atrial Fibrillation: A Nationwide Case-Control Study," *International Journal of Cardiology* 168, no. 1 (2013): 312.

73. Morten Schmidt et al., "Non-steroidal Anti-Inflammatory Drug Use and Risk of Atrial Fibrillation or Flutter: Population Based Case-Control Study," *British Medical Journal* 343, no. 7814 (2011): 82.

74. Chao et al., "The Association Between the Use of Non-steroidal Anti-Inflammatory Drugs and Atrial Fibrillation," 314.

75. Morten Schmidt and Kenneth J. Rothman, "Mistaken Inference Caused by Reliance on and Misinterpretation of a Significance Test," *International Journal of Cardiology* 177, no. 3 (2014): 1089–1090.

76. Lisse et al., "Gastrointestinal Tolerability and Effectiveness of Rofecoxib Versus Naproxen," 543.

77. Alex Berenson, "Evidence in Vioxx Suits Shows Intervention by Merck Officials," *New York Times*, April 24, 2005.

78. Shaoni Bhattacharya, "Up to 140,000 Heart Attacks Linked to Vioxx," *New Scientist*, January 25, 2005.

79. Fisher, "The Statistical Method in Psychical Research," 191.

80. Ziliak and McCloskey, *The Cult of Statistical Significance*, 29.

81. Valentin Amrhein, Sander Greenland, and Blake McShane, "Scientists Rise Up Against Statistical Significance," *Nature* 567, no. 7748 (2019): 305–307.

82. Wasserstein and Lazar, "The ASA Statement on *p*-Values," 129–133.

83. Ronald L. Wasserstein, Allen L. Schirm, and Nicole A. Lazar, "Moving to a World Beyond '$p < 0.05$,'" *American Statistician* 73, supp. 1 (2019): 2.

84. Wasserstein, Schirm, and Lazar, "Moving to a World Beyond '$p < 0.05$,'" 2.

85. Jacky G. Goetz et al., "Biomechanical Remodeling of the Microenvironment by Stromal Caveolin-1 Favors Tumor Invasion and Metastasis," *Cell* 146, no. 1 (2011): 148–163.

86. Fritz Strack, Leonard L. Martin, and Sabine Stepper, "Inhibiting and Facilitating Conditions of the Human Smile: A Nonobtrusive Test of the Facial Feedback Hypothesis," *Journal of Personality and Social Psychology* 54, no. 5 (1988): 768.

87. Monya Baker, "Is There a Reproducibility Crisis?," *Nature* 533 (2016): 452–454.

7. THE WAY OUT

1. Constance Reid, *Neyman—from Life* (New York: Springer-Verlag, 1982), 229.

2. bell hooks, *Writing Beyond Race: Living Theory and Practice* (New York: Routledge, 2013), 4.

3. Edwin T. Jaynes, *Probability Theory: The Logic of Science*, ed. G. Larry Bretthorst (Cambridge: Cambridge University Press, 2003), 506.

4. Ronald A. Fisher, *Statistical Methods and Scientific Inference* (Edinburgh: Oliver and Boyd, 1956), 16–17.

5. Jaynes, *Probability Theory*, 478.

6. We note here in passing that a similar problem arises with Fisher's idea of sampling from a "hypothetical infinite population." Since there is no uniform probability measure on the subsets of an infinite population, it is impossible to do such a sample, even an imaginary one.

7. Charles Lambdin, "Significance Tests As Sorcery: Science Is Empirical—Significance Tests Are Not," *Theory and Psychology* 22, no. 1 (2012): 73.

8. David Trafimow, "Editorial," *Basic and Applied Social Psychology* 36, no. 1 (2014): 1–2.

9. David Trafimow and Michael Marks, "Editorial," *Basic and Applied Social Psychology* 37, no. 1 (2015): 1–2.

10. Chris Woolston, "Psychology Journal Bans *P* Values," *Nature* 519, no. 7541 (2015): 9.

11. Ronald L. Wasserstein and Nicole Lazar, "The ASA Statement on *p*-Values: Context, Process, and Purpose," *American Statistician* 70, no. 2 (2016): 130.

12. Wasserstein and Lazar, "The ASA Statement on *p*-Values," 130.

13. Daniel J. Benjamin et al., "Redefine Statistical Significance." *Nature Human Behaviour* 2, no. 1 (2018): 6–10.

14. Ronald A. Fisher, *The Design of Experiments* (London: Oliver and Boyd, 1935), 16.

15. Stephen M. Stigler, *The History of Statistics: The Measurement of Uncertainty Before 1900* (Cambridge, MA: Belknap Press, 1986), 69–77.

16. F. J. Anscombe, "Graphs in Statistical Analysis," *American Statistician* 27, no. 1 (1973): 17–21.

17. Ward Edwards, Harold Lindman, and Leonard J. Savage, "Bayesian Statistical Inference for Psychological Research," *Psychological Review* 70, no. 3 (1963): 217.

18. Stanislav Andreski, *Social Sciences as Sorcery* (New York: St. Martin's Press, 1972), 110.

19. Andreski, *Social Sciences as Sorcery*, 103–104.

20. Ronald A. Fisher, "Inverse Probability and the Use of Likelihood," *Mathematical Proceedings of the Cambridge Philosophical Society* 28, no. 3 (1932): 259.

21. Harry M. Collins, "Son of Seven Sexes: The Social Destruction of a Physical Phenomenon," *Social Studies of Science* 11, no. 1 (1981): 33–62.

22. Harry M. Collins and Trevor J. Pinch. *The Golem: What You Should Know About Science*, 2nd ed. (Cambridge: Cambridge University Press, 1998).

23. Sextus Empiricus, *Outlines of Pyrrhonism*, trans. R. G. Bury. (London: Heinemann, 1933), 195.

BIBLIOGRAPHY

Amrhein, Valentin, Sander Greenland, and Blake McShane. "Scientists Rise Up Against Statistical Significance." *Nature* 567, no. 7748 (2019): 305–307.

Andreski, Stanislav. *Social Sciences as Sorcery*. New York: St. Martin's Press, 1972.

Anscombe, F. J. "Graphs in Statistical Analysis." *American Statistician* 27, no. 1 (1973): 17–21.

Bakan, David. "The Test of Significance in Psychological Research." *Psychological Bulletin* 66, no. 6 (1966): 423–437.

Baker, Graham J. "Christianity and Eugenics: The Place of Religion in the British Eugenics Education Society and the American Eugenics Society, c. 1907–1940." *Social History of Medicine* 27, no. 2 (2014): 281–302.

Baker, Monya. "Is There a Reproducibility Crisis?" *Nature* 533 (2016): 452–454.

Bargh, John A., Mark Chen, and Lara Burrows. "Automaticity of Social Behavior: Direct Effects of Trait Construct and Stereotype Activation on Action." *Journal of Personality and Social Psychology* 71, no. 2 (1996): 230.

Bartlett, M. S. "The Information Available in Small Samples." *Mathematical Proceedings of the Cambridge Philosophical Society* 32, no. 4 (1936): 560–566.

Bayes, Thomas. "LII. An Essay Towards Solving a Problem in the Doctrine of Chances. By the Late Rev. Mr. Bayes, F.R.S. Communicated by Mr. Price, in a Letter to John Canton, A.M.F.R.S." *Philosophical Transactions of the Royal Society of London* 53 (1763): 370–418.

Begley, C. Glenn, and Lee M. Ellis. "Raise Standards for Preclinical Cancer Research." *Nature* 483, no. 7391 (2012): 531–533.

Bem, Daryl J. "Feeling the Future: Experimental Evidence for Anomalous Retroactive Influences on Cognition and Affect." *Journal of Personality and Social Psychology* 100, no. 3 (2011): 407–425.

Bem, Daryl J. "Writing the Empirical Journal Article." In *The Compleat Academic: A Practical Guide for the Beginning Social Scientist*, ed. Mark P. Zanna and John M. Darley, 171–201. Hove, East Sussex, UK: Psychology Press, 1987.

Bem, Daryl J., and Charles Honorton. "Does Psi Exist? Replicable Evidence for an Anomalous Process of Information Transfer." *Psychological Bulletin* 115, no. 1 (1994): 4–18.

Benjamin, Daniel J., James O. Berger, Magnus Johannesson, Brian A. Nosek, E.-J. Wagenmakers, Richard Berk, Kenneth A. Bollen et al. "Redefine Statistical Significance." *Nature Human Behaviour* 2, no. 1 (2018): 6–10.

Bennett, J. H. *Statistical Inference and Analysis: Selected Correspondence of R. A. Fisher*. Oxford: Oxford University Press, 1990.

Berenson, Alex. "Evidence in Vioxx Suits Shows Intervention by Merck Officials." *New York Times*. April 24, 2005.

Berkson, Joseph. "Some Difficulties of Interpretation Encountered in the Application of the Chi-Square Test." *Journal of the American Statistical Association* 33, no. 203 (1938): 526–536.

Berkson, Joseph. "Tests of Significance Considered as Evidence." *Journal of the American Statistical Association* 37, no. 219 (1942): 325–335.

Bernoulli, Jacob. *On the Law of Large Numbers*. Trans. Oscar Sheynin. Berlin, 2005. http://www.sheynin.de/download/bernoulli.pdf.

Bhattacharya, Shaoni. "Up to 140,000 Heart Attacks Linked to Vioxx." *New Scientist*, January 25, 2005.

Black, Edwin. *War Against the Weak: Eugenics and America's Campaign to Create a Master Race*. New York: Four Walls Eight Windows, 2003.

Boole, George. *An Investigation of the Laws of Thought: On Which Are Founded the Mathematical Theories of Logic and Probabilities*. London: Walton and Maberly, 1854.

Boole, George. "XII. On the Theory of Probabilities." *Philosophical Transactions of the Royal Society of London* 152 (1862): 225–252.

Boring, Edwin G. "Mathematical vs. Scientific Significance." *Psychological Bulletin* 16, no. 10 (1919): 335–338.

Box, Joan Fisher. *R. A. Fisher, the Life of a Scientist*. New York: Wiley, 1978.

Buchanan, Mark. "Statistics: Conviction by Numbers." *Nature* 445, no. 7125 (2007): 254–256.

Buehler, Robert J., and Alan P. Feddersen. "Note on a Conditional Property of Student's *t*." *Annals of Mathematical Statistics* 34, no. 3 (1963): 1098–1100.

Button, Katherine S., John P. A. Ioannidis, Claire Mokrysz, Brian A. Nosek, Jonathan Flint, Emma S. J. Robinson, and Marcus R. Munafò. "Power Failure: Why Small Sample Size Undermines the Reliability of Neuroscience." *Nature Reviews Neuroscience* 14, no. 5 (2013): 365–376.

Buttrick, Nick, Anup Gampa, Lilian Hummer, and Brian Nosek. "Replication of Analytic Thinking Promotes Religious Disbelief." Center for Open Science. Accessed April 28, 2020. https://osf.io/r4dve/.

Cacioppo, John T., Stephanie Cacioppo, Gian C. Gonzaga, Elizabeth L. Ogburn, and Tyler J. VanderWeele. "Marital Satisfaction and Break-Ups Differ Across On-Line and Off-Line Meeting Venues." *Proceedings of the National Academy of Sciences* 110, no. 25 (2013): 10135–10140.

Camerer, Colin F., Anna Dreber, Eskil Forsell, Teck-Hua Ho, Jürgen Huber, Magnus Johannesson, Michael Kirchler et al. "Evaluating Replicability of Laboratory Experiments in Economics." *Science* 351, no. 6280 (2016): 1433–1436.

Camerer, Colin F., Anna Dreber, Felix Holzmeister, Teck-Hua Ho, Jürgen Huber, Magnus Johannesson, Michael Kirchler et al. "Evaluating the Replicability of Social Science Experiments in *Nature* and *Science* Between 2010 and 2015." *Nature Human Behaviour* 2, no. 9 (2018): 637–644.

Campbell, John P. "Some Remarks from the Outgoing Editor." *Journal of Applied Psychology* 67, no. 6 (1982): 691–700.

Campbell, Lewis, and William Garnett. *The Life of James Clerk Maxwell: With a Selection from His Correspondence and Occasional Writings and a Sketch of His Contributions to Science*. New York: Macmillan, 1882.

Carey, Benedict. "Journal's Paper on ESP Expected to Prompt Outrage." *New York Times*, January 5, 2011.

Carney, Dana R., Amy J. C. Cuddy, and Andy J. Yap. "Power Posing: Brief Nonverbal Displays Affect Neuroendocrine Levels and Risk Tolerance." *Psychological Science* 21, no. 10 (2010): 1363–1368.

Chao, Tze-Fan, Chia-Jen Liu, Su-Jung Chen, Kang-Ling Wang, Yenn-Jiang Lin, Shih-Lin Chang, Li-Wei Lo et al. "The Association Between the Use of Non-steroidal Anti-Inflammatory Drugs and Atrial Fibrillation: A Nationwide Case-Control Study." *International Journal of Cardiology* 168, no. 1 (2013): 312–316.

Chesterton, Gilbert Keith. *Eugenics and Other Evils*. London: Cassell, 1922.

Christopher, Milbourne. *ESP, Seers, and Psychics*. New York: Crowell, 1970.

Cicero, Marcus Tullius, and Richard D. McKirahan. *De Natura Deorum. I.* Bryn Mawr Latin Commentaries. Bryn Mawr, PA: Thomas Library, Bryn Mawr College, 1997. Cohen, Jacob. "The Earth Is Round ($p < .05$)." *American Psychologist* 49, no. 12 (1994): 997–1003.

Colhoun, Helen M., Paul M. McKeigue, and George Davey Smith. "Problems of Reporting Genetic Associations with Complex Outcomes." *The Lancet* 361, no. 9360 (2003): 865–872.

Collins, Harry M. "Son of Seven Sexes: The Social Destruction of a Physical Phenomenon." *Social Studies of Science* 11, no. 1 (1981): 33–62.

Collins, Harry M., and Trevor J. Pinch. *The Golem: What You Should Know About Science*. 2nd ed. Cambridge: Cambridge University Press, 1998.

Corbett, Edward P. J., W. Rhys Roberts, and Ingram Bywater. *The Rhetoric and the Poetics of Aristotle*. New York: Modern Library, 1984.

Cournot, Antoine Augustin. *Exposition de la théorie des chances et des probabilités*. Paris: L. Hachette, 1843.

de Moivre, Abraham. *The Doctrine of Chances: Or, A Method of Calculating the Probability of Events in Play*. London: W. Pearson, 1718.

Diaconis, Persi, and David Freedman. "The Persistence of Cognitive Illusions." *Behavioral and Brain Sciences* 4, no. 3 (1981): 333–334.

Diaconis, Persi, and Brian Skyrms. *Ten Great Ideas About Chance*. Princeton, NJ: Princeton University Press, 2018.

Donkin, William Fishburn. "XLVII. On Certain Questions Relating to the Theory of Probabilities." *London, Edinburgh, and Dublin Philosophical Magazine and Journal of Science*, 4th ser., 1, no. 5 (1851): 353–368.

Doyen, Stéphane, Olivier Klein, Cora-Lise Pichon, and Axel Cleeremans. "Behavioral Priming: It's All in the Mind, but Whose Mind?" *PlOS One* 7, no. 1 (2012).

Eddy, David M. "Probabilistic Reasoning in Clinical Medicine: Problems and Opportunities." In *Judgment Under Uncertainty: Heuristics and Biases*, ed. Daniel Kahneman, Paul Slovic, and Amos Tversky, 249–267. Cambridge: Cambridge University Press, 1982.

Edwards, Ward, Harold Lindman, and Leonard J. Savage. "Bayesian Statistical Inference for Psychological Research." *Psychological Review* 70, no. 3 (1963): 193–242.

Ellenberg, Jordan. *How Not to Be Wrong: The Power of Mathematical Thinking*. New York: Penguin, 2015.

Ellis, Robert Leslie. *On the Foundations of the Theory of Probabilities*. London: John W. Parker, 1843.

Ellis, Robert Leslie. "Remarks on an Alleged Proof of the "Method of Least Squares," Contained in a Late Number of the Edinburgh Review." *London, Edinburgh, and Dublin Philosophical Magazine and Journal of Science*, 3rd ser., 37, no. 251 (1850): 321–328.

Engber, Daniel. "Daryl Bem Proved ESP Is Real Which Means Science Is Broken." *Slate*, May 17, 2017. Accessed April 28, 2020. https://slate.com/health-and-science/2017/06/daryl-bem -proved-esp-is-real-showed-science-is-broken.html

Engdahl, F. William. *Seeds of Destruction: The Hidden Agenda of Genetic Manipulation*. Montreal: Global Research, 2007.

Everitt, Brian, and Anders Skrondal. *The Cambridge Dictionary of Statistics*. Cambridge: Cambridge University Press, 2002.

Falk, Ruma, and Charles W. Greenbaum. "Significance Tests Die Hard: The Amazing Persistence of a Probabilistic Misconception." *Theory and Psychology* 5, no. 1 (1995): 75–98.

Feynman, Richard P., Robert B. Leighton, and Matthew Sands. *Six Easy Pieces: Essentials of Physics Explained by Its Most Brilliant Teacher*. Reading, MA: Addison-Wesley, 1995.

Fisher, Ronald A. "The Bearing of Genetics on Theories of Evolution." *Science Progress in the Twentieth Century (1919–1933)* 27, no. 106 (1932): 273–287.

Fisher, Ronald A. "The Concepts of Inverse Probability and Fiducial Probability Referring to Unknown Parameters." *Proceedings of the Royal Society of London. Series A, Containing Papers of a Mathematical and Physical Character* 139, no. 838 (1933): 343–348.

Fisher, Ronald A. *Contributions to Mathematical Statistics*. New York: Wiley, 1950.

Fisher, Ronald A. "Dangers of Cigarette-Smoking." *British Medical Journal* 2, no. 5039 (1957): 297.

Fisher, Ronald A. *The Design of Experiments*. London: Oliver and Boyd, 1935. 6th ed., 1951. 7th ed., 1960. 9th ed., 1971.

Fisher, Ronald A. *The Genetical Theory of Natural Selection*. Oxford: Clarendon Press, 1930.

Fisher, Ronald A. "Inverse Probability." *Mathematical Proceedings of the Cambridge Philosophical Society* 26, no. 4 (1930): 528–535.

Fisher, Ronald A. "Inverse Probability and the Use of Likelihood." *Mathematical Proceedings of the Cambridge Philosophical Society* 28, no. 3 (1932): 257–261.

Fisher, Ronald A. *Natural Selection, Heredity, and Eugenics: Including Selected Correspondence of R. A. Fisher with Leonard Darwin and Others*. Ed. J. H. Bennett. Oxford: Clarendon Press, 1983.

Fisher, Ronald A. "A Note on Fiducial Inference." *Annals of Mathematical Statistics* 10, no. 4 (1939): 383–388.

Fisher, Ronald A. "On the Mathematical Foundations of Theoretical Statistics." *Philosophical Transactions of the Royal Society of London. Series A, Containing Papers of a Mathematical or Physical Character* 222 (1922): 309–368.

Fisher, Ronald A. "On the 'Probable Error' of a Coefficient of Correlation Deduced from a Small Sample." *Metron* 1 (1921): 1–32.

Fisher, Ronald A. "Probability Likelihood and Quantity of Information in the Logic of Uncertain Inference." *Proceedings of the Royal Society of London. Series A, Containing Papers of a Mathematical and Physical Character* 146, no. 856 (1934): 1–8.

Fisher, Ronald A. "Some Hopes of a Eugenist." *Eugenics Review* 5, no. 4 (1914): 309.

Fisher, Ronald A. "The Statistical Method in Psychical Research." *Proceedings of the Society for Psychical Research* 39 (1929): 189–192.

Fisher, Ronald A. "Statistical Methods and Scientific Induction." *Journal of the Royal Statistical Society: Series B (Methodological)* 17, no. 1 (1955): 69–78.

Fisher, Ronald A. *Statistical Methods and Scientific Inference*. Edinburgh: Oliver and Boyd, 1956. 3rd ed. New York: Hafner, 1973.

Fisher, Ronald A. *Statistical Methods for Research Workers*. Biological Monographs and Manuals No. 3. Edinburgh, London: Oliver and Boyd, 1925. 5th ed. 1932.

Fisher, Ronald A. "Student." *Annals of Eugenics* 9, no. 1 (1939): 1–9.

Fox, Craig R., and Jonathan Levav. "Partition-Edit-Count: Naive Extensional Reasoning in Judgment of Conditional Probability." *Journal of Experimental Psychology: General* 133, no. 4 (2004): 626.

Freedman, Leonard P., Iain M. Cockburn, and Timothy S. Simcoe. "The Economics of Reproducibility in Preclinical Research." *PLOS Biology* 13, no. 6 (2015): e1002165.

Fries, Jakob Friedrich. *Versuch einer Kritik der Principien der Wahrscheinlichkeitsrechnung*. Braunschweig: F. Vieweg u. sohn, 1842.

Galton, Francis. *Hereditary Genius: An Inquiry Into Its Laws and Consequences*. London: Macmillan, 1869. 2nd ed. 1892.

Galton, Francis. "Hereditary Talent and Character." *Macmillan's Magazine* 12 (1865): 157–166, 318–327.

Galton, Francis. Letter to the editor. *The Times*, June 5, 1873.

Galton, Francis. *Memories of My Life*. New York: Dutton, 1909.

Galton, Francis. *Natural Inheritance*. London: Macmillan, 1889.

Galton, Francis, *Probability, the Foundation of Eugenics: The Herbert Spencer Lecture Delivered on June 5, 1907*. Oxford: Clarendon Press, 1907.

Galton, Francis. "Regression Towards Mediocrity in Hereditary Stature." *Journal of the Anthropological Institute of Great Britain and Ireland* 15 (1886): 246–263.

Galton, Francis. "Statistics by Intercomparison, with Remarks on the Law of Frequency of Error." *London, Edinburgh, and Dublin Philosophical Magazine and Journal of Science* 49, no. 322 (1875): 33–46.

Gardner, Martin. *The 2nd Scientific American Book of Mathematical Puzzles and Diversions*. New York: Simon and Schuster, 1961.

Gelman, Andrew. *Bayesian Data Analysis*. 3rd ed. Boca Raton, FL: CRC Press, 2014.

Gelman, Andrew and Eric Loken. "The Statistical Crisis in Science." *American Scientist* 102, no. 6 (2014): 460-65. Gervais, Will M., and Ara Norenzayan. "Analytic Thinking Promotes Religious Disbelief." *Science* 336, no. 6080 (2012): 493–496.

Gervais, Will M., and Ara Norenzayan. "Author Comment." Center for Open Science. Accessed April 28, 2020. https://osf.io/q64td/.

Gigerenzer, Gerd. *The Empire of Chance: How Probability Changed Science and Everyday Life*. Cambridge: Cambridge University Press, 1989.

Gigerenzer, Gerd. "The Superego, the Ego, and the Id in Statistical Reasoning." In *A Handbook for Data Analysis in the Behavioral Sciences: Methodological Issues*, ed. G. Keren and C. Lewis, 311–339. Hillsdale, NJ: Erlbaum, 1993.

Gillies, Donald. *Philosophical Theories of Probability*. New York: Routledge, 2000.

Goetz, Jacky G., Susana Minguet, Inmaculada Navarro-Lérida, Juan José Lazcano, Rafael Samaniego, Enrique Calvo, Marta Tello et al. "Biomechanical Remodeling of the Microenvironment by Stromal Caveolin-1 Favors Tumor Invasion and Metastasis." *Cell* 146, no. 1 (2011): 148–163.

Gönen, Mithat, Wesley O. Johnson, Yonggang Lu, and Peter H. Westfall. "The Bayesian Two-Sample *t* Test." *American Statistician* 59, no. 3 (2005): 252–257.

Hacking, Ian. "Karl Pearson's History of Statistics." *British Journal for the Philosophy of Science* 32, no. 2 (1981): 177–183.

Hagger, Martin S., Nikos L. D. Chatzisarantis, Hugo Alberts, Calvin O. Anggono, Cedric Batailler, Angela R. Birt, Ralf Brand et al. "A Multilab Preregistered Replication of the Ego-Depletion Effect." *Perspectives on Psychological Science* 11, no. 4 (2016): 546–573.

Haldane, J. B. S. "Karl Pearson, 1857–1957." *Biometrika* 44, no. 3/4 (1957): 303–313.

Hansel, Charles Edward Mark. *The Search for Psychic Power: ESP and Parapsychology Revisited.* Buffalo, NY: Prometheus Books, 1989.

Hill, Ray. "Multiple Sudden Infant Deaths—Coincidence or Beyond Coincidence?" *Paediatric and Perinatal Epidemiology* 18, no. 5 (2004): 320–326.

Hitler, Adolf, and Ralph Manheim. *Mein Kampf.* Boston: Houghton Mifflin, 1971.

Hogben, Lancelot Thomas. *Genetic Principles in Medicine and Social Science.* New York: Knopf, 1932.

Hogben, Lancelot Thomas. *Principles of Evolutionary Biology.* Cape Town: Juta, 1927.

hooks, bell. *Writing Beyond Race: Living Theory and Practice.* New York: Routledge, 2013.

Hoyle, Fred. *Mathematics of Evolution.* Memphis, TN: Acorn Enterprises, 1999.

Hume, David. *An Enquiry Concerning Human Understanding: A Critical Edition.* Ed. Tom L. Beauchamp. Oxford: Oxford University Press, 2000.

Inge, William R. "Some Moral Aspects of Eugenics." *Eugenics Review* 1 (1909–1910): 26–36.

Ioannidis, John P. A. "Contradicted and Initially Stronger Effects in Highly Cited Clinical Research." *JAMA* 294, no. 2 (2005): 218–228.

Ioannidis, John P. A. "Why Most Published Research Findings Are False." *PLOS Medicine* 2, no. 8 (2005): e124.

Ioannidis, John P. A., T. D. Stanley, and Hristos Doucouliagos. "The Power of Bias in Economics Research." *Economic Journal* 127, no. 605 (2017): F236–F265.

Jaynes, Edwin T. *Probability Theory: The Logic of Science.* Ed. G. Larry Bretthorst. Cambridge: Cambridge University Press, 2003.

Jaynes, Edwin T. "The Well-Posed Problem." *Foundations of Physics* 3, no. 4 (1973): 477–492.

Jeffreys, Harold. "On the Theory of Errors and Least Squares." *Proceedings of the Royal Society of London. Series A, Containing Papers of a Mathematical and Physical Character* 138, no. 834 (1932): 48–55.

Jeffreys, Harold. "Probability, Statistics, and the Theory of Errors." *Proceedings of the Royal Society of London. Series A, Containing Papers of a Mathematical and Physical Character* 140, no. 842 (1933): 523–535.

Jeffreys, Harold. "Probability and Scientific Method." *Proceedings of the Royal Society of London. Series A, Containing Papers of a Mathematical and Physical Character* 146, no. 856 (1934): 9–16.

Jeffreys, Harold. *Theory of Probability.* Oxford: Clarendon Press, 1939. 3rd ed. 1961.

Keynes, J. M. *A Treatise on Probability.* London: Macmillan, 1962.

Kolmogorov, Andreï Nikolaevich. *Foundations of the Theory of Probability.* Trans. Nathan Morrison. With an added bibliography by Albert T. Bharucha-Reid. 2nd English ed. Mineola, NY: Dover, 2018.

Kurt, Will. *Bayesian Statistics the Fun Way: Understanding Statistics and Probability with Star Wars, LEGO, and Rubber Ducks*. San Francisco: No Starch Press, 2019.

Lambdin, Charles. "Significance Tests as Sorcery: Science Is Empirical—Significance Tests Are Not." *Theory and Psychology* 22, no. 1 (2012): 67–90.

Lane, Arthur Henry. *The Alien Menace: A Statement of the Case*. 5th ed. London: Boswell, 1934.

Laplace, Pierre-Simon. *Essai philosophique sur les probabilités*. Paris: Bachelier, 1825.

Laplace, Pierre-Simon. "Mémoire sur les probabilités." *Mémoires de l'Académie Royale des sciences de Paris* 1778. Trans. Richard J. Pulskamp (1781): 227–332.

Laplace, Pierre-Simon. *Théorie analytique des Probabilités*. 2 vols. Paris: Courcier Imprimeur, 1812.

Lisse, Jeffrey R., Monica Perlman, Gunnar Johansson, James R. Shoemaker, Joy Schechtman, Carol S. Skalky, Mary E. Dixon et al. "Gastrointestinal Tolerability and Effectiveness of Rofecoxib Versus Naproxen in the Treatment of Osteoarthritis: A Randomized, Controlled Trial." *Annals of Internal Medicine* 139, no. 7 (2003): 539–546.

Lowery, George. "Study Showing That Humans Have Some Psychic Powers Caps Daryl Bem's Career." *Cornell Chronicle*, December 6, 2010.

Lowry, Oliver H., Nira J. Rosebrough, A. Lewis Farr, and Rose J. Randall. "Protein Measurement with the Folin Phenol Reagent." *Journal of Biological Chemistry* 193 (1951): 265–275.

Lykken, David T. "Statistical Significance in Psychological Research." *Psychological Bulletin* 70, no. 3 (1968): 151–159.

MacKenzie, Donald. "Statistical Theory and Social Interests: A Case-Study." *Social Studies of Science* 8, no. 1 (1978): 35–83.

MacKenzie, Donald A. *Statistics in Britain, 1865–1930: The Social Construction of Scientific Knowledge*. Edinburgh: Edinburgh University Press, 1981.

Matthews, Robert A. J. "Why *Should* Clinicians Care About Bayesian Methods?" *Journal of Statistical Planning and Inference* 94, no. 1 (2001): 43–58.

Mazumdar, Pauline. *Eugenics, Human Genetics and Human Failings*. Florence: Routledge, 1992.

McElreath, Richard. *Statistical Rethinking: A Bayesian Course with Examples in R and Stan*. Boca Raton, FL: CRC Press, 2018.

Meadow, Roy. *ABC of Child Abuse*. Ed. Roy Meadow. 3rd ed. London: BMJ Publishing Group, 1997.

Meehl, Paul E. "Theoretical Risks and Tabular Asterisks: Sir Karl, Sir Ronald, and the Slow Progress of Soft Psychology." *Journal of Consulting and Clinical Psychology* 46, no. 4 (1978): 806–834.

Meehl, Paul E. "Theory-Testing in Psychology and Physics: A Methodological Paradox." *Philosophy of Science* 34, no. 2 (1967): 103–115.

Meehl, Paul E. "Why Summaries of Research on Psychological Theories Are Often Uninterpretable." *Psychological Reports* 66, no. 1 (1990): 195–244.

Mill, John Stuart. *A System of Logic, Ratiocinative and Inductive, Being a Connected View of the Principles of Evidence and the Methods of Scientific Investigation*. Vol. 2. London: John W. Parker, 1843.

Mischel, Walter, and Ebbe B. Ebbesen. "Attention in Delay of Gratification." *Journal of Personality and Social Psychology* 16, no. 2 (1970): 329.

Mowbray, Miranda, and Dieter Gollmann. "Electing the Doge of Venice: Analysis of a 13th Century Protocol." In *20th IEEE Computer Security Foundations Symposium*, 295–310. Los Alamitos, CA: IEEE, 2007.

"A New Bem Theory." Statistical Modeling, Causal Inference, and Social Science. Accessed April 28, 2020. https://andrewgelman.com/2013/08/25/a-new-bem-theory.

Newman, Dennis. "The History of Statistics in the 17th and 18th Centuries, Against the Changing Background of Intellectual, Scientific and Religious Thought: Lectures by Karl Pearson, 1921–1933." *Journal of the Royal Statistical Society: Series A (General)* 143, no. 1 (1980): 78–79.

Neyman, Jerzy. *Lectures and Conferences on Mathematical Statistics and Probability*. Washington, DC: Graduate School, U.S. Department of Agriculture, 1952.

Norton, Bernard J. "Karl Pearson and Statistics: The Social Origins of Scientific Innovation." *Social Studies of Science* 8, no. 1 (1978): 3–34.

Nosek, Brian A., Jeffrey R. Spies, and Matt Motyl. "Scientific Utopia: II. Restructuring Incentives and Practices to Promote Truth Over Publishability." *Perspectives on Psychological Science* 7, no. 6 (2012): 615–631.

Open Science Collaboration. "Estimating the Reproducibility of Psychological Science." *Science* 349, no. 6251 (2015): aac4716.

Pearson, Egon S. *Karl Pearson: An Appreciation of Some Aspects of His Life and Work*. Cambridge: Cambridge University Press, 1938.

Pearson, Egon S. "Studies in the History of Probability and Statistics. XX: Some Early Correspondence Between W. S. Gosset, R. A. Fisher and Karl Pearson, with Notes and Comments." *Biometrika* 55, no. 3 (1968): 445–457.

Pearson, Karl. "III. Contributions to the Mathematical Theory of Evolution." *Philosophical Transactions of the Royal Society of London* 185 (1894): 71–110.

Pearson, Karl. *The Grammar of Science*. London: W. Scott, 1892. 2nd ed. 1900.

Pearson, Karl. *The Life, Letters and Labours of Francis Galton*. Vol. 1. Cambridge: Cambridge University Press, 1924.

Pearson, Karl. *National Life from the Standpoint of Science: An Address Delivered at Newcastle November 19, 1900*. London: Alan and Charles Black, 1901.

Pearson, Karl. "On the Inheritance of the Mental and Moral Characters in Man: II." *Biometrika* 3 (1904): 131–160.

Pearson, Karl, and Margaret Moul. "The Problem of Alien Immigration Into Great Britain, Illustrated by an Examination of Russian and Polish Jewish Children: Part I." *Annals of Eugenics* 1, no. 1 (1925): 5–54.

Pearson, Karl, and Margaret Moul. "The Problem of Alien Immigration Into Great Britain, Illustrated by an Examination of Russian and Polish Jewish Children: Part II." *Annals of Eugenics* 2, no. 1–2 (1927): 111–244.

Pearson, Karl, and Egon S. Pearson. *The History of Statistics in the 17th and 18th Centuries Against the Changing Background of Intellectual, Scientific, and Religious Thought: Lectures by Karl Pearson Given at University College, London, During the Academic Sessions, 1921–1933*. London: C. Griffin, 1978.

Peile, James Hamilton Francis. "Eugenics and the Church." *Eugenics Review* 1, no. 3 (1909): 163.

Peng, Roger. "The Reproducibility Crisis in Science: A Statistical Counterattack." *Significance* 12, no. 3 (2015): 30–32.

People v. Collins, 68 Cal. 2d 319, 438 P.2d 33, 66 Cal. Rptr. 497 (1968).

Pius XI. "Casti Connubii." *Acta Apostolicae Sedis* 22 (1930): 539–592.

Poisson, Siméon Denis. *Recherches sur la probabilité des jugements en matière criminelle et en matière civile*. Paris: Bachelier, 1837.

Porter, Theodore M. *Karl Pearson: The Scientific Life in a Statistical Age*. Princeton, NJ: Princeton University Press, 2004.

Porter, Theodore M. *The Rise of Statistical Thinking, 1820–1900*. Princeton, NJ: Princeton University Press, 1986.

R v. Clark, [2003] EWCA Crim 1020.

Ramsey, Frank P. "Truth and Probability." In *Readings in Formal Epistemology*, ed. Horacio Arló-Costa, Victor F. Hendricks, and Johan van Benthem, 21–45. Cham, Switz.: Springer, 2016.

Ranehill, Eva, Anna Dreber, Magnus Johannesson, Susanne Leiberg, Sunhae Sul, and Roberto A. Weber. "Assessing the Robustness of Power Posing: No Effect on Hormones and Risk Tolerance in a Large Sample of Men and Women." *Psychological Science* 26, no. 5 (2015): 653–656.

Reid, Constance. *Neyman—from Life*. New York: Springer-Verlag, 1982.

"Report of Committee for Legalizing Eugenic Sterilization." *Postgraduate Medical Journal* 6, no. 61 (1930): 13.

"Reproducibility Project: Cancer Biology." Ed. Roger J. Davis et al. eLife. Accessed April 28, 2020. https://elifesciences.org/collections/9b1e83d1/reproducibility-project-cancer-biology/.

Resnick, Brian. "More Social Science Studies Just Failed to Replicate. Here's Why This Is Good." *Vox*. August 27, 2018. Accessed April 28, 2020. https://www.vox.com/science-and-health/2018/8/27/17761466/psychology-replication-crisis-nature-social-science

Rothenberg, Thomas J., Franklin M. Fisher, and Christian Bernhard Tilanus. "A Note on Estimation from a Cauchy Sample." *Journal of the American Statistical Association* 59, no. 306 (1964): 460–463.

Royal Statistical Society. "Royal Statistical Society Concerned by Issues Raised in Sally Clark Case." News release, October 23, 2001. Accessed April 28, 2020. http://www.inference.org.uk/sallyclark/RSS.html

Rozeboom, William W. "The Fallacy of the Null-Hypothesis Significance Test." *Psychological Bulletin* 57, no. 5 (1960): 416–428.

Sandys, John Edwin, ed. *The Rhetoric of Aristotle*. Cambridge: Cambridge University Press, 1909.

Savage, Leonard J. "The Foundations of Statistics Reconsidered." In *Proceedings of the Fourth Berkeley Symposium on Mathematical Statistics and Probability, Volume 1: Contributions to the Theory of Statistics*, 575–586. Berkeley, CA: University of California Press, 1961.

Savage, Leonard J. "On Rereading R. A. Fisher." *Annals of Statistics* 4, no. 3 (1976): 441–500.

Schmidt, Morten, Christian F. Christiansen, Frank Mehnert, Kenneth J. Rothman, and Henrik Toft Sorensen. "Non-steroidal Anti-Inflammatory Drug Use and Risk of Atrial Fibrillation or Flutter: Population Based Case-Control Study." *British Medical Journal* 343, no. 7814 (2011): 82.

Schmidt, Morten, and Kenneth J. Rothman. "Mistaken Inference Caused by Reliance on and Misinterpretation of a Significance Test." *International Journal of Cardiology* 177, no. 3 (2014): 1089–1090.

Schneps, Leila, and Coralie Colmez. *Math on Trial: How Numbers Get Used and Abused in the Courtroom*. New York: Basic Books, 2013.

Schweber, Silvan S. "The Origin of the *Origin* Revisited." *Journal of the History of Biology* 10, no. 2 (1977): 229–316.

Seidenfeld, Teddy. "R. A. Fisher's Fiducial Argument and Bayes' Theorem." *Statistical Science* 7, no. 3 (1992): 358–368.

Sextus Empiricus. *Outlines of Pyrrhonism*. Trans. R. G. Bury. London: Heinemann, 1933.

Sheynin, Oscar B. "Studies in the History of Probability and Statistics. XXI. On the Early History of the Law of Large Numbers." *Biometrika* 55, no. 3 (1968): 459–467.

Simmons, Joseph P., Leif D. Nelson, and Uri Simonsohn. "False-Positive Psychology: Undisclosed Flexibility in Data Collection and Analysis Allows Presenting Anything as Significant." *Psychological Science* 22, no. 11 (2011): 1359–1366.

Smith, George, and L. G. Wickham Legg. *The Dictionary of National Biography, 1931–1940: With an Index Covering the Years 1901–1940 in One Alphabetical Series*. London: Oxford University Press, 1949.

Soal, S. G. "Experimental Evidence for Extra-Sensory Perception." *Nature* 185 (1960): 950–951.

Soal, S. G., and Frederick Bateman. *Modern Experiments in Telepathy*. New Haven, CT: Yale University Press, 1954.

"Social Biology." *New Statesman and Nation*, December 26, 1931, 816–817.

Sokal, Alan D. "Transgressing the Boundaries: Towards a Transformative Hermeneutics of Quantum Gravity." *Social Text* 46/47 (1996): 217–252.

Soper, H. E., A. W. Young, B. M. Cave, Alice Lee, and Karl Pearson. "On the Distribution of the Correlation Coefficient in Small Samples. Appendix II to the Papers of 'Student' and R. A. Fisher." *Biometrika* 11, no. 4 (1917): 328–413.

Spottiswoode, William. "On Typical Mountain Ranges: An Application of the Calculus of Probabilities to Physical Geography." *Journal of the Royal Geographical Society of London* 31 (1861): 149–154.

Sripada, Chandra, Daniel Kessler, and John Jonides. "Methylphenidate Blocks Effort-Induced Depletion of Regulatory Control in Healthy Volunteers." *Psychological Science* 25, no. 6 (2014): 1227–1234.

"Statistiek in het strafproces." *NOVA/Den Haag Vandaag*. Petra Greeven and Marcel Hammink. Hilversum, Netherlands: VARA/NOS, November 4, 2003.

Stigler, Stephen M. "Fisher in 1921." *Statistical Science* 20, no. 1 (2005): 32–49.

Stigler, Stephen M. *The History of Statistics: The Measurement of Uncertainty Before 1900*. Cambridge, MA: Belknap Press, 1986.

Stigler, Stephen M. "Studies in the History of Probability and Statistics. XXXIII Cauchy and the Witch of Agnesi: An Historical Note on the Cauchy Distribution." *Biometrika* 61, no. 2 (1974): 375–380.

Stigler, Stephen M. "The True Title of Bayes's Essay." *Statistical Science* 28, no. 3 (2013): 283–288.

Stoppard, Tom. *Rosencrantz and Guildenstern Are Dead*. New York: Grove, 1967.

Strack, Fritz, Leonard L. Martin, and Sabine Stepper. "Inhibiting and Facilitating Conditions of the Human Smile: A Nonobtrusive Test of the Facial Feedback Hypothesis." *Journal of Personality and Social Psychology* 54, no. 5 (1988): 768.

Student. "Probable Error of a Correlation Coefficient." *Biometrika* 6, no. 2–3 (1908): 302–310.

Tierney, John. "Behind Monty Hall's Doors: Puzzle, Debate, and Answer?" *New York Times*, July 21, 1991.

Trafimow, David. "Editorial." *Basic and Applied Social Psychology* 36, no. 1 (2014): 1–2.

Trafimow, David, and Michael Marks. "Editorial." *Basic and Applied Social Psychology* 37, no. 1 (2015): 1–2.

Turnbull, Craig. *A History of British Actuarial Thought*. Cham, Switz.: Springer International Publishing, 2017.

UNESCO. *The Race Question in Modern Science: Race and Science*. New York: Columbia University Press, 1961.

Utts, Jessica. "Replication and Meta-Analysis in Parapsychology." *Statistical Science* 6, no. 4 (1991): 363–378.

Venn, John. *The Logic of Chance: An Essay on the Foundations and Province of the Theory of Probability, with Especial Reference to Its Logical Bearings and Its Application to Moral and Social Science, and to Statistics*. London: Macmillan, 1866. 3rd ed. 1888..

vos Savant, Marilyn. "Ask Marilyn." *Parade*, September 9, 1990.

vos Savant, Marilyn. *The Power of Logical Thinking: Easy Lessons in the Art of Reasoning . . . and Hard Facts About Its Absence in Our Lives*. New York: Macmillan, 1997.

Wagenmakers, Eric-Jan, Titia Beek, Laura Dijkhoff, Quentin F. Gronau, A. Acosta, R. B. Adams Jr., D. N. Albohn et al. "Registered Replication Report: Strack, Martin, & Stepper (1988)." *Perspectives on Psychological Science* 11, no. 6 (2016): 917–928.

Wagenmakers, Eric-Jan, Don van den Bergh, Maarten Marsman, Johnny van Doorn, and Alexander Ly. "Supplement: Bayesian Analyses for 'Evaluating Replicability of Social Science Experiments in Nature and Science.' " Center for Open Science. Accessed April 28, 2020. https://osf.io/nsxgj/.

Wagenmakers, Eric-Jan, Ruud Wetzels, Denny Borsboom, and Han L. J. Van Der Maas. "Why Psychologists Must Change the Way They Analyze Their Data: The Case of Psi: Comment on Bem (2011)." *Journal of Personality and Social Psychology* 100, no. 3 (2011): 426–432.

Wasserstein, Ronald L., and Nicole Lazar. "The ASA Statement on p-Values: Context, Process, and Purpose." *American Statistician* 70, no. 2 (2016): 129–133.

Wasserstein, Ronald L, Allen L. Schirm, and Nicole A. Lazar. "Moving to a World Beyond '$p < 0.05$.'" *American Statistician* 73, supp. 1 (2019): 1–19.

Watts, Tyler W., Greg J. Duncan, and Haonan Quan. "Revisiting the Marshmallow Test: A Conceptual Replication Investigating Links Between Early Delay of Gratification and Later Outcomes." *Psychological Science* 29, no. 7 (2018): 1159–1177.

Weiss, Sheila Faith. "After the Fall: Political Whitewashing, Professional Posturing, and Personal Refashioning in the Postwar Career of Otmar Freiherr Von Verschuer." *Isis* 101, no. 4 (2010): 722–58.

"Which Statistics Test Should I Use?" Social Science Statistics. Accessed April 28, 2020. https://www.socscistatistics.com/tests/what_stats_test_wizard.aspx.

"Who Is the Greatest Biologist of All Time?" Edge. Accessed April 28, 2020. https://www.edge.org/conversation/who-is-the-greatest-biologist-of-all-time.

Woodhouse, Jayne. "Eugenics and the Feeble-Minded: The Parliamentary Debates of 1912–14." *History of Education* 11, no. 2 (1982): 127–137.

Woolston, Chris. "Psychology Journal Bans *P* Values." *Nature* 519, no. 7541 (2015): 9.

Yandell, Benjamin. *The Honors Class: Hilbert's Problems and Their Solvers*. Boca Raton, FL: CRC Press, 2001.

Zabell, Sandy L. "R. A. Fisher and Fiducial Argument." *Statistical Science* 7, no. 3 (1992): 369–387.

Zabell, Sandy L. "R. A. Fisher on the History of Inverse Probability." *Statistical Science* 4, no. 3 (1989): 247–256.

Ziliak, Stephen T., and Deirdre Nansen McCloskey. *The Cult of Statistical Significance: How the Standard Error Costs Us Jobs, Justice, and Lives*. Ann Arbor: University of Michigan Press, 2008.

INDEX

Printed in the USA
CPSIA information can be obtained
at www.ICGtesting.com
JSHW021437221024
72172JS00004B/37